ELECTRONIC STRUCTURE CALCULATIONS FOR SOLIDS AND MOLECULES: THEORY AND COMPUTATIONAL METHODS

Electronic structure problems are studied in condensed matter physics and theoretical chemistry to provide important insights into the properties of matter. This graduate level textbook describes the main theoretical approaches and computational techniques for studying the behavior of electrons in molecules and solids, from the simplest approximations to the most sophisticated methods.

The first part describes in detail the various theoretical approaches to calculating the electronic structure of solids and molecules, including density functional theory (DFT) and chemical methods based on Hartree–Fock theory. The basic approximations, adiabatic, classical nuclei, local density, and gradient corrections, are thoroughly discussed, and an in-depth overview of recent advances and alternative approaches in DFT is given.

The second part discusses the different practical methods proposed in order to solve the electronic structure problem computationally. This is developed in parallel for DFT and Hartree–Fock approaches, including a detailed discussion of basis sets and practical aspects of the main methods in use, including plane waves, Gaussian basis sets, and augmentation methods. The final chapters address the issues of diagonalization, self-consistency, and first-principles molecular dynamics simulations.

At variance with most books in this area, this textbook addresses graduate students in both physics and chemistry, and is intended to improve communication between the two communities. It also serves as a reference for researchers coming into the field.

JORGE KOHANOFF is Reader in Applied Mathematics and Theoretical Physics at Queen's University Belfast. Following study at the University of Buenos Aires and SISSA in Trieste, he worked with IBM in Zurich, Switzerland, and gained his Ph.D. from ETH-Zurich. He then spent some time as a post-doctoral research assistant in France, before leading the Electronic Structure and Computational Physics Programme at the International Centre for Theoretical Physics, Trieste, Italy. He has published over 50 papers, including contributions in computational methods and applications in electronic structure, statistical mechanics, and Car–Parrinello molecular dynamics simulations.

ELECTRONIC STRUCTURE CALCULATIONS FOR SOLIDS AND MOLECULES:

Theory and Computational Methods

JORGE KOHANOFF

School of Mathematics and Physics, Queen's University Belfast

CAMBRIDGE UNIVERSITY PRESS
Cambridge, New York, Melbourne, Madrid, Cape Town, Singapore, São Paulo

Cambridge University Press
The Edinburgh Building, Cambridge CB2 2RU, UK

Published in the United States of America by Cambridge University Press, New York

www.cambridge.org
Information on this title: www.cambridge.org/9780521815918

© J. Kohanoff 2006

This publication is in copyright. Subject to statutory exception
and to the provisions of relevant collective licensing agreements,
no reproduction of any part may take place without
the written permission of Cambridge University Press.

First published 2006

Printed in the United Kingdom at the University Press, Cambridge

A catalog record for this publication is available from the British Library

ISBN-13 978-0-521-81591-8 hardback
ISBN-10 0-521-81591-6 hardback

Cambridge University Press has no responsibility for the persistence or accuracy of URLs for external
or third-party internet websites referred to in this publication, and does not guarantee that any
content on such websites is, or will remain, accurate or appropriate.

A Clarisa, Anna, Elena, Tuta y Blanca

Contents

Preface		*page* xi
List of symbols		xvi
List of acronyms		xx
	Part I Theory	**1**
1	**The problem of the structure of matter**	**3**
	1.1 Adiabatic approximation	6
	1.2 Classical nuclei approximation	10
2	**The electronic problem**	**15**
	2.1 Screening: Gouy–Chapman and Debye–Hückel analysis	17
	2.2 The pair correlation function	19
	2.3 Many-body theory of electronic systems	22
3	**Quantum many-body theory: chemical approaches**	**28**
	3.1 The Hartree–Fock approximation	31
	3.2 Post-Hartree–Fock methods	37
4	**Density functional theory**	**51**
	4.1 Thomas–Fermi theory	51
	4.2 Modern density functional theory	56
	4.3 Kinetic correlation: the adiabatic connection	65
	4.4 Some observations about Kohn–Sham theory	70
5	**Exchange and correlation in DFT: approximations and their performances**	**75**
	5.1 The local density approximation	77
	5.2 Gradient expansions	85
	5.3 Non-locality: the weighted density approximation	94

	5.4	Hybrid HF–KS approaches	96
	5.5	Exact exchange: the optimized effective potential method	97
	5.6	Orbital-dependent correlation functionals	103
	5.7	Van der Waals (dispersion) interactions	108
	5.8	Green's function approach: the GW approximation	110
	5.9	Strong correlations: LDA+U and LDA+DMFT	112
	5.10	Summary of exchange-correlation functionals	113

Part II Computational methods — 121

6 Solving the electronic problem in practice — 123
- 6.1 Kohn–Sham and Hartree–Fock equations — 123
- 6.2 Condensed phases: Bloch's theorem and periodic boundary conditions — 128

7 Atomic pseudopotentials — 143
- 7.1 Pseudopotential theory — 144
- 7.2 Construction of pseudopotentials — 148
- 7.3 Separable form of atomic pseudopotentials — 164
- 7.4 Ultrasoft pseudopotentials — 169
- 7.5 Some practical aspects of pseudopotentials — 174

8 Basis sets — 178
- 8.1 Periodic systems — 181
- 8.2 Plane waves — 182
- 8.3 Other floating basis sets — 188
- 8.4 Atom-centered basis sets — 189
- 8.5 Mixed basis sets — 202
- 8.6 Augmented basis sets — 202

9 Electronic structure methods — 217
- 9.1 Multiple scattering methods: the KKR approach — 218
- 9.2 All-electron methods based on augmentation spheres — 220
- 9.3 The pseudopotential plane wave method (PPW) — 228
- 9.4 Atom-centered basis sets — 248
- 9.5 Gaussian basis sets — 259

10 Simplified approaches to the electronic problem — 270
- 10.1 Tight-binding methods — 272
- 10.2 Semiempirical approaches in quantum chemistry — 286
- 10.3 Relation between tight-binding and semiempirical methods — 294
- 10.4 Many-body classical potentials — 296

10.5	Classical force fields	298
10.6	Hybrid QM-MM methods	300
10.7	Orbital-free density functional approaches	304

11 Diagonalization and electronic self-consistency — **311**

11.1	Diagonalization	312
11.2	Self-consistency: mixing schemes	316
11.3	Direct minimization of the electronic energy functional	318

12 First-principles molecular dynamics (Car–Parrinello) — **323**

12.1	Density functional molecular dynamics	324
12.2	The Car–Parrinello Lagrangian	325
12.3	The Car–Parrinello equations of motion	328
12.4	Orthonormalization	330
12.5	Pre-conditioning	332
12.6	Performance of CPMD	333

Index — 339

Preface

Very often when a postgraduate student begins a research activity in a new field, she or he is presented with a partial view of the big picture, according to the arena where the research group carries out its activity. For the student this implies narrowing the focus to a class of systems such as molecules, surfaces, liquids, defects, or magnetic systems, and to concentrate on a particular theoretical approach and one or a few specific computational techniques out of the many possibilities available. All this is perfectly understandable and reasonable because it is very difficult to absorb rapidly the knowledge accumulated during many decades, since the pioneering work of Hartree, Fock, Slater, Thomas, Fermi, Bloch, Dirac, and Wigner in the twenties and thirties, until the most recent developments in areas such as electronic correlation. This knowledge is built up during many years of practice in the field, participating in schools, conferences, and workshops that promote exchange and collaboration between the different subareas, discussing advantages and disadvantages of different approaches with colleagues around the world, and keeping up with the latest developments in the literature.

Although many excellent books and reviews are available in areas such as density functional theory, solid state physics, quantum chemistry, electronic structure methods, and Car–Parrinello simulations, it is desirable to have all this reference material condensed in a single book that may be used by a fresh graduate student, by a postdoc who is moving into the field, or by a researcher with experience in one or a few subareas who wants to broaden her or his horizons. In this way the researcher can become familiar with the basic ideas and the most common theoretical approaches and approximations and, at the same time, learn about the various computational methods devised to solve the electronic structure problem in practice. A recent book by Richard Martin (*Electronic Structure: Basic Theory and Practical Methods*, Cambridge University Press, 2004) fills that gap especially in the area of solid state physics. The present book covers many of those issues, too. However, it is specifically oriented to both physicists and chemists, by describing methods used by the two communities to study solids, liquids, and molecules, and trying to make connections whenever possible.

This book is the result of the experience accumulated in practice and through organization and participation in electronic structure activities. Four institutions were fundamental in this respect: the Scuola Internazionale Superiore di Study Avanzati (SISSA) in Trieste, where I first met some of the main practitioners in the field like Michele Parrinello and Roberto Car, the International Centre for Theoretical Physics (ICTP) in Trieste, where I spent five years (1994–1999) organizing electronic structure and computational physics activities, the Centre Européen de Calcul Atomique et Moléculaire (CECAM) in Lyon, where I attended several workshops, and Queen's University Belfast (Northern Ireland), which gave me the possibility of devoting part of my time to the writing of this book. The germ of this book can be traced back to an invitation from Abhijit Mookerjee to lecture in a School he organized at the S. N. Bose Centre in Calcutta (India) in 1998. In previous years I had the chance to organize several activities at ICTP, where many lecturers kindly provided notes for the participants. I then thought it would be a good idea to try and put together this kind of material in a systematic and consistent fashion, and set off to prepare a set of notes for the Calcutta Workshop. Along the years I kept on revising and enriching these lecture notes, which were used as reference material for postgraduate courses. They were also on-line for the advantage of scientists around the world. It was my feeling, and it was also suggested by Mookerjee and other colleagues who read the notes, like Mariana Weissman and Pablo Ordejón, that it could be a good idea to use them as a starting point for a book. The chance eventually materialized when Mike Finnis mentioned these notes to Eoin O'Sullivan from Cambridge University Press, and he suggested that I put forward a proposal for this book.

I started in the field of electronic structure calculations in 1990 guided by Michele Parrinello and Giulia Galli at IBM Zurich. Through conversations with other members of the group I rapidly learnt about the diversity of the field, and realized that different groups tend to choose a specific theoretical approach according to their goals, expertise, and background. Usually, physicists prefer DFT and chemists favor Hartree–Fock-derived methods, although in recent times there have been numerous crossings from one to the other side, and the very existence of hybrid HF-DFT methods, and the combination of configuration interaction approaches with DFT, proves that the interaction between these two communities is very beneficial. By including both types of approach in this book, my intention was to make it useful to both communities. Although clearly biased towards DFT methods, it gives physicists a general idea of what type of calculations are carried out by chemists, and vice versa. At least as important is making theoretical approaches into practical schemes for calculating electronic properties of materials from first principles. This means devising an appropriate framework for implementing the Hartree–Fock or DFT methodologies in a computer. Here is where the various groups depart from each other. The field becomes more technical, and precise mathematical and numerical techniques acquire a prominent role. The second part of this book is an attempt to give a broad picture of the different methods available, but it also goes in detail into some areas such as the pseudopotential plane wave method and the Car–Parrinello approach.

The first part of this book, Chapters 1 to 5, contains theoretical material that covers from the very basic approximations up to the latest developments in the field of electronic correlation. Chapter 1 begins with an introduction to the problem of the structure of matter, discusses the adiabatic and classical nuclei approximations – which are often confused – and ends with the statement of the electronic structure problem. Chapter 2 is a conceptual introduction to the general problem of calculating the electronic structure of a many-electron system. The physical origin of correlation is qualitatively discussed and introduced at a simple level in the case of classical systems, and is then extended to electronic systems. *Ab initio* approaches used in the theoretical chemistry community are described in Chapter 3, starting from the Hartree–Fock approximation, and moving onto various ways of including correlation such as configuration interaction and Møller–Plesset perturbation theory. Density functional theory is introduced and discussed in detail in Chapter 4, while Chapter 5 is devoted to exchange and correlation within DFT. It begins with the local density and generalized gradient approximations (LDA and GGA), and covers up to the most recent meta-GGA and hybrid HF-DFT approaches. Next, this chapter focuses on topics somehow out of the main stream, presenting the state-of-the-art in DFT. This chapter intends to be a reference that condenses the present understanding of the most popular (and unpopular) approximations.

The second part of this book is devoted to practical methods. Chapter 6 begins by setting the general framework and discussing the main technical issues that arise in an actual calculation: the choice of a basis set for representing the electronic orbitals, and the choice between all-electron and pseudopotential methods. It then reviews some basic concepts from solid state physics, like the unit cell, Bloch's theorem, Brillouin zone, and how aperiodic systems can be treated in periodic cell calculations. Due to the enormous role played in the past two decades, Chapter 7 is devoted to the development of pseudopotential theory in great detail, and discusses the many issues that arise, such as the choice of atomic configuration, ghost states, and non-linear core corrections. At the end of this chapter there is a set of practical hints on how to construct pseudopotentials. Chapter 8 is concerned with basis sets. After setting the general framework for molecular and crystalline systems, it describes in detail plane wave and atom-centered basis sets, specializing in the case of Gaussians. Other approaches are also mentioned, and a final section is devoted to augmentation schemes such as the APW and MTO basis sets, together with their linearized versions. Chapter 9 reviews in more or less detail the different electronic structure methods generally used, from the KKR approach to all-electron methods based on augmentation spheres, and the pseudopotential plane wave approach. Then it focuses on methods based on atom-centered basis sets, specifically discussing some technical aspects related to the use of Gaussian basis sets. Chapter 10 uses the framework set in the previous chapters to discuss simplified approaches to the electronic problem such as tight-binding schemes normally used in solid state physics, and semiempirical schemes commonly used in the theoretical chemistry community like extended Hückel, CNDO, and MNDO. A connection between these two parallel worlds is attempted. Simpler approaches like bond order potentials and empirical force fields are also reviewed, hybrid QM-MM methods are described, and a final section is devoted to simplified, orbital-free density functionals. Chapter 11 is concerned with the issues of diagonalization and self-consistency, and the final chapter is devoted to the Car–Parrinello extended Lagrangian method for first-principles molecular dynamics simulations.

I would have liked to include some other topics in this book, such as quantum Monte Carlo methods, density functional perturbation theory, electronic excitations, and quantum nuclear effects, for the sake of completeness. Unfortunately they remained outside the book, but readers will surely have no difficulty in locating the appropriate literature.

Most of what is in this book I learnt from reading and listening to the main actors along the years. It would take several pages to name them all, with the almost certainty of forgetting to mention someone. I want to thank, however, a few people who helped me with parts of the manuscript. First is my long-time friend Andrés Saul, who patiently read the first four chapters of the manuscript and helped me with the tight-binding section. Next is Sasha Lozovoi, who checked Chapters 6 and 7 and provided me with useful data. Then is my good friend Nikitas Gidopoulos. We wrote together a chapter on DFT in a recent handbook, which constituted the basis for Chapters 4 and 5. Mike Finnis critically read the section on tight-binding methods, and Tony Paxton had to suffer me bothering him so many times trying to understand how muffin-tin orbitals are constructed. Thanks also to Bob Jones and David Vanderbilt for kindly providing figures for reproduction. Last, but not least, I want to thank my three mentors: Erio Tosatti, Michele Parrinello, and Ruth Lynden-Bell. I am largely indebted to the three of them for their generous guidance and friendship.

I would also like to mention some people who involuntarily participated in this book by sharing their experience on various subjects with me. In no particular order, these are the people I first met in SISSA like Mario Tosi, Stefano de Gironcoli, Stefano Baroni, Raffaele Resta, Roberto Car, Annabella Selloni, Giorgio Pastore, Furio Ercolessi, and Guido Chiarotti. Then there are Giulia Galli, Pedro Serena, Franco Buda, Wanda Andreoni, Peter Blöchl, François Gygi, Jürg Hutter, Andrew Fisher, Michiel Sprik, Ursula Röthlisberger, Dominik Marx, Mark Tuckerman, Ali Alavi, Mike Klein, and Kari Laasonen, whom I met during my years at IBM Zurich. Jean-Pierre Hansen, Hong Xu, Enrico Smargiassi, and Carlo Pierleoni were some of my colleagues in Lyon, and the following people I met through the years, in different places and times: David Ceperley, Charusita Chakravarty, Pietro Ballone, Daniel Laría, Rob Coalson, Michael Methfessel, David Vanderbilt, Sandro Scandolo, Darío Estrín, Ruben Weht, Ricardo Migoni, Sergio Koval, Ezequiel Leiva, Barbara Montanari, Mariana Weissman, Stuart Murdock, Carlo Cavazzoni, Saverio Moroni, Pablo Ordejón, Jose Soler, Emilio Artacho, Daniel Sánchez-Portal, Hardy Gross, Tchavdar Todorov, Ron Cohen, Detlef Hohl, Mike Gillan, Graeme Ackland, Cristián Sánchez, Raul Cachau, Mario Del Popolo, Tristan Youngs, Peijun Hu, Paolo Giannozzi, Andy Görling, Stewart Clark, Paul Madden, Richard Martin, Alfredo Caro, Jorge Sofo, Gleb Gribakin, Jose Luis Martins, Renata Wentzcovitch, Abdallah Qteish, and my students Simone Selenu, Carlos Pinilla, and Iván Scivetti, who had to suffer (or enjoy?) my disappearance during the final stages of the writing of this book. Special thanks go to my first Ph.D. student, Giuseppe Colizzi, who forced me to study pseudopotential theory properly, and provided me with the figures that are shown in the corresponding chapter. Finally, I can never be grateful enough to Clarisa and my daughters Anna and Elena for their patience and encouragement. Now it is time to make up for all those boring weekends watching me working on this book!

Symbols

Operators

$\hat{\mathcal{H}}$	Hamiltonian operator
$\hat{h}_e$	electronic Hamiltonian
$\hat{T}$	electronic kinetic energy operator
$\hat{T}_n$	nuclear kinetic energy operator
$\hat{U}_{ee}$	electron–electron interaction
$\hat{V}_{ne}$	electron–nuclear interaction
$\hat{V}_{nn}$	nuclear–nuclear interaction
$\hat{V}_{ext}$	external interaction acting on electrons
$\hat{h}_1(i)$	core Hamiltonian for electron i
$\hat{v}_2(i,j)$	electron–electron interaction between electrons i and j
$\hat{\mathcal{F}}$	Fock operator
$\hat{\mathcal{J}}_i$	Coulomb operator
$\hat{\mathcal{K}}_i$	exchange operator
$\hat{H}_{KS}$	Kohn–Sham one-electron Hamiltonian operator
$\hat{V}_{PS}$	pseudopotential electron–nuclear interaction
$\hat{I}$	identity operator
$\hat{P}_l$	projection operator onto angular momentum l subspace
$\hat{S}$	overlap operator

Energy contributions

T	electronic kinetic energy
U_{ee}	electron–electron interaction energy
E_X	exchange energy
E_C	correlation energy
E_{XC}	exchange and correlation energy
$\tilde{E}_{XC}$	coupling constant averaged exchange and correlation energy
E_H	Hartree (direct Coulomb) energy
V_{ne}	electron–nuclear interaction energy
V_{nn}	nuclear–nuclear interaction energy

V_{ext}	interaction energy of electrons with external sources
E_{KS}	Kohn–Sham energy
T_{R}	non-interacting electronic kinetic energy
K_{e}	fictitious kinetic energy in CPMD

Integrals in Hartree–Fock theory

E_{ii}	one-electron integrals
J_{ij}	Coulomb integrals
K_{ij}	exchange integrals

Potentials

$v_{\text{ext}}(\mathbf{r})$	external potential acting on electrons
$v_{\text{H}}(\mathbf{r})$	Hartree electrostatic potential
$\mu_{\text{XC}}(\mathbf{r})$	exchange-correlation potential
$v_{\text{KS}}(\mathbf{r})$	Kohn–Sham potential
$v_{\text{PS}}(\mathbf{r})$	pseudopotential

System specifications and constants

P	number of particles (atoms or nuclei)
M_I	mass of particle I
Z_I	charge of particle I
$\mathbf{R}_I$	coordinates of particle I
$\mathbf{R}$	coordinates of all the nuclei
N	number of electrons
m	mass of the electron
e	charge of the electron
$\mathbf{r}_i$	coordinates of electron i
$\mathbf{r}$	coordinates of all the electrons
$\mathbf{x}_i$	combined spatial and spin coordinates of electron i
$\hbar$	Planck's constant
ϵ_{F}	Fermi energy

Wave functions and eigenvalues

$\Psi_n(\mathbf{R}, \mathbf{r})$, $|\Psi_n\rangle$	total (electron–nuclear) wave function
$\mathcal{E}_n$	total (electron–nuclear) energy
$\Theta_n(\mathbf{R})$, $|\Theta_n\rangle$	nuclear wave function
$\Phi_n(\mathbf{R}, \mathbf{r})$, $|\Phi_n\rangle$	electronic wave function
$E_n(\mathbf{R})$	electronic eigenstates
$\varphi_i(\mathbf{r})$, $|\varphi\rangle$	one-electron orbitals
ε_i	one-electron eigenvalues
f_i	occupation numbers
$\psi_{\mathbf{k}}(\mathbf{r})$, $|\psi_{\mathbf{k}}\rangle$	one-electron wave function of momentum $\mathbf{k}$
$\varepsilon(\mathbf{k})$	electronic dispersion relation
$\phi_\alpha(\mathbf{r})$, $|\phi_\alpha\rangle$	generic basis function
M	size of the basis set

Notation for crystalline systems

$\mathbf{a}_i$	lattice vectors
$\mathbf{T}$	lattice translations
$\mathbf{b}_i$	reciprocal lattice vectors
$\mathbf{G}$	reciprocal lattice translations
$\mathbf{k}$	wave vector in the first Brillouin zone
Γ-point	Brillouin zone center
$N_\mathbf{k}$	number of wave vectors in the first Brillouin zone
$\omega_\mathbf{k}$	weight of wave vector $\mathbf{k}$ in BZ sums
Ω	volume of unit cell or supercell
E_{cut}	plane wave energy cutoff
$S(\mathbf{G})$	structure factor

Density functionals

$f[\rho]$	functional
$\delta f[\rho]/\delta\rho(x)$	functional derivative
$\epsilon_{\text{XC}}[\rho]$	exchange-correlation energy density
$F_{\text{XC}}[\rho]$	exchange-correlation enhancement factor
$\tau[\rho]$	kinetic energy density

Density and response functions

$\rho(\mathbf{r})$	electronic density
$\zeta(\mathbf{r})$	spin-polarization or magnetization density
$\rho_1(\mathbf{r},\mathbf{r}')$	one-body density matrix
$\rho_2(\mathbf{r},\mathbf{r}')$	two-body density matrix
$g(\mathbf{r},\mathbf{r}')$	pair correlation function
$\tilde{g}(\mathbf{r},\mathbf{r}')$	coupling-constant-averaged pair correlation function
$\rho_{\text{XC}}(\mathbf{r},\mathbf{r}')$	exchange-correlation hole
$\chi_\sigma(\mathbf{r},\mathbf{r}')$	linear response function for spin σ
$G_\sigma(\mathbf{r},\mathbf{r}';E)$	Green's function for spin σ and energy E
$\Sigma(\mathbf{r},\mathbf{r}';E)$	self-energy
$W(\mathbf{r},\mathbf{r}';E)$	screened interaction

Atomic quantities

n, l, m	principal, angular, and magnetic quantum numbers
$\Phi_{\text{AE}}^{nlm}(r)$	all-electron atomic wave function
$Y_{lm}(\theta,\phi)$, $|lm\rangle$	spherical harmonic function
$P_l(\cos\theta)$	Legendre polynomial
$R_{nl}(r)$, $\chi^l(r)$	radial component of the atomic wave function
$u^l(r)$	$r\chi^l(r)$
$\Phi_{\text{PS}}^{lm}(r)$	pseudo-atomic wave function
$|\zeta^{lm}\rangle$, $|\xi^{lm}\rangle$, $|\beta^{lm}\rangle$	pseudopotential projector functions
$\eta_l(\varepsilon)$	phase shift
ζ_{nl}	exponents of Slater-type and Gaussian-type orbitals
$G_{ijk}(\mathbf{r},\alpha,\mathbf{R}_I)$	primitive Cartesian Gaussian

Notation for atom-centered and augmented basis sets

$\mathbf{H}$, $H_{\alpha\beta}$	Hamiltonian matrix
$\mathbf{F}$, $F_{\alpha\beta}$	Fock matrix
$\mathbf{S}$, $S_{\alpha\beta}$	overlap matrix
$T^{IJ}_{\mu\nu}$	kinetic energy matrix elements
$U^{IJ}_{\mu\nu}$	nuclear attraction matrix elements
$h^{IJ}_{\mu\nu}$	one-electron matrix elements
$\langle\mu\gamma\|\nu\delta\rangle$, $\langle\phi^I_\mu\phi^K_\gamma\|\phi^J_\nu\phi^L_\delta\rangle$	two-electron integrals
$O^{IJ}_{\mu\nu}(\mathbf{r})$	overlap distribution or differential overlap
ε^I_α	on-site energies
$t^{IJ}_{\alpha\beta}$, $\beta^{IJ}_{\alpha\beta}$	hopping or resonance integrals
$C_{LL'L''}$	Gaunt coefficients
$B_{JL',IL}$	structure constants
R_{MT}	muffin-tin radius

Acronyms

Quantum chemistry

CASSCF	complete active space self-consistent field
CI	configuration interaction
CC	coupled clusters
ECP	effective core potential
HF	Hartree–Fock
RPA	random phase approximation
MCSCF	multi-configuration self-consistent field
MRCI	multi-reference configuration interaction
MP	Møller–Plesset perturbation theory
S,D,T,Q	single, double, triple, and quadruple excitations
SCF	self-consistent field

Density functional theory

DFT	density functional theory
DFPT	density functional perturbation theory
DMFT	dynamical mean-field theory
EXX	exact exchange
GGA	generalized gradient approximation
KS	Kohn–Sham
LDA	local density approximation
LSDA	local spin density approximation
MBPT	many-body perturbation theory
OEP	optimized effective potential
SIC	self-interaction correction
TFD	Thomas–Fermi–Dirac
WDA	weighted density approximation
XC	exchange-correlation

Plane wave and related methods

AE	all-electron
BZ	Brillouin zone
FFT	fast Fourier transform

KB	Kleinman–Bylander
NLCC	non-linear core corrections
OPW	orthogonalized plane waves
PAW	projected augmented waves
PBC	periodic boundary conditions
PPW	pseudopotential plane wave
PS	pseudopotential
PW	plane waves
US	ultrasoft

Atom-centered methods

AO	atomic orbital
BSSE	basis set superposition error
CGTO	contracted Gaussian-type orbitals
GTO	Gaussian-type orbitals
HOMO	highest occupied molecular orbital
LCAO	linear combination of atomic orbitals
LUMO	lowest unoccupied molecular orbital
MO	molecular orbital
PGTO	primitive Gaussian-type orbitals
STO	Slater-type orbitals
SZ,DZ,DZP,TZP, etc.	single-zeta, double zeta+polarization, etc.

Augmentation and related methods

APW	augmented plane waves
ASA	atomic sphere approximation
FP	full-potential
KKR	Korringa–Kohn–Rostoker
LAPW	linearized augmented plane waves
LCMTO	linear combination of muffin-tin orbitals
LMTO	linear muffin-tin orbitals
MT	muffin-tin
MTA	muffin-tin approximation
MTO	muffin-tin orbitals

Simplified methods

AM1	Austin model 1
CNDO	complete neglect of differential overlap
EAM	embedded atom model
INDO	intermediate neglect of differential overlap
MINDO	modified intermediate neglect of differential overlap
MNDO	modified neglect of differential overlap
NDDO	neglect of diatomic differential overlap
PM3	parametric model 3
QM – MM	quantum mechanics – molecular mechanics
SCCT	self-consistent charge transfer

SCTB		self-consistent tight-binding
TB		tight-binding
ZDO		zero differential overlap

***Ab initio* molecular dynamics**

BOMD		Born–Oppenheimer molecular dynamics
CPMD		Car–Parrinello molecular dynamics
DFMD		density functional molecular dynamics
DKSO		dynamical Kohn–Sham orbitals

Part I
Theory

1

The problem of the structure of matter

The description of the physical and chemical properties of matter is a central issue that has occupied the minds of scientists since the age of the ancient Greeks. In their route to dissect matter down to what cannot be divided any further, they coined the term *atom*, the *indivisible*. Matter became then a collection of atoms. More than twenty centuries had to pass until the development of a more precise concept of atom, thanks, amongst others, to the systematic studies of Mendeleyev and the establishment in 1869 of the periodic table of the elements (Mendeleyev, 1869). The discovery of the electron in 1897 and the first modern model of the atomic structure by Sir Joseph Thomson were soon refined by his student, Sir Ernest Rutherford, who in 1910 showed that an atom was made of a positively charged small nucleus and a number of negatively charged electrons that neutralize the nuclear charge. Much in the spirit of planetary systems, and drawing from the analogy between gravitational and electrostatic interactions, scientists in the beginning of the twentieth century built an image of the atom that consisted of a number Z of electrons – of elementary charge $-e$ – orbiting around the nucleus of charge Ze.

A number of experimental observations, though, were incompatible with this idea of orbiting electrons. In particular, according to the successful electromagnetic theory, charged electrons in orbital (radially accelerated) motion should radiate energy, thus decelerating and eventually collapsing onto the nucleus. Clearly, such a picture would imply that matter is essentially unstable, in flagrant contradiction with our everyday experience of the very existence of matter. It was this kind of incompatibility that motivated the idea that matter at such small scale does not obey the laws of classical mechanics and electromagnetism, but a different set of laws, whose body became known as *quantum mechanics* (Bohr, 1913). To solve the problem of electron radiation, Bohr postulated the existence of certain peculiar orbits for which the electron would not radiate. These orbits correspond to specific energies and radii, and the promotion from one orbit to another could only happen through the absorption or emission of a *quantum* of energy. One of the appealing aspects of this theory was its conceptual compatibility with Planck's theory of black body (electromagnetic) radiation, which also required discrete energy levels. The emission of light of very precise frequency by atoms could then be explained as the decay of an electron from a higher to a lower energy level, by emitting a quantum of radiation whose frequency was proportional to the energy difference between these levels. These ebullient new ideas gave a tremendous momentum to the field, and, in a few decades, the mathematical apparatus, the language of quantum mechanics, was largely developed. Schrödinger's equation was published in 1926 (Schrödinger, 1926), and it was soon applied to multi-electronic atoms and to polyatomic systems such as molecules (Heitler and London, 1927) and solids (Bloch, 1928). It is the main goal of this book to describe the different approaches to the description of matter at the atomic scale, i.e. in terms of atomic nuclei and electrons.

In general terms, we can imagine a piece of matter as a collection of interacting atoms, sometimes under the influence of an external field. This ensemble of particles may be in the gas phase (molecules, clusters) or in a condensed phase (bulk solids, surfaces, wires). It could be in a solid, liquid or amorphous phase, either homogeneous or heterogeneous (molecules in solution, interfaces, adsorbates on surfaces). However, at this scale, we can unambiguously describe all these systems as a set of atomic nuclei and electrons interacting via coulombic, electrostatic forces. Formally, we can write the Hamiltonian of such a system in the following general form:

$$\hat{\mathcal{H}} = -\sum_{I=1}^{P} \frac{\hbar^2}{2M_I} \nabla_I^2 - \sum_{i=1}^{N} \frac{\hbar^2}{2m} \nabla_i^2 + \frac{e^2}{2} \sum_{I=1}^{P} \sum_{J \neq I}^{P} \frac{Z_I Z_J}{|\mathbf{R}_I - \mathbf{R}_J|}$$
$$+ \frac{e^2}{2} \sum_{i=1}^{N} \sum_{j \neq i}^{N} \frac{1}{|\mathbf{r}_i - \mathbf{r}_j|} - e^2 \sum_{I=1}^{P} \sum_{i=1}^{N} \frac{Z_I}{|\mathbf{R}_I - \mathbf{r}_i|}, \quad (1.1)$$

where $\mathbf{R} = \{\mathbf{R}_I, \ I = 1, \ldots, P\}$ is a set of P nuclear coordinates, and $\mathbf{r} = \{\mathbf{r}_i, \ i = 1, \ldots, N\}$ is a set of N electronic coordinates. Z_I and M_I are the nuclear charges and masses, respectively. Since the electrons are fermions, the total electronic wave function must be antisymmetric, i.e. it should change sign whenever the coordinates of any two electrons are exchanged. Different nuclear species are distinguishable, but nuclei of the same species also obey a specific statistics according to the nuclear spin. They are fermions for half-integer nuclear spin (e.g. H, ^{3}He) and bosons for integer spin (e.g. D, ^{4}He, H_2). All the ingredients are well known and, in principle, all the properties can be derived by solving the time-independent Schrödinger equation:

$$\hat{\mathcal{H}} \Psi_n(\mathbf{R}, \mathbf{r}) = \mathcal{E}_n \Psi_n(\mathbf{R}, \mathbf{r}), \quad (1.2)$$

where $\mathcal{E}_n$ are the energy eigenvalues and $\Psi_n(\mathbf{r}, \mathbf{R})$ are the corresponding eigenstates, or wave functions, which must be antisymmetric with respect to exchange of electronic coordinates in $\mathbf{r}$, and symmetric or antisymmetric with respect to exchange of nuclear variables in $\mathbf{R}$.

In practice, this problem is almost impossible to treat within a full quantum mechanical framework. Only in a few cases, such as hydrogenoid atoms or the H_2^+ molecule, a complete analytic solution is available. Exact numerical solutions are also limited to a few cases, mostly atoms and very small molecules. There are several features that contribute to this difficulty, but the most important is that this is a multi-component many-body system, and the two-body nature of the Coulomb interaction makes the above Schrödinger equation not separable.

In order to fix ideas, let us confine ourselves to the case of an atom with Z electrons, and focus on the electronic wave function. First of all, to respect the antisymmetry of the wave function against electron exchange, we can, in principle, write such a wave function as an antisymmetrized product of one-electron wave functions (a so-called Slater determinant). This assumes, however, some kind of separability of the Schrödinger equation, implying that the probability of finding an electron at some point in space is essentially independent of where the other electrons are located. The repulsive electron–electron interaction is quite at odds with this picture, because an electron located at point $\mathbf{r}$ in space precludes other electrons from approaching this location, much in the spirit of an *exclusion zone*. Hence, the probability of finding an electron at $\mathbf{r}$ depends on the location of the other $Z - 1$ electrons. This phenomenon is known as *correlation*, and it implies that the exact many-body wave function should contain factors depending on two electronic coordinates. Therefore, the image in terms of one-electron wave functions can be somewhat crude in many cases.

This means that the full Schrödinger equation cannot be easily decoupled into a set of equations, so that, in general, we have to deal with $3(P+N)$ coupled degrees of freedom. The usual choice is to resort to a few reasonable and well-controlled approximations, which encompass a wide variety of problems of interest. Of course, there are systems where the hypotheses leading to these approximations are violated, and these are by no means uninteresting. They require, however, a much larger theoretical and computational effort, and thus, historically, precedence has been given to the *easier* systems.

In the first part of this book we shall develop the theory starting from two major approximations: the adiabatic separation of the nuclear and electronic degrees of freedom, and the classical treatment of atomic nuclei. We shall then discuss in detail the different approaches to tackling the electronic problem, as emerged from two, often contrasting, but mostly cooperating, communities: *chemists* and *physicists*, trying to establish a common language between them. This will be done in the spirit of facilitating the understanding of chemical approaches by physicists, and vice versa. In fact, similar approaches are sometimes adopted by the two communities using different names, as is the case of tight-binding and semiempirical methods. In the second part, we shall concentrate on the different computational methods proposed to cast the problem of the structure of matter in a form amenable to numerical treatment, devoting our attention also to specific algorithms.

1.1 Adiabatic approximation

A first observation is that the time scale associated with the motion of nuclei is usually much slower than that associated with electrons. In fact, the most unfavorable case of a single proton already corresponds to a mass ratio of 1:1836, i.e. less than 1%. Within a classical picture we could say that, under typical conditions, the velocity of the electron is much larger than that of the heavy particle (the proton). In this spirit, Born and Oppenheimer (1927) proposed a scheme for separating the motion of nuclei from that of the electrons. The original work studied the time-independent Hamiltonian (1.1) perturbatively in the mass ratio $\kappa = (m/M)^{1/4}$. The influence of the nuclei on the electronic wave functions becomes apparent when considering a new set of nuclear variables defined by $\mathbf{R} = \mathbf{R}_0 + \kappa \mathbf{u}$, where $\mathbf{u}$ represents the displacement of the nuclei with respect to their equilibrium positions $\mathbf{R}_0$. Using this change of variables and expanding the $\mathbf{R}$-dependent terms in the potential in powers of $\mathbf{u}$, they obtained an expansion of the Hamiltonian in powers of κ. By keeping terms up to fourth order in κ, they showed that no mixing of different electronic stationary states happened due to the interaction with the nuclei. Therefore, under appropriate conditions that are discussed below, *the electrons do not undergo transitions between stationary states*. This is called the *adiabatic approximation*. The reason for the name is based more on dynamical, rather than stationary, arguments. We will first present a semiclassical picture which serves to fix the ideas, and then the quantum mechanical derivation from time-dependent perturbation theory. This will allow us to inspect the limits of validity of the adiabatic approximation.

Let us first analyze the case of a molecular system, where the electronic spectrum is discrete. The arguments here have to be adapted to the case of infinite systems such as solids or liquids, which exhibit a continuum spectrum of excitations. In a molecule there are basically three types of motion: electronic, nuclear vibrations, and nuclear rotations, each one corresponding to a typical time (or energy) scale. The energy scale of the electronic motion is given by the separation between successive electronic eigenstates, which is of the same order of magnitude as the ground state energy. This quantity is of the order of

$E_e \approx \hbar^2/(ma^2)$, where a is a typical interatomic distance in the molecule. This distance a gives an indication of the order of magnitude of the amplitude of the electronic motion. Typical distances are of the order of a few Å, and the associated energies are in the order of 1 eV. The energy scale of nuclear vibrations is given, in the harmonic approximation, by the quantum $E_v = \hbar\omega$, where ω is the frequency of the vibrational motion. The order of magnitude of such a frequency can be obtained by thinking that the energy of a classical oscillatory motion of amplitude a is $M\omega^2 a^2$. With a being a typical interatomic distance, a motion of this amplitude would cause the extraction of an atom from the molecule. Since atoms in molecules bind mostly via shared electrons (chemical bonding), this energy can be identified with that of breaking a chemical bond, which is of the same order of magnitude as the electronic energy E_e. Hence, $M\omega^2 a^2 \approx E_e$, and replacing a in terms of E_e from the first expression leads to $E_v = \hbar\omega \approx (m/M)^{1/2} E_e$. Rotational energies are the smallest of them all. They are related to the angular momentum of the molecule, which is quantized in levels separated by $\approx \hbar$. The energy is $E_r = L^2/I$, where I is the moment of inertia of the molecule, which is of the order of Ma^2. Therefore, $E_r \approx \hbar^2/Ma^2 = (m/M)\hbar^2/ma^2 = (m/M)E_e$. In summary, there is the following relation between these three energy scales:

$$E_r \approx \kappa^2 E_v \approx \kappa^4 E_e. \tag{1.3}$$

Therefore, if we take a value of $\kappa \approx 0.066$ (corresponding to the N_2 molecule), we have that rotational energies are about two orders of magnitude smaller than vibrational energies, and these in turn are two orders of magnitude smaller than electronic energies. For example, the first electronic excitation energy in the N_2 molecule is 7.5 eV, the vibrational excitation energy is about 300 meV, and the rotational level separation is around 0.5 meV.

In extended systems the electronic spectrum is continuous and, in principle, so is the excitation spectrum. Here we have to distinguish two cases. For insulating and semiconducting systems the smallest electronic excitation is given by the energy gap, which, as in the case of molecules, falls in the region of a few eV ($E_g \approx 1.1$ eV for silicon and ≈ 4 eV for diamond). For metallic systems the electronic excitations form a continuum, so that, formally, $E_e = 0$ and, strictly speaking, the adiabatic approximation should not be applicable. We shall discuss this at the end of this section, but, for the moment being, let us analyze the situation in which, for some reason, the electronic energy levels are more narrowly spaced than what has been assumed above. In that case the electronic energy scale becomes comparable to that of the nuclear vibrational motion, the adiabatic separation breaks down, and nuclei and electrons have to be treated in a unified framework. Examples of this behavior are when the electronic gap closes due to some external factor, e.g. pressure, doping, or temperature. States can be formed where electrons are intimately coupled to a nuclear vibration, thus giving rise to exotic entities like *polarons*. Thermal agitation also modifies the spacing between electronic levels but in a dynamical way. Occasionally, two or more levels may become so close that the nuclear motion promotes an electronic transition from one to another level, thus violating the hypotheses of the adiabatic approximation. If these crossings involve electron–nuclear interactions that occur during a limited period of time, then they can be studied using asymptotic techniques that view the non-adiabatic region in terms of transition probabilities, like the Landau–Zener approach (Messiah, 1961). The same phenomenon happens in atomic and molecular collisions. A more general approach requires the dynamical treatment of the electronic motion coupled to the nuclear motion, but described at the quantum-mechanical level. Practical schemes to achieve this goal within a harmonic description of the nuclear motion (phonons) have only very recently become available (see, e.g., Horsfield *et al.*, 2004).

The electrons can then be thought of as *instantaneously following the motion of the nuclei, while remaining always in the same stationary state of the electronic Hamiltonian*. This stationary state will vary in time because of the electrostatic coupling of the two sets of degrees of freedom, but, if the electrons were, e.g., in the (many-electron) ground state, they would remain there forever. In other words, as the nuclei follow their dynamics, the electrons instantaneously adjust their wave function according to the nuclear wave function. This approximation ignores the possibility of having non-radiative transitions between different electronic eigenstates. Transitions can only arise through the coupling with an external electromagnetic field, but this issue will not be addressed in the following.

All this can be cast in a formal mathematical framework by proposing a solution to Eq. (1.2) of the following form:

$$\Psi(\mathbf{R}, \mathbf{r}, t) = \sum_n \Theta_n(\mathbf{R}, t) \Phi_n(\mathbf{R}, \mathbf{r}), \tag{1.4}$$

where $\Theta_n(\mathbf{R}, t)$ are wave functions describing the evolution of the nuclear subsystem in each one of the *adiabatic* electronic eigenstates $\Phi_n(\mathbf{R}, \mathbf{r})$. These satisfy the time-independent Schrödinger equation

$$\hat{h}_e \Phi_n(\mathbf{R}, \mathbf{r}) = E_n(\mathbf{R}) \Phi_n(\mathbf{R}, \mathbf{r}), \tag{1.5}$$

where the electronic Hamiltonian is:

$$\hat{h}_e = \hat{T} + \hat{U}_{ee} + \hat{V}_{ne} = \hat{\mathcal{H}} - \hat{T}_n - \hat{V}_{nn}. \tag{1.6}$$

Here, $\hat{T}$ is the electronic kinetic operator, $\hat{U}_{ee}$ is the electron–electron interaction, $\hat{V}_{ne}$ the electron–nuclear interaction, $\hat{T}_n$ the nuclear kinetic operator, and $\hat{V}_{nn}$ the inter-nuclear interaction.

1.1 Adiabatic approximation

In this partial differential equation on the **r** variables, the $3P$ nuclear coordinates **R** enter as parameters. This expansion, which is always mathematically possible, is called the expansion in the *adiabatic basis*, because $\Phi_n(\mathbf{R}, \mathbf{r})$ are solutions of the time-independent electronic Schrödinger equation, corresponding to a particular nuclear configuration. Equation (1.5) has to be solved for all nuclear configurations **R** where the nuclear wave function is non-vanishing.

By replacing the above ansatz into the full Schrödinger equation we obtain:

$$\left[i\hbar\frac{\partial}{\partial t} + \sum_{I=1}^{P} \frac{\hbar^2}{2M_I} \nabla_I^2 - V_{nn} - E_q(\mathbf{R}) \right] \Theta_q(\mathbf{R}, t) =$$

$$-\sum_n \sum_{I=1}^{P} \frac{\hbar^2}{2M_I} \langle \Phi_q | \nabla_I^2 | \Phi_n \rangle \Theta_n(\mathbf{R}, t)$$

$$-2 \sum_n \sum_{I=1}^{P} \frac{\hbar^2}{2M_I} \vec{\nabla}_I \Theta_n(\mathbf{R}, t) \cdot \langle \Phi_q | \vec{\nabla}_I | \Phi_n \rangle, \tag{1.7}$$

which constitutes a set (infinite, in principle) of coupled partial differential equations containing off-diagonal terms. Here we have used Dirac's bra-ket notation to indicate matrix elements of the type

$$\langle \Phi_q | \vec{\nabla}_I | \Phi_n \rangle = \int \Phi_q^*(\mathbf{R}, \mathbf{r}) \vec{\nabla}_I \Phi_n(\mathbf{R}, \mathbf{r}) \, d\mathbf{r}. \tag{1.8}$$

Therefore, the reduction of the full wave function to an expression of the type

$$\Psi(\mathbf{R}, \mathbf{r}, t) = \Theta_n(\mathbf{R}, t) \Phi_n(\mathbf{R}, \mathbf{r}) \tag{1.9}$$

is not completely correct, because, even if the system was initially *prepared* in a *pure* state like the above one, the off-diagonal terms will mix (excite) the different electronic eigenstates along the temporal evolution. These are precisely the non-radiative transitions alluded to above. If this is the case, then the dynamics is said to be *non-adiabatic*. However, if the off-diagonal terms can be neglected, then an expression like (1.9) is valid because the nuclear dynamics has no means to cause electronic transitions, and the electrons remain always in the same (n) adiabatic state (ground or excited). In this case, the dynamics is said to be *adiabatic*.

The necessary condition for neglecting the non-adiabatic couplings is that

$$\left| \sum_{I=1}^{P} \frac{\hbar^2}{M_I} \langle \Theta_q | \vec{\nabla}_I | \Theta_n \rangle \cdot \langle \Phi_q | \vec{\nabla}_I | \Phi_n \rangle \right| \ll |E_q(\mathbf{R}) - E_n(\mathbf{R})| \tag{1.10}$$

or, equivalently,

$$\frac{m}{M} \left| \frac{\hbar \Omega_v}{E_q(\mathbf{R}) - E_n(\mathbf{R})} \right| \ll 1, \tag{1.11}$$

where Ω_v is the maximum frequency of rotation of the electronic wave function due to the nuclear motion, and the energies in the denominator correspond to the electronic adiabatic eigenstates (the energy gap if $q = 1$ and $n = 0$). Notice that the mass ratio m/M is always smaller than 5×10^{-4}, thus justifying the adiabatic approximation unless a very small gap occurs, as for open-shell, conical intersections or Jahn–Teller systems. The case of lighter particles such as muons would be different. Typical electronic excitations are of the order of 1 eV, while typical nuclear excitations (phonons) are of the order of 0.01 eV. This indicates that there is a clear separation of energy (and consequently time) scales. There are situations in which this approximation is not adequate, but they are rather exceptional cases.

The arguments in favor of an adiabatic treatment of metallic systems are more subtle. In principle, it can be argued that the adiabatic approximation breaks down because the energy gap is zero and electronic excitations of vanishing energy are possible. However, since typical temperatures (between room temperature and a few thousand degrees) are usually much lower than the electronic Fermi temperature, excitations are confined to a narrow region around the Fermi surface, and most properties are little affected by neglecting non-adiabatic contributions due to these few electrons. In fact, the usual treatment of electronic transport phenomena in metals begins with the adiabatic description and introduces non-adiabatic terms (in the form of electron–phonon interactions) afterwards, perturbatively. In terms of the ratio of energy scales, it can also be realized that the relevant excitations in metals at small wave numbers are not electron–hole pairs, which, besides being very few, carry a small oscillator strength (Migdal's theorem). The relevant energy scale is actually dictated by the plasmon (collective charge excitation), which is again typically of the order of a few eV.

1.2 Classical nuclei approximation

Therefore, according to the adiabatic approximation, the total wave function can be written in the form of Expression (1.9), where $\Theta_n(\mathbf{R}, t)$ is the nuclear wave function. At room temperature the thermal wavelength is $\lambda_T = (\hbar^2/2Mk_BT)^{1/2}$, which, for hydrogen at room temperature, is of the order of 0.2 Å. Regions of space separated by more than λ_T do not exhibit quantum phase coherence. Interatomic distances are normally of the order of 1 Å, and then the total nuclear wave function can be considered as an incoherent superposition of individual nuclear wave packets:

$$\Theta_n(\mathbf{R}, t) = \prod_{I=1}^{P} \Theta_n^{(I)}(\mathbf{R}, \mathbf{R}^{(I)}(t), t), \qquad (1.12)$$

3.2 Post-Hartree–Fock methods

Its eigenstates and energies can be expanded in powers of the parameter λ as follows:

$$\Phi_i = \Phi_i^{(0)} + \lambda \Phi_i^{(1)} + \lambda^2 \Phi_i^{(2)} + \lambda^3 \Phi_i^{(3)} + \cdots,$$
$$E_i = E_i^{(0)} + \lambda E_i^{(1)} + \lambda^2 E_i^{(2)} + \lambda^3 E_i^{(3)} + \cdots. \quad (3.48)$$

By replacing (3.48) into (3.47) and equating the various powers of λ (Messiah, 1961), equations are obtained for the different contributions. The first order correction to the energies is simply the expectation value of the perturbation in the corresponding unperturbed eigenstates:

$$E_i^{(1)} = \langle \Phi_i^{(0)} | \Delta \hat{\mathcal{H}} | \Phi_i^{(0)} \rangle. \quad (3.49)$$

The first order correction to the eigenfunctions, which is assumed to be orthogonal to the corresponding unperturbed eigenstates, is obtained as an infinite sum over the entire spectrum of the unperturbed Hamiltonian:

$$\Phi_i^{(1)} = \sum_{j \neq i} \frac{\langle \Phi_j^{(0)} | \Delta \hat{\mathcal{H}} | \Phi_i^{(0)} \rangle}{E_i^{(0)} - E_j^{(0)}} \Phi_j^{(0)}. \quad (3.50)$$

The second order correction to the energies involves the first order correction to the eigenstates:

$$E_i^{(2)} = \langle \Phi_i^{(0)} | \Delta \hat{\mathcal{H}} | \Phi_i^{(1)} \rangle = \sum_{j \neq i} \frac{|\langle \Phi_j^{(0)} | \Delta \hat{\mathcal{H}} | \Phi_i^{(0)} \rangle|^2}{E_i^{(0)} - E_j^{(0)}}. \quad (3.51)$$

Expressions for the second and higher order corrections to the eigenstates and energies become increasingly complex, but can be derived recursively by equating the different powers of λ, and can be found in standard textbooks in quantum mechanics (Messiah, 1961). The important fact is that corrections to every order can be written in terms of unperturbed eigenstates and energies. Whether or not a few terms in the expansion will suffice to converge to the solution of the perturbed problem will depend on how large the perturbation is, and on the convergence properties of the expansion.

Perturbations in quantum many-body theory

The problem of interest in the present context is that of many interacting quantum particles. Being a central problem in theoretical physics for such a long time, it has been attacked from different angles. One of the traditional approaches (Fetter and Walecka, 1971) begins by solving exactly the problem of *non-interacting* particles, which in the present situation are electrons (fermions). Next, the Coulomb interaction between electrons is introduced as a perturbation, which can be done in a systematic way by means of diagrammatic techniques such as the one developed by Feynman. This is beyond the scope of this book, and will not be pursued here, but there are, however, some important general remarks.

First of all, the electron–electron interaction is by no means a small term. Therefore, it cannot be expected that a few terms in the perturbative expansion will solve the problem. This leads to the idea of considering infinite sums of perturbative terms of certain classes (re-summations), which are expected to be the most relevant ones. One such case is the Hartree–Fock approximation, where only two types of diagram are retained, direct Coulomb and exchange. However, HF considers an infinite sum of perturbative terms corresponding to many virtual interactions of these two types. The random phase approximation (RPA) is another possible re-summation of diagrams that begins from the non-interacting electron system as a reference.

A system of non-interacting electrons, however, is not necessarily the best possible starting point for which the exact solution is available. In fact, the Hartree–Fock approximation is a numerically tractable scheme, and can be solved exactly or at least to an excellent extent using various computational approaches.

Møller–Plesset theory

Another possibility is to take the sum of Fock operators,

$$\hat{\mathcal{H}}_0 = \sum_{i=1}^{N} \hat{\mathcal{F}}_i = \sum_{i=1}^{N} \left[\hat{h}_i + \sum_{j=1}^{N} (\hat{\mathcal{J}}_{ij} - \hat{\mathcal{K}}_{ij}) \right], \tag{3.52}$$

as the reference Hamiltonian for a perturbative expansion. This has the enormous advantage that the unperturbed eigenstates are the well-known, easy-to-calculate *Hartree–Fock* determinants, which are far closer to the exact solution than the non-interacting system. In this case the perturbation takes the form:

$$\Delta \hat{\mathcal{H}} = \frac{1}{2} \sum_{i \neq j} \hat{v}_2(i, j) - \sum_{ij} (\hat{\mathcal{J}}_{ij} - \hat{\mathcal{K}}_{ij}). \tag{3.53}$$

This theory was developed in the thirties by Møller and Plesset (1934), and accordingly is called Møller–Plesset (MP) perturbation theory. The unperturbed energies are just the sum of the eigenvalues of the Fock operator over all the occupied states:

$$E_{\text{MP}}^{(0)} = \sum_{i=1}^{N} \varepsilon_i, \tag{3.54}$$

3.2 Post-Hartree–Fock methods

and the first order energy, i.e. the sum of the unperturbed energy and the first order correction, is the Hartree–Fock energy:

$$E_{\text{MP1}} = E_{\text{MP}}^{(0)} + E_{\text{MP}}^{(1)} = \sum_{i=1}^{N} \varepsilon_i + \frac{1}{2} V_{\text{ee}} - V_{\text{ee}} = \sum_{i=1}^{N} \varepsilon_i - \frac{1}{2} V_{\text{ee}} = E_{\text{HF}}, \quad (3.55)$$

where

$$V_{\text{ee}} = \langle \Phi_k^{(0)} | \sum_{i \neq j} \hat{v}_2(i,j) | \Phi_k^{(0)} \rangle \quad (3.56)$$

is the electrostatic energy calculated as the expectation value of the two-body Coulomb operator in the unperturbed Hartree–Fock state $\Phi_k^{(0)}$. The HF ground state corresponds to $k = 0$.

The first non-trivial correction is to second order (MP2), which involves matrix elements of $\Delta\hat{\mathcal{H}}$ between the Hartree–Fock reference state, e.g. the HF ground state, and all the excited states constructed as in a CI expansion (see Section 3.2.1). A very useful property of this form of the perturbation is that, since $\Delta\hat{\mathcal{H}}$ is a two-body operator, the only excited states contributing a non-zero matrix element are the double excitations. Therefore, the first energy correction involving correlation is:

$$E_{\text{MP}}^{(2)} = \sum_{\mu < \nu = N_{\text{occ}}+1}^{ND} \sum_{i < j=1}^{N_{\text{occ}}} \frac{|\langle \Phi_k^{(0)} | \Delta\hat{\mathcal{H}} | \Phi_k^{ij\mu\nu} \rangle|^2}{E_k^{(0)} - E_k^{ij\mu\nu}}, \quad (3.57)$$

where electrons in the unperturbed many-body state k that occupy the orbitals i and j are promoted to the empty orbitals μ and ν.

The sum of this term to the HF energy, $E_{\text{MP2}} = E_{\text{MP}}^{(0)} + E_{\text{MP}}^{(1)} + E_{\text{MP}}^{(2)}$, is known as the MP2 energy. The MP2 approximation consists of terminating the perturbative expansion for the energy at this level while correcting the wave functions to first order. The scaling of the MP2 method with the number of basis functions (M) is of the order M^5. The pre-factor, however, is rather small because only matrix elements between the HF determinant and double excitations have to be calculated. For modest-size systems (around 10 atoms), the cost of the MP2 correction is comparable to that of an HF calculation, although the more unfavorable scaling takes its toll for larger systems.

Higher orders in the perturbation expansion can also be included, and may still be less expensive than CI calculations. The energy at the MP3 level can be calculated using the first order corrected wave functions, by virtue of the $2n+1$-theorem. Therefore, MP3 also involves only double excitations, but now the doubles interact between themselves. The scaling is of the order M^6, and it typically recovers 90–95% of the correlation energy. This is consistent with the observation that double excitations are energetically the most important ones, as mentioned in the context of CI calculations. The next level in perturbation theory is MP4, which involves single, double, triple, and quadruple excitations, and scales like M^7.

The MP2 approximation is the way of choice in quantum chemical calculations whenever low-level electronic correlation is required. It is the least expensive correlated method, and rather accurate as it typically recovers around 80–90% of the correlation energy. The remainder, however, is quite important in certain situations, particularly when the system consists of weakly interacting fragments. In fact, MP2 does not reproduce the correct dissociation limit. This requires at least an MP4 calculation, as is shown in Fig. 3.1, which shows the dissociation curve of the H_2 molecule at various theory levels. The cost of an MP4 calculation is similar to a CISD calculation, and it accounts typically for 95–98% of the correlation energy. Higher perturbative orders have not been explored extensively. At variance with the CI expansion, perturbative methods are not variational, so that the energy may be lower than the exact energy.

The main limitation of the MP method is that it relies on an HF reference state. If HF provides a poor description of the system, then the correction terms are larger and the perturbation expansion requires more terms to be accurate. In

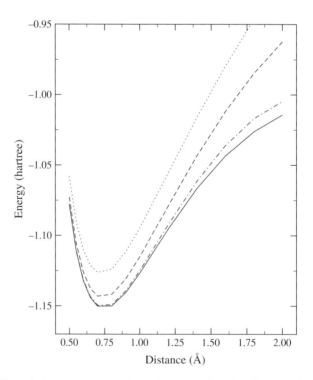

Fig. 3.1 Dissociation curve for H_2 at different levels of approximation: HF (dotted line), MP2 (dashed line), MP4 (dot-dashed line), and CISD (solid line). The result at the CCSD level is indistinguishable from the CISD.

fact, in some cases the expansion may not even converge. This will happen, e.g. when the unperturbed wave function has a large multi-reference character. In that case, a multi-reference MP scheme can be devised, in the form of a CASMP2 expansion around a CASSCF reference.

Re-summation perturbative methods: coupled clusters

As mentioned before, quantum many-body perturbation expansions can be constructed to include an infinite number of terms of a certain class, while discarding all the terms of other type. This idea was proposed and exploited by Bartlett (1989) and it is known by the name of *coupled clusters* (CC). In this method the wave function is written as

$$\Phi_{cc} = e^{\hat{T}} \Phi_0, \tag{3.58}$$

where Φ_0 is the HF wave function and $\hat{T} = \hat{T}_1 + \hat{T}_2 + \cdots + \hat{T}_N$ is an operator acting on the HF wave function that has the effect of generating all the excited Slater determinants up to a certain order. In fact, single excitations correspond to the first term,

$$\hat{T}_1 \Phi_k^{(0)} = \sum_{\mu=N_{occ}+1}^{ND} \sum_{i=1}^{N_{occ}} t_i^\mu \Phi_k^{i\mu}, \tag{3.59}$$

while the second term,

$$\hat{T}_2 \Phi_k^{(0)} = \sum_{\mu<\nu=N_{occ}+1}^{ND} \sum_{i<j=1}^{N_{occ}} t_{ij}^{\mu\nu} \Phi_k^{ij\mu\nu}, \tag{3.60}$$

includes double excitations, and so on. The coefficients t are the equivalent of the coefficient C in the CI expansion.

By expanding the exponential,

$$e^{\hat{T}} = \hat{I} + \hat{T}_1 + \left(\hat{T}_2 + \frac{1}{2}\hat{T}_1^2\right) + \left(\hat{T}_3 + \hat{T}_2\hat{T}_1 + \frac{1}{6}\hat{T}_1^3\right) + \cdots, \tag{3.61}$$

it can be seen that the CC wave function includes all the excited Slater determinants. For example, the second order term will involve genuine (connected) doubles through $\hat{T}_2$, and disconnected doubles through $\hat{T}_1^2$, much in the spirit of connected and disconnected diagrams in perturbative expansions in field theories.

Using the above wave function, the Schrödinger equation reads

$$\hat{\mathcal{H}}\left(e^{\hat{T}}\Phi_0\right) = E_{cc}\left(e^{\hat{T}}\Phi_0\right), \tag{3.62}$$

and the corresponding energy is:

$$E_{cc} = \langle \Phi_0 | \hat{\mathcal{H}} e^{\hat{T}} | \Phi_0 \rangle. \tag{3.63}$$

Interestingly, using the fact that the Hamiltonian contains only one- and two-body operators, it turns out that the CC energy is given *exactly* in terms of singles and doubles, in the following form:

$$E_{cc} = E_{HF} + \sum_{\mu<\nu=N_{occ}+1}^{ND} \sum_{i<j=1}^{N_{occ}} \left(t_{ij}^{\mu\nu} + t_i^{\mu} t_j^{\nu} - t_i^{\nu} t_j^{\mu} \right) \langle \Phi_0 | \Delta \hat{\mathcal{H}} | \Phi^{ij\mu\nu} \rangle, \quad (3.64)$$

where the coefficients t_i^{μ} and $t_{ij}^{\mu\nu}$ are determined by solving a hierarchy of CI-type equations derived from the expansion of the exponential. Keeping all orders in this hierarchy is equivalent to full CI, and it is practically impossible. Truncated expansions are used, instead, in the same spirit of CI calculations. Thus, choosing $\hat{T} = \hat{T}_2$ leads to the CCD level of approximation, and using $\hat{T} = \hat{T}_1 + \hat{T}_2$ corresponds to the CCSD approximation. The computational effort involved in these calculations scales like M^6, i.e. as an MP3 calculation. When the $\hat{T}_3$ operator is also included the scaling becomes M^8, and it is more demanding than a CISDT calculation. CCSD, however, is as costly as CISD, but the re-summation involved makes it more accurate. In fact, CCSD is as accurate as MP4, but at a significantly lower cost. Expressions for the CC equations can be found in the original articles (Bartlett, 1989) and in quantum chemistry books (Szabo and Ostlund, 1989; Jensen, 1999).

3.2.4 Explicitly correlated methods

All the methods described above are based on one-electron orbitals. In fact, these are easier to deal with, but they ignore the fact that the position of an electron is correlated to the position of all the other electrons. This is recovered in a painful way by constructing combinations of products of one-electron orbitals (CI) or via perturbative expansions (MP or CC). A more efficient approach would be to try and build the correlation directly into the trial wave function. Helgaker *et al.* discuss this issue in detail for the He atom (Helgaker *et al.*, 2000, Chapter 7). Here we shall only mention the main ideas. First, we notice that the exact wave function, as a function of $r_{12} = |\mathbf{r}_1 - \mathbf{r}_2|$, exhibits a cusp at the origin ($r_{12} = 0$), meaning a discontinuous first derivative. The same happens as a function of the electron–nuclear distance (nuclear cusp). By carrying out a CI expansion for the He atom in terms of Slater-type orbitals (see Section 8.4.4), it can be shown that a reasonable ansatz is the Hylleraas function (Hylleraas, 1929):

$$\Phi_H(r_1, r_2, r_{12}) = e^{-\zeta(r_1+r_2)} \sum_{ijk} C_{ijk} (r_1^i r_2^j + r_2^i r_1^j) r_{12}^k, \quad (3.65)$$

or

$$\Phi_H(s, t, u) = e^{-\zeta s} \sum_{ijk} c_{ijk} s^i t^{2j} u^k, \quad (3.66)$$

with $s = r_1 + r_2$, $t = r_1 - r_2$, and $u = r_{12}$, and the coefficients c_{ijk} are variational parameters. The Hylleraas method is very accurate, and only a few terms in the expansion are required. Unfortunately it is only applicable to atomic systems with a few electrons.

An alternative approach is to introduce an explicit dependence on the interelectronic distance such that the cusp condition is verified. A general approach is to propose a wave function of the form $\Phi_{R12} = \gamma(\{r_{ij}\})\Phi_{CI}$, where γ is an appropriate correlating function. Possible expressions are

$$\gamma = 1 + \beta \sum_{i>j} r_{ij}, \tag{3.67}$$

$$\gamma = \prod_{i>j}(1 + \beta r_{ij}), \tag{3.68}$$

$$\gamma = e^{\beta \sum_{i>j} r_{ij}}. \tag{3.69}$$

Correlating functions of this type are also used in variational quantum Monte Carlo calculations, in particular *Jastrow-type factors* as in (3.69).

A minimal approach for a two-electron system would be to supplement the CI expansion with a single term proportional to r_{12} and to the dominant HF configuration $\Phi_{R12} = \Phi_{CI} + c_{12} r_{12} \Phi_{HF}$, where the coefficients of the original CI expansion can be retained or further optimized. The addition of a single linear term reduces the error in the energy by two orders of magnitude. This method can be applied not only to the CI wave function, but to any correlated calculation like MP or CC theory. In general they are called R12 methods, e.g., CI–R12 (Klopper *et al.*, 1991). Although explicitly correlated methods are potentially more accurate than the usual one-electron approaches, they have not yet reached the efficiency required to become widely adopted as a standard tool.

References

Bartlett, R. J. (1989). Coupled-cluster approach to molecular structure and spectra: a step toward predictive quantum chemistry. *J. Phys. Chem.* **93**, 1697–1708.

Davidson, E. R. (1975). The iterative calculation of a few of the lowest eigenvalues and corresponding eigenvectors of large real-symmetric matrices. *J. Comput. Phys.* **17**, 87–94.

Fetter, A. and Walecka, J. D. (1971). *Quantum Theory of Many-Particle Systems*. New York, McGraw-Hill.

Fock, V. (1930). Näherungsmethode zur Losung des quantenmechanischen Mehrkörperprobleme. *Z. Phys.* **61**, 126–148.

Hartree, D. R. (1928). The wave mechanics of an atom with a non-coulomb central field. I. Theory and methods. *Proc. Cambridge Phil. Soc.* **24**, 89–110.

Helgaker, T., Jørgensen, P., and Olsen, J. (2000). *Molecular Electronic-Structure Theory*. Chichester, Wiley.

Hylleraas, E. A. (1929). Calculation of the energy of helium. *Z. Phys.* **54**, 347–366.

Jensen, F. (1999). *Introduction to Computational Chemistry*. Chichester, Wiley.
Klopper, W., Röhse, R., and Kutzelnigg, W. (1991). CID and CEPA calculations with linear $r12$ terms. *Chem. Phys. Lett.* **178**, 455–461.
Koopmans, T. A. (1933). Über die Zuordnung von Wellenfunktionen und Eigenwerten zu den einzelnen Elektronen eines Atoms. *Physica* **1**, 104–113.
Messiah, A. (1961). *Quantum Mechanics*. Amsterdam, North Holland.
Møller, C. and Plesset, M. S. (1934). Note on an approximation treatment for many-electron systems. *Phys. Rev.* **46**, 618–622.
Pople, J. A. and Beveridge, D. L. (1970). *Approximate Molecular Orbital Theory*. Advanced Chemistry, London, McGraw-Hill.
Slater, J. C. (1928). The self-consistent field and the structure of atoms. *Phys. Rev.* **32**, 339–348.
    (1930). Note on Hartree's method. *Phys. Rev.* **35**, 210–211.
Szabo, A. and Ostlund, N. S. (1989). *Modern Quantum Chemistry*. New York, McGraw-Hill.

4
Density functional theory

Parallel to the approaches described in the previous chapter, a different line of thought drove L. H. Thomas and E. Fermi to propose, at about the same time as Hartree (1927–1928), that the full electronic density was the fundamental variable of the many-body problem. From this idea they derived a differential equation for the density without resorting to one-electron orbitals (Thomas, 1927; Fermi, 1928). The original *Thomas–Fermi* approximation was actually too crude, mainly because the approximation used for the kinetic energy of the electrons was unable to sustain bound states. However, it set up the basis for the later development of *density functional theory* (DFT), which has been the way of choice in electronic structure calculations in condensed matter physics during the past twenty years, and, recently, it also became accepted by the quantum chemistry community because of its computational advantages compared to post-Hartree–Fock methods of comparable quality.

This chapter is organized as follows: we first give an account of Thomas–Fermi theory and then present a modern approach to DFT that takes into account the formal properties of density functionals. We then move into the core of DFT by stating the basic theorems and developing the mathematical framework of orbital-based (Kohn–Sham) DFT. Next, we describe the most common approximations to exchange and correlation. A detailed analysis of these approximations and a general idea of which kind of systems can be safely treated within them is deferred to Chapter 5.

4.1 Thomas–Fermi theory

Thomas (1927), and independently Fermi (1928), gave a prescription for calculating the energy of an electronic system exclusively in terms of the electronic density. A thorough account of the historical perspective can be found in several articles and books (see, e.g., March, 1992). In their original work Thomas and

Fermi proposed an expression for the total electronic energy where the kinetic, exchange, and correlation contributions were taken from the homogeneous electron gas, for which good approximations were known. The idea was to construct the same quantities for the inhomogeneous system as

$$E_\alpha[\rho] = \int \rho(\mathbf{r}) \varepsilon_\alpha[\rho(\mathbf{r})]\, d\mathbf{r}, \tag{4.1}$$

where $\varepsilon_\alpha[\rho(\mathbf{r})]$ is the energy *density* of contribution α (kinetic, exchange, and correlation), calculated *locally* at the value assumed by the density at every point in space. This was the first time that the *local density approximation*, or LDA, was proposed. In the above expression the square brackets indicate a functional dependence of the energy and energy density on the electronic density.

For a homogeneous electron gas the electronic density is related to the Fermi energy (ϵ_F) by (see, e.g., Kittel, 1996)

$$\rho = \frac{1}{3\pi^2}\left(\frac{2m}{\hbar^2}\right)^{3/2} \epsilon_F^{3/2}, \tag{4.2}$$

and the kinetic energy is $T = 3\rho\epsilon_F/5$, so that the kinetic energy density is:

$$t[\rho] = \frac{3}{5}\frac{\hbar^2}{2m}(3\pi^2)^{2/3}\rho^{2/3}. \tag{4.3}$$

Therefore, the LDA kinetic energy is written

$$T_{TF} = C_k \int \rho(\mathbf{r})^{5/3}\, d\mathbf{r}, \tag{4.4}$$

with $C_k = 3(3\pi^2)^{2/3}/10 (= 2.871)$ hartree (the atomic unit of energy, equivalent to twice the ionization energy of the hydrogen atom, i.e. 27.21 eV). Exchange can be introduced into this picture in this same *local* spirit quite straightforwardly by considering Slater's expression for the homogeneous electron gas (Dirac, 1930; Slater, 1951)

$$E_X[\rho] = -C_X \int \rho(\mathbf{r})^{4/3}\, d\mathbf{r}, \tag{4.5}$$

with $C_X = 3(3/\pi)^{1/3}/4 (= 0.739)$ hartree. When exchange is treated at this level of approximation, the theory is called *Thomas–Fermi–Dirac* (TFD). Correlation can also be easily included by using any local approximation to the homogeneous electron gas, for instance the one proposed by Wigner (1938):

$$E_C[\rho] = -0.056 \int \frac{\rho(\mathbf{r})^{4/3}}{0.079 + \rho(\mathbf{r})^{1/3}}\, d\mathbf{r}, \tag{4.6}$$

where all the numerical constants are given in atomic units.

4.1 Thomas–Fermi theory

By replacing the above approximations into the general expression for the energy of an inhomogeneous electronic system given in (2.30), at the end of Chapter 2, we obtain TFD's energy expression:

$$E_{\text{TFD}}[\rho] = C_k \int \rho(\mathbf{r})^{5/3} d\mathbf{r} + \int \rho(\mathbf{r}) v_{\text{ext}}(\mathbf{r}) \, d\mathbf{r} + \frac{1}{2} \int \int \frac{\rho(\mathbf{r})\rho(\mathbf{r}')}{|\mathbf{r}-\mathbf{r}'|} d\mathbf{r} \, d\mathbf{r}'$$
$$- C_X \int \rho^{4/3}(\mathbf{r}) \, d\mathbf{r} + E_C[\rho]. \tag{4.7}$$

It can be seen that the only dependence of E_{TFD} on the electronic variables is through the electronic density. In that sense it is said that it is a *functional of the density*. Assuming intuitively some variational principle, we can search for the density $\rho(\mathbf{r})$ that minimizes $E_{\text{TFD}}[\rho]$, subject to the constraint that the total integrated charge be equal to the number of electrons: $\int \rho(\mathbf{r}) d\mathbf{r} = N$. Since the variation is not with respect to a parameter but a function, i.e. the density $\rho(\mathbf{r})$, the minimization assumes the form of the search for a function in three-dimensional space that makes the energy stationary with respect to any kind of density variations.

4.1.1 Functional derivation

The functional derivation idea is a generalization of the concept of partial derivation of a function of several variables to the case of a continuum of variables; in this case the value of the density at every point in space. If we want to minimize a function of several variables we have to look for stationary points, i.e. points where the partial derivatives vanish and the Hessian is positive definite. The generalization to the continuum implies that we should look for densities such that the derivative of the energy functional with respect to the value of the density at every point in space vanishes. This procedure is called functional derivation.

Suppose that $F[f]$ is a functional of $f(x)$, and we modify the function f by adding a small contribution $f(x) \to f(x) + \xi g(x)$. The change in the functional F is given by

$$\Delta F = F[f + \xi g] - F[f]. \tag{4.8}$$

If we discretize the variable x on a grid of points x_i with $i = 1, \ldots, n$, then the function f will be represented by its values on the grid $f_i = f(x_i)$. This can be visualized in Fig. 4.1, where a random variation of the values of the function at each grid point has been indicated with vertical bars. This generates another set of values $h_i = f_i + \xi g_i$.

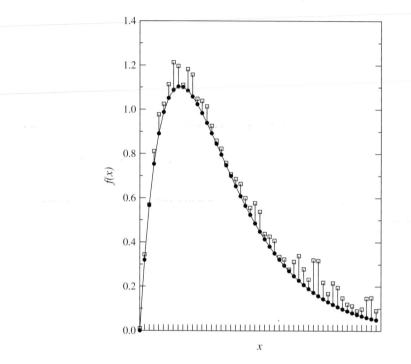

Fig. 4.1 A generic function $f(x)$ (solid line) discretized on a regular grid x_i (filled circles). After a variation of the values of the function at each grid point we have $h_i = f_i + \xi g_i$ (open squares). The vertical bars indicate the change in the values of the function.

With this discretization the functional $F[f]$ can be thought of as a function of the n variables f_i ($i = 1, \ldots, n$). Therefore, a small change in F can be written as

$$\Delta F = \sum_{i=1}^{n} \left(\frac{\partial F}{\partial f_i}\right) \Delta f_i = \xi \sum_{i=1}^{n} \left(\frac{\partial F}{\partial f_i}\right) g_i, \tag{4.9}$$

which for $\xi \to 0$ becomes

$$\lim_{\xi \to 0} \frac{\Delta F}{\xi} = \sum_{i=1}^{n} \left(\frac{\partial F}{\partial f_i}\right) g_i = \nabla F \cdot \mathbf{g}. \tag{4.10}$$

Going back to the continuum implies that we should consider grid points infinitesimally close, so that the derivative is carried out with respect to the value of the function at point x, i.e. $f(x)$. In order to make a distinction with respect to partial derivation, this derivative is indicated by $\delta F[f]/\delta f(x)$ and named *functional derivative*. In this limit the sum over grid points becomes an integral, and the change in the functional F becomes

$$\lim_{\xi \to 0} \frac{\Delta F}{\xi} = \int_{-\infty}^{\infty} \left(\frac{\delta F}{\delta f(x)}\right) g(x)\, dx. \tag{4.11}$$

4.1 Thomas–Fermi theory

A set of functional derivation rules analogous to those of usual derivation can be derived. Here we indicate a few cases that are important for our purposes.

$$\frac{\delta}{\delta f(x)}\left(\int g[f]dx\right) = \frac{\partial g[f]}{\partial f(x)},$$

$$\frac{\delta}{\delta f(x)}\left(\int g[f]f(x)dx\right) = \frac{\partial g[f]}{\partial f(x)}f(x) + g[f(x)],$$

$$\frac{\delta}{\delta f(x)}\left(\frac{1}{2}\int\int g(x,x')f(x)f(x')dxdx'\right) = \int g(x,x')f(x')dx',$$

where the function $f(x)$ is identified with the density $\rho(\mathbf{r})$. The second expression is useful for calculating the functional derivative of local density approximation terms such as the kinetic, exchange, and correlation terms in the Thomas–Fermi–Dirac approach, and the third expression is useful for Coulomb-type integrals when identifying $g(x, x')$ with the Coulomb potential $1/|\mathbf{r}-\mathbf{r}'|$.

The extension to higher-order functional derivatives is straightforward, and it is also possible to carry out a Taylor expansion in the functional derivatives as

$$F[f + \xi g] = F[f] + \xi \int \frac{\delta F[f]}{\delta f(x)} g(x) dx$$

$$+ \frac{\xi^2}{2}\int\int \frac{\delta^2 F[f]}{\delta f(x)\delta f(x')} g(x)g(x') dx dx' + \cdots. \quad (4.12)$$

In the case of Coulomb-type integrals we have:

$$\frac{\delta^2}{\delta f(x)\delta f(x')}\left(\frac{1}{2}\int\int g(x,x')f(x)f(x')dxdx'\right) = g(x,x'). \quad (4.13)$$

4.1.2 The Thomas–Fermi–Dirac equation

We now apply the rules of functional derivation to the Thomas–Fermi functional (4.7), subject to the constraint of a fixed number of electrons. This constraint is introduced via Lagrange multipliers, as is usually done in multi-dimensional constrained minimization problems. From the expression

$$\frac{\delta}{\delta \rho(\mathbf{r})}\left(E_{\text{TFD}}[\rho] - \mu \int \rho(\mathbf{r}) \, d\mathbf{r}\right) = 0 \quad (4.14)$$

we obtain

$$\mu = \frac{5}{3}C_k\rho(\mathbf{r})^{2/3} + v_{\text{ext}}(\mathbf{r}) + \int \frac{\rho(\mathbf{r}')}{|\mathbf{r}-\mathbf{r}'|} d\mathbf{r} - \frac{4}{3}C_X\rho(\mathbf{r})^{1/3} + \mu_c[\rho(\mathbf{r})], \quad (4.15)$$

with μ the chemical potential and $\mu_c[\rho] = \delta E_c[\rho]/\delta\rho(\mathbf{r})$ the correlation potential. This equation can be inverted to obtain the density as a *unique* function of the external potential. Actually, this form is not easy to invert directly due to the integral term and the non-integer powers of the density. In practice it is solved using iterative methods.

While the idea of thinking of the inhomogeneous system as locally homogeneous for the kinetic energy term is sensible for condensed phases, in particular simple metals, it represents a rather severe approximation for atomic and molecular systems. The total energies are poor, the density profile diverges at the nucleus and does not decay exponentially at long distances, and the atomic shell structure is absent. Improving over these features requires a better treatment of the kinetic functional. There is a class of *orbital-free* density functionals that have been developed along the years with this aim. These will be discussed in Section 10.7, but now we turn to the more usual approaches in modern DFT, where the kinetic energy is treated to a better accuracy at the expense of re-introducing one-particle orbitals.

4.2 Modern density functional theory

The Thomas–Fermi approach was developed in the hopes that the energy can in fact be written exclusively in terms of the electronic density. This idea, however, was intuitive at the time, but a proof that this was the case had to wait more than thirty years. In 1964, Hohenberg and Kohn (1964) formulated and proved a theorem that put on solid mathematical grounds the former ideas. The theorem is divided into two parts.

4.2.1 The Hohenberg–Kohn theorem

Theorem 1 The external potential is univocally determined by the electronic density, besides a trivial additive constant.

Proof We first suppose the opposite to hold, i.e. that the external potential is not univocally determined by the density. In this case we should be able to find two potentials v, v' such that their ground state density ρ is the same. Let Φ and $E_0 = \langle \Phi | \hat{H} | \Phi \rangle$ be the ground state wave function and ground state energy of the Hamiltonian $\hat{H} = \hat{T} + \hat{V}_{\text{ext}} + \hat{U}_{\text{ee}}$. Let also Φ' and $E_0' = \langle \Phi' | \hat{H}' | \Phi' \rangle$ be the ground state wave function and ground state energy of the Hamiltonian $\hat{H}' = \hat{T} + \hat{V}'_{\text{ext}} + \hat{U}_{\text{ee}}$. According to Rayleigh–Ritz's variational principle we have:

$$E_0 < \langle \Phi' | \hat{H} | \Phi' \rangle = \langle \Phi' | \hat{H}' | \Phi' \rangle + \langle \Phi' | \hat{H} - \hat{H}' | \Phi' \rangle$$
$$= E_0' + \int \rho(\mathbf{r}) \left[v_{\text{ext}}(\mathbf{r}) - v'_{\text{ext}}(\mathbf{r}) \right] d\mathbf{r}, \quad (4.16)$$

where we have used the fact that different Hamiltonians necessarily correspond to different ground states $\Phi \neq \Phi'$. This is straightforward to prove, since the potential is a multiplicative operator. Now we can simply exchange the roles of Φ and Φ' ($\hat{H}$ and $\hat{H}'$), and readily obtain:

$$E_0' < \langle \Phi | \hat{H}' | \Phi \rangle = \langle \Phi | \hat{H} | \Phi \rangle + \langle \Phi | \hat{H}' - \hat{H} | \Phi \rangle$$
$$= E_0 - \int \rho(\mathbf{r}) \left[v_{\text{ext}}(\mathbf{r}) - v_{\text{ext}}'(\mathbf{r}) \right] d\mathbf{r}. \quad (4.17)$$

Adding these two inequalities, it turns out that $E_0 + E_0' < E_0' + E_0$, which is absurd. Therefore, there cannot be $v_{\text{ext}}(\mathbf{r}) \neq v_{\text{ext}}'(\mathbf{r})$ that correspond to the same electronic density for the ground state, unless they differ by a trivial additive constant.

Corollary Since $\rho(\mathbf{r})$ univocally determines $v_{\text{ext}}(\mathbf{r})$, it also determines the ground state wave function Φ, which should be obtained by solving the full many-body Schrödinger equation.

Theorem 2 Let $\tilde{\rho}(\mathbf{r})$ be a non-negative density normalized to N. We define the variational energy E_v, which is a functional of the density because of the previous theorem, in the following way:

$$E_v[\tilde{\rho}] = F[\tilde{\rho}] + \int \tilde{\rho}(\mathbf{r}) v_{\text{ext}}(\mathbf{r}) d\mathbf{r}, \quad (4.18)$$

with

$$F[\tilde{\rho}] = \langle \Phi[\tilde{\rho}] | \hat{T} + \hat{U}_{\text{ee}} | \Phi[\tilde{\rho}] \rangle. \quad (4.19)$$

Here $\Phi[\tilde{\rho}]$ is the ground state of a potential which has $\tilde{\rho}$ as its ground state density, so that $E_0 = E_v[\rho]$ verifies

$$E_0 < E_v[\tilde{\rho}] \quad (4.20)$$

for any $\tilde{\rho} \neq \rho$, and is thus the ground state energy.

Proof Quite simply, we have

$$\langle \Phi[\tilde{\rho}] | \hat{H} | \Phi[\tilde{\rho}] \rangle = F[\tilde{\rho}] + \int \tilde{\rho}(\mathbf{r}) v_{\text{ext}}(\mathbf{r}) d\mathbf{r}$$
$$= E_v[\tilde{\rho}] \geq E_v[\rho] = E_0 = \langle \Phi[\rho] | \hat{H} | \Phi[\rho] \rangle. \quad (4.21)$$

The inequality follows from Rayleigh–Ritz's variational principle for the wave function, but applied to the electronic density. Therefore, the variational principle states that

$$\delta \left\{ E_v[\rho] - \mu \left(\int \rho(\mathbf{r}) d\mathbf{r} - N \right) \right\} = 0, \quad (4.22)$$

which leads to a generalization of the Thomas–Fermi equation:

$$\mu = \frac{\delta E_v[\rho]}{\delta \rho} = v_{\text{ext}}(\mathbf{r}) + \frac{\delta F[\rho]}{\delta \rho}. \tag{4.23}$$

The knowledge of $F[\rho]$ implies the knowledge of the solution of the full many-body Schrödinger equation. It has to be remarked that $F[\rho]$ is a *universal* functional, which does not depend explicitly on the external potential. It depends only on the electronic density. In the Hohenberg–Kohn formulation, $F[\rho] = \langle \Phi | \hat{T} + \hat{U}_{ee} | \Phi \rangle$, where Φ is the ground state many-body wave function. These two theorems form the mathematical basis of *density functional theory*, or DFT.

4.2.2 Constrained search formulation

In the Hohenberg–Kohn theorem the electronic density determines the external potential. However, it is also required that the density corresponds to some ground state antisymmetric wave function. While this is a necessary condition for the *true* density ρ, it may not be the case for other trial densities $\tilde{\rho}$. In fact, unacceptable densities can easily be obtained in a variational search strategy if this is not done carefully.

With this observation in mind, two decades later Levy (1982) reformulated DFT in such a way that the antisymmetric origin of the density is guaranteed. Levy used the constrained search method, which was then widely applied by several authors in similar contexts. The main idea is to redefine the universal functional $F[\rho]$ given by Expression (4.19) in the following way:

$$F[\rho] = \min_{\Phi \to \rho} \left\{ \langle \Phi | \hat{T} + \hat{U}_{ee} | \Phi \rangle \right\}, \tag{4.24}$$

where ρ is any non-negative density such that

$$\int \rho(\mathbf{r}) d\mathbf{r} = N \quad \text{and} \quad \int |\nabla \rho^{1/2}(\mathbf{r})|^2 d\mathbf{r} < \infty, \tag{4.25}$$

with the additional constraint that the density should arise from an antisymmetric wave function. The search is thus constrained to the subspace of all the antisymmetric Φ that give rise to the *same* density ρ, thus eliminating the conceptual difficulty of possible unphysical densities.

4.2.3 Ground and excited states in DFT

Using DFT one can determine the electronic ground state density and energy exactly, provided that $F[\rho]$ is known. A common misconception is that DFT is a ground state theory, and that the question of excited states cannot be addressed within it. This is actually an incorrect statement. In fact, since the density determines the potential univocally, by solving the full many-body Schrödinger equation one can determine univocally the many-body wave functions, *ground* and *excited* states. The problem is practical rather than conceptual, because of the inherent difficulty in solving the full many-body problem.

A practical scheme for determining the ground state was devised by Kohn and Sham (1965), and will be discussed at length in the next subsection. For excited states there are a number of extensions and generalizations of Kohn–Sham theory, which have been used with different degrees of success. The main roads explored up to date are ensemble DFT (Theophilou, 1979), time-dependent density functional theory (TDDFT) (Petersilka *et al.*, 1996), Green's function methods and in particular the GW approximation (Hedin, 1965), and generalizations of the Kohn–Sham formalism (Görling, 1996). This topic will not be addressed any further in this book, and the interested reader is referred to the original literature.

4.2.4 The Kohn–Sham equations

The general features of the electron–electron interaction U_{ee} have already been discussed in Section 2.3. There, it was argued that a good strategy is to separate the classical electrostatic energy (the so-called Hartree term) from the exchange and correlation contributions, as in Eqs. (2.30) and (2.31). This strategy is useful because it divides the generally unknown electron–electron interaction energy into pieces of decreasing importance from the energetic point of view: Hartree, exchange, and correlation. More importantly, the Hartree term, which is by far the largest contribution, is just the classical electrostatic energy, which is known exactly. The second largest, i.e. the exchange term, is also well-known and, in principle, can also be calculated exactly as in Hartree–Fock theory. In practice, however, for computational reasons most of the time this term is approximated. Eventually, all the ignorance about the electronic many-body problem is displaced to the smallest contribution, namely the correlation term. Dealing with correlation is, in fact, the biggest difficulty. This is an active field of research, which has produced significant improvements over the past decades, and will be discussed at length in Chapter 5. Here we just mention that reasonably good approximations exist for a large class of systems of interest.

The main problem at this stage is with the kinetic energy $T = \langle \Phi | \hat{T} | \Phi \rangle$ because its explicit expression in terms of the electronic density is not known. According to (2.23) the exact calculation of the kinetic energy term requires the knowledge of the Laplacian of the one-body density matrix, which is not related to the density in an obvious manner. The Thomas–Fermi expression is local in the density, which is a severe limitation because this model does not hold bound states, and also the atomic shell structure is absent. The main problem with this approach is that the kinetic operator is inherently non-local. This can be easily seen by considering the first-order finite difference expression of the Laplacian of a function f in a discrete one-dimensional grid:

$$\left(\nabla^2 f\right)_{x_i} = \frac{f(x_{i+1}) + f(x_{i-1}) - 2f(x_i)}{\Delta^2}, \qquad (4.26)$$

where $\{x_i\}$ are the grid points and Δ is the spacing between them. More accurate expressions that extend beyond nearest neighbors are available, and they show that the contribution from more distant neighbors becomes increasingly smaller. This analysis tells us that the kinetic operator, although non-local, is quite near-sighted. Therefore, semi-local gradient expansions such as those that will be discussed in Section 10.7 are not unlikely to be successful.

A more general approach was suggested by Kohn and Sham (1965). It starts from the observation that a system of non-interacting electrons is exactly described by an antisymmetric wave function of the Slater determinant type, made of one-electron orbitals. As in Hartree–Fock theory, for such a wave function the kinetic energy can be easily obtained in terms of the one-electron orbitals (see Section 3.1). In this case the ground state density matrix $\rho_1(\mathbf{r}, \mathbf{r}')$ is given by

$$\rho_1(\mathbf{r}, \mathbf{r}') = \sum_{i=1}^{\infty} f_i \varphi_i(\mathbf{r}) \varphi_i^*(\mathbf{r}'), \qquad (4.27)$$

where $\varphi_i(\mathbf{r})$ are the one-electron orbitals and f_i are the occupation numbers corresponding to these orbitals. Then, according to (2.23), the exact expression for the kinetic energy of non-interacting electrons is

$$T = -\frac{\hbar^2}{2m} \sum_{i=1}^{\infty} f_i \langle \varphi_i | \nabla^2 | \varphi_i \rangle. \qquad (4.28)$$

Kohn and Sham's idea was that, if one can find a system of non-interacting electrons that produces the same electronic density of the interacting system, then the kinetic energy of the non-interacting system can be calculated exactly via (4.28). Of course, this is not the exact kinetic energy of the interacting system. The missing fraction is due to the fact that the true many-body wave function is not a Slater determinant. There is then a correlation contribution to the kinetic energy that is not taken into account, which must be included in the correlation energy term. The relative magnitude of the non-interacting and correlation contributions to the kinetic energy will be discussed below.

In the following we assume that the equivalent non-interacting system, i.e. a system of non-interacting electrons whose ground state density coincides with that

4.2 Modern density functional theory

of the interacting system, does exist. This system will be called the *non-interacting reference system* of density $\rho(\mathbf{r})$, and is described by the Hamiltonian

$$\hat{\mathcal{H}}_R = \sum_{i=1}^{N}\left[-\frac{\hbar^2}{2m}\nabla_i^2 + v_R(\mathbf{r}_i)\right], \quad (4.29)$$

with N the number of electrons. Here, the potential $v_R(\mathbf{r})$, or *reference potential*, is such that the ground state density of $\hat{H}_R$ equals $\rho(\mathbf{r})$. If that is the case, Hohenberg–Kohn's theorem ensures that the ground state energy equals the energy of the interacting system.

This Hamiltonian has no electron–electron interactions. Therefore, its eigenstates can be expressed in the form of Slater determinants

$$\Phi(\mathbf{r}) = \frac{1}{\sqrt{N_s!}} \mathrm{SD}\left[\varphi_1(\mathbf{r}_1)\varphi_2(\mathbf{r}_2)\cdots\varphi_{N_s}(\mathbf{r}_{N_s})\right], \quad (4.30)$$

where we have chosen the closed shell situation, where the occupation numbers are 2 for $i \leq N_s$ and 0 for $i > N_s$, with $N_s = N/2$ the number of doubly occupied orbitals. For simplicity we are ignoring a possible spin dependence. This would arise, e.g., in magnetic or open shell systems. Spin-dependent expressions, however, are straightforward extensions to the above and will be presented in Section 4.2.5. Within this assumption, the density reads

$$\rho(\mathbf{r}) = 2\sum_{i=1}^{N_s}|\varphi_i(\mathbf{r})|^2, \quad (4.31)$$

while the kinetic term is

$$T_R[\rho] = -\frac{\hbar^2}{m}\sum_{i=1}^{N_s}\langle\varphi_i|\nabla^2|\varphi_i\rangle. \quad (4.32)$$

The single-particle orbitals $\varphi_i(\mathbf{r})$ are the N_s lowest-energy eigenfunctions of the one-electron Hamiltonian

$$\hat{H}_{KS} = -\frac{\hbar^2}{2m}\nabla^2 + v_R(\mathbf{r}), \quad (4.33)$$

which are obtained by solving the one-electron Schrödinger equation

$$\hat{H}_{KS}\,\varphi_i(\mathbf{r}) = \varepsilon_i\,\varphi_i(\mathbf{r}). \quad (4.34)$$

Using $T_R[\rho]$, the universal density functional can be re-written in the following form:

$$F[\rho] = T_R[\rho] + \frac{1}{2}\int\int\frac{\rho(\mathbf{r})\rho(\mathbf{r}')}{|\mathbf{r}-\mathbf{r}'|}d\mathbf{r}\,d\mathbf{r}' + \tilde{E}_{XC}[\rho], \quad (4.35)$$

which defines a modified exchange and correlation energy $\tilde{E}_{XC}$, different from the E_{XC} given by (2.31) in that it accounts also for the kinetic correlation ignored in $T_R[\rho]$.

By substituting this expression for F into the total energy functional $E_v[\rho] = F[\rho] + \int \rho(\mathbf{r})v_{ext}(\mathbf{r})d\mathbf{r}$, we finally obtain the Kohn–Sham (KS) functional:

$$E_{KS}[\rho] = T_R[\rho] + \int \rho(\mathbf{r})v_{ext}(\mathbf{r})\,d\mathbf{r} + \frac{1}{2}\int\int \frac{\rho(\mathbf{r})\rho(\mathbf{r}')}{|\mathbf{r}-\mathbf{r}'|} d\mathbf{r}\,d\mathbf{r}'$$
$$+ \tilde{E}_{XC}[\rho]. \qquad (4.36)$$

In this way the energy functional is expressed in terms of the N_s orbitals that minimize the non-interacting electronic kinetic energy under the fixed density constraint. The one-electron orbitals are usually called the *Kohn–Sham orbitals*.

The Kohn–Sham orbitals are the result of a mathematical construction devised in order to simplify the problem. In principle they do not have an obvious meaning by themselves, but only in that they are used to construct the density according to (4.31). In practice, however, it is customary to think of them as single-particle physical eigenstates. On the other hand, it is also usual to hear that the Kohn–Sham orbitals are meaningless, and cannot be identified as single-particle eigenstates corresponding to electronic excitations. The truth is actually in between. A rigorous perturbative treatment due to Görling (1996) shows that Kohn–Sham eigenvalue differences are a well-defined approximation to excitation energies, to order zero in the electron–electron interaction. The first-order correction is due exclusively to exchange, while correlation enters at second and higher orders.

The Kohn–Sham orbitals satisfy the one-electron Kohn–Sham equations (4.34), but so far we do not know what the reference potential v_R is (or the Kohn–Sham potential v_{KS}, as it is also known). What we know is that v_R is a potential that ensures that the density of the non-interacting reference system is the same as the true density of the interacting system. It should then be possible to determine it by minimizing the KS functional (4.36) with respect to the density, under the constraint that this density integrates to N particles. The variational principle is the same as for the Thomas–Fermi functional (4.14), but now applied to Kohn–Sham's functional

$$\frac{\delta}{\delta\rho(\mathbf{r})}\left(E_{KS}[\rho] - \mu\int\rho(\mathbf{r})\,d\mathbf{r}\right) = 0. \qquad (4.37)$$

Using the rules of functional derivation, we obtain the following equation for the minimizing ground state density:

$$\frac{\delta T_R[\rho]}{\delta\rho(\mathbf{r})} + v_{ext}(\mathbf{r}) + \int \frac{\rho(\mathbf{r}')}{|\mathbf{r}-\mathbf{r}'|}d\mathbf{r}' + \frac{\delta \tilde{E}_{XC}[\rho]}{\delta\rho(\mathbf{r})} = \mu. \qquad (4.38)$$

4.2 Modern density functional theory

The functional derivative $\delta T_R[\rho]/\delta\rho(\mathbf{r})$ can be readily obtained by considering the non-interacting Hamiltonian $\hat{H}_R$ of (4.29). Since the particles in the reference system only interact with the reference potential, and not between themselves, this Hamiltonian corresponds to the energy functional

$$E_{v_R}[\tilde{\rho}] = T_R[\tilde{\rho}] + \int \tilde{\rho}(\mathbf{r})v_R(\mathbf{r})\,d\mathbf{r}, \tag{4.39}$$

whose ground state energy is the same as that of the interacting system because they share the same electronic density. Therefore, in general $E_{v_R}[\tilde{\rho}] \geq E_0$ and the equality is verified only for the ground state density ρ. This means that the functional derivative of $E_{v_R}[\tilde{\rho}]$ must vanish for the ground state density. Applying the variational principle (4.37) to $E_{v_R}[\tilde{\rho}]$, we obtain

$$\frac{\delta T_R[\rho]}{\delta\rho(\mathbf{r})} + v_R(\mathbf{r}) = \mu_R, \tag{4.40}$$

where μ_R is the chemical potential of the non-interacting system, which should coincide with that of the interacting system μ. Otherwise, if the interacting and the equivalent non-interacting reference system were put into contact, there would be charge flow from one to the other.

By comparing (4.38) and (4.40), we obtain the following expression for the reference potential:

$$v_R(\mathbf{r}) = v_{\text{ext}}(\mathbf{r}) + \int \frac{\rho(\mathbf{r}')}{|\mathbf{r}-\mathbf{r}'|}\,d\mathbf{r}' + \mu_{\text{XC}}[\rho](\mathbf{r}) \tag{4.41}$$

with

$$\mu_{\text{XC}}[\rho](\mathbf{r}) = \frac{\delta\tilde{E}_{\text{XC}}[\rho]}{\delta\rho(\mathbf{r})}. \tag{4.42}$$

Notice that, as in Hartree and Hartree–Fock theories, the reference potential depends on the solutions of the one-electron Schrödinger equation (the Kohn–Sham orbitals) through the electronic density. Therefore, this equation has to be solved self-consistently, making sure that the density used to construct the reference potential coincides with that obtained from the solutions of the equation via (4.31).

4.2.5 Extension to spin-polarized systems

In order to extend Kohn–Sham theory to spin-polarized systems it is sufficient to consider the electronic density as composed by two independent spin densities,

$\rho = \rho_\uparrow + \rho_\downarrow$. Each one of these densities is constructed with the Kohn–Sham spin orbitals, which satisfy the self-consistent Kohn–Sham equations

$$\left\{-\frac{\hbar^2}{2m}\nabla^2 + v_{R,s}(\mathbf{r})\right\} \varphi_{i,s}(\mathbf{r}) = \varepsilon_{i,s}\, \varphi_{i,s}(\mathbf{r}), \quad (4.43)$$

where the subindex s indicates the spin component ($\uparrow$ or $\downarrow$). The reference potentials $v_{R,s}$ are obtained as an extension of Eqs. (4.41) and (4.42) to the spin-dependent case

$$v_{R,s}(\mathbf{r}) = v_{\text{ext}}(\mathbf{r}) + \int \frac{\rho(\mathbf{r}')}{|\mathbf{r}-\mathbf{r}'|}\, d\mathbf{r}' + \mu_{XC,s}[\rho_\uparrow, \rho_\downarrow](\mathbf{r}), \quad (4.44)$$

with

$$\mu_{XC,s}[\rho_\uparrow, \rho_\downarrow](\mathbf{r}) = \frac{\delta \tilde{E}_{XC}[\rho_\uparrow, \rho_\downarrow]}{\delta \rho_s(\mathbf{r})}. \quad (4.45)$$

Now the exchange-correlation energy and potential depend on the spin densities, which are constructed via the Kohn–Sham spin orbitals according to

$$\rho_s(\mathbf{r}) = \sum_{i=1}^{N_s} |\varphi_{i,s}(\mathbf{r})|^2, \quad (4.46)$$

with N_s the number of occupied spin orbitals with spin projection s.

Generally, exchange-correlation functionals are given in terms of the total electronic density ρ and spin-polarization (or magnetization) density ζ, which are defined as

$$\rho(\mathbf{r}) = \rho_\uparrow(\mathbf{r}) + \rho_\downarrow(\mathbf{r}) \quad (4.47)$$

and

$$\zeta(\mathbf{r}) = \rho_\uparrow(\mathbf{r}) - \rho_\downarrow(\mathbf{r}). \quad (4.48)$$

The density given by expressions (4.46) and (4.47) involves two different numbers of electrons $N_\uparrow$ and $N_\downarrow$, such that the sum $N = N_\uparrow + N_\downarrow$ is the total number of electrons in the system. Notice that it is not necessary that $N_\uparrow = N_\downarrow$. In fact, magnetic and open-shell systems have unbalanced spin projections. The values of N_s have to be determined according to the single-particle eigenvalues, by asking for the lowest N to be occupied. This defines a Fermi energy ε_F such that the occupied eigenstates have $\varepsilon_{i,s} < \varepsilon_F$. This approach is known as *spin density functional theory* (SDFT). In the case of non-spin-polarized systems $\rho_\uparrow(\mathbf{r}) = \rho_\downarrow(\mathbf{r})$, and SDFT reduces to the simpler case of DFT with double occupancy of the single-particle orbitals. In SDFT the total energy is written as

$$E_{KS}[\rho_\uparrow,\rho_\downarrow] = T_R[\rho_\uparrow,\rho_\downarrow] + \int \rho(\mathbf{r})v_{ext}(\mathbf{r})d\mathbf{r} + \frac{1}{2}\int\int \frac{\rho(\mathbf{r})\rho(\mathbf{r}')}{|\mathbf{r}-\mathbf{r}'|}d\mathbf{r}\,d\mathbf{r}'$$
$$+\tilde{E}_{XC}[\rho_\uparrow,\rho_\downarrow], \qquad (4.49)$$

with

$$T_R[\rho_\uparrow,\rho_\downarrow] = -\frac{\hbar^2}{2m}\sum_{s=1}^{2}\sum_{i=1}^{N_s}\langle\varphi_{i,s}|\nabla^2|\varphi_{i,s}\rangle. \qquad (4.50)$$

The similitude between the Kohn–Sham and Hartree equations (3.14) is remarkable. In fact, the solution of the Kohn–Sham equations must be obtained by an iterative procedure in the same way as for the Hartree and Hartree–Fock equations. As in these methods, the total energy cannot be written simply as the sum of the eigenvalues $\varepsilon_{i,s}$. Double counting terms have to be subtracted:

$$E_{KS}[\rho_\uparrow,\rho_\downarrow] = \sum_{s=1}^{2}\sum_{i=1}^{N_s}\varepsilon_{i,s} - \frac{1}{2}\int\int\frac{\rho(\mathbf{r})\rho(\mathbf{r}')}{|\mathbf{r}-\mathbf{r}'|}d\mathbf{r}\,d\mathbf{r}'$$
$$+\left\{\tilde{E}_{XC}[\rho_\uparrow,\rho_\downarrow] - \int \rho(\mathbf{r})\mu_{XC}[\rho_\uparrow,\rho_\downarrow](\mathbf{r})\,d\mathbf{r}\right\}, \qquad (4.51)$$

and a similar expression for the unpolarized case, where the sum over spin components is replaced by a factor of 2, and the spin-polarized expressions for the exchange-correlation potential are replaced by the unpolarized ones.

Spin-density functional theory assumes that the projection of the total spin of the system is a good quantum number. There are cases, however, when this is not the case. Non-collinear magnetism can appear in disordered or low-symmetry structures, and also in the form of spiral phases such as γ-Fe, and in Fe clusters (Oda *et al.*, 1998). The theory of non-collinear magnetism, developed by von Barth and Hedin (1972), is based on a 2×2 spinorial representation of the spin density

$$\rho_{\alpha\beta}(\mathbf{r}) = \frac{1}{2}\rho(\mathbf{r})\delta_{\alpha\beta} + \frac{1}{2}\sum_{i=1}^{3}\zeta_i(\mathbf{r})\sigma^i_{\alpha\beta}, \qquad (4.52)$$

where σ^i are the Pauli spin matrices and $\zeta_i(\mathbf{r})$ are the Cartesian components of the spin polarization vector.

4.3 Kinetic correlation: the adiabatic connection

So far, no approximations to the electronic problem have been introduced. DFT is an exact theory for the ground state and, in principle, it can also be extended to excited states. What has been done is to confine all the ignorance about the many-electron problem to the smallest possible energetic contribution represented by the term $\tilde{E}_{XC}[\rho]$, while the remaining energy terms are well-known. Exchange and correlation functionals will be discussed in the following chapter, but now we come back to the question of how good an approximation is $T_R[\rho]$ to the exact $T[\rho]$, and how $E_{XC}[\rho]$ has to be modified to account for the correlation missing in the non-interacting kinetic energy.

The two quantities $T_R[\rho]$ and $T[\rho]$ are the expectation values of the kinetic operator, but in different states. The non-interacting $T_R[\rho]$ corresponds to the expectation value of the kinetic operator in its own ground state $|\Phi_T\rangle$, i.e.

$$T_R[\rho] = \min_{\Phi_T \to \rho} \left\{ \langle \Phi_T | \hat{T} | \Phi_T \rangle \right\}. \tag{4.53}$$

This is because in this representation there are no explicit electron–electron interactions, but only the interaction with the *external* reference potential. This latter, though, includes the electron–electron interaction in the form of a self-consistent density dependence.

On the other side, the interacting $T[\rho]$ corresponds to the ground state of the full Hamiltonian $|\Phi_0\rangle$, including the electron–electron interaction:

$$T[\rho] = \langle \Phi_0 | \hat{T} | \Phi_0 \rangle, \tag{4.54}$$

where $\Phi_0 \neq \Phi_T$ is not the ground state of the kinetic operator. Therefore, for variational reasons it must be that $T_R[\rho] \leq T[\rho]$, and thus the correlation term $\tilde{E}_C[\rho]$ contains a positive contribution arising from kinetic correlations.

The original definition (2.31) of the exchange-correlation energy, which does not contain kinetic contributions, can be used only if the exact expression for the kinetic energy is known. However, this is not the case. In Kohn–Sham theory the non-interacting expression for the kinetic energy is used, and then the exchange-correlation term is redefined as

$$\tilde{E}_{XC}[\rho] = E_{XC}[\rho] + T[\rho] - T_R[\rho]. \tag{4.55}$$

The kinetic contribution to the exchange term is given by Pauli's principle, and this is already contained in $T_R[\rho]$ and in the density when adding up the contributions of the N_s, or N, lowest eigenstates according to (4.31) and (4.32), or (4.46) and (4.50) for spin polarized systems. Therefore, the exchange term is not modified by the introduction of the non-interacting reference system.

One way to address the issue of the kinetic contribution to the correlation energy is to introduce the concept of *adiabatic connection*. The main idea, which

4.3 Kinetic correlation: the adiabatic connection

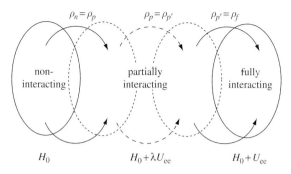

Fig. 4.2 Pictorial representation of the adiabatic connection between the non-interacting reference system and the fully interacting system. These are connected by a continuous sequence of partially interacting states described by Hamiltonian $\hat{\mathcal{H}}_\lambda[\rho]$. The parameter λ varies from 0 to 1, but the density ρ remains always invariant.

is shown schematically in Fig. 4.2, is to switch the electron–electron interaction gradually from the non-interacting reference system towards the fully interacting system. This is done by introducing a switching parameter λ that multiplies the interaction term in the Hamiltonian (Langreth and Perdew, 1977):

$$\hat{\mathcal{H}}_\lambda[\rho] = \hat{T}[\rho] + \hat{V}_{\text{ext}}[\rho] + \lambda \hat{U}_{\text{ee}}[\rho], \tag{4.56}$$

so that $\lambda = 0$ corresponds to the non-interacting system and $\lambda = 1$ represents the fully interacting system. The switching is done in such a way that the electronic density ρ is unchanged all along the connecting path.

This construction leads to the anticipated modification of the correlation term that includes kinetic correlations. This latter can be obtained by averaging the pair correlation function $g(\mathbf{r}, \mathbf{r}')$ over the strength of the electron–electron interaction λ (Jones and Gunnarsson, 1989), i.e.

$$\tilde{E}_{\text{XC}}[\rho] = \frac{1}{2} \int \int \frac{\rho(\mathbf{r})\rho(\mathbf{r}')}{|\mathbf{r} - \mathbf{r}'|} \left[\tilde{g}(\mathbf{r}, \mathbf{r}') - 1\right] d\mathbf{r} \, d\mathbf{r}', \tag{4.57}$$

where

$$\tilde{g}(\mathbf{r}, \mathbf{r}') = \int_0^1 g_\lambda(\mathbf{r}, \mathbf{r}') \, d\lambda \tag{4.58}$$

and $g_\lambda(\mathbf{r}, \mathbf{r}')$ is the pair correlation function corresponding to the Hamiltonian (4.56), for varying values of λ. Separating exchange and correlation contributions:

$$\tilde{g}(\mathbf{r}, \mathbf{r}') = g_{\text{X}}(\mathbf{r}, \mathbf{r}') + \tilde{g}_{\text{C}}(\mathbf{r}, \mathbf{r}'), \tag{4.59}$$

we have that the exchange contribution is

$$g_X(\mathbf{r},\mathbf{r}') = 1 - \frac{|\rho_1(\mathbf{r},\mathbf{r}')|^2}{\rho(\mathbf{r})\rho(\mathbf{r}')}, \quad (4.60)$$

with $\rho_1(\mathbf{r},\mathbf{r}')$ the one-body density matrix, which in general is a non-diagonal operator. This expression is the same as in (2.29), but now the density and density matrix are evaluated using the Kohn–Sham orbitals instead of the Hartree–Fock ones. The spin-dependent expression is obtained simply by replacing the numerator in the second term of (4.60) with $\sum_{s=1}^{2}|\rho_s(\mathbf{r},\mathbf{r}')|^2$.

For the homogeneous electron gas the expression for the density matrix is well known and the exchange contribution to $\tilde{g}(\mathbf{r},\mathbf{r}')$ assumes the following analytic closed form:

$$g_X(\mathbf{r},\mathbf{r}') = g_X(|\mathbf{r}-\mathbf{r}'|,\rho) = 1 - \frac{9}{2}\left[\frac{j_1(k_F|\mathbf{r}-\mathbf{r}'|)}{k_F|\mathbf{r}-\mathbf{r}'|}\right]^2, \quad (4.61)$$

where $j_1(x) = [\sin(x) - x\cos(x)]/x^2$ is the spherical Bessel function of order one and k_F is the Fermi momentum, which is related to the uniform density by $k_F = (3\pi^2\rho)^{1/3}$.

Figure 4.3 shows the shape of the non-oscillatory part of the pair distribution function $g(r)$ and its coupling constant average $\tilde{g}(r)$ for the unpolarized (left panel) and fully polarized (right panel) uniform electron gas of density parameter $r_s = 2$, typical of metallic systems. It is clear that the kinetic contribution to the pair correlation function is a rather small correction, at least for the homogeneous electron gas. This result is not very sensitive to the density, as shown in Perdew and Wang (1992).

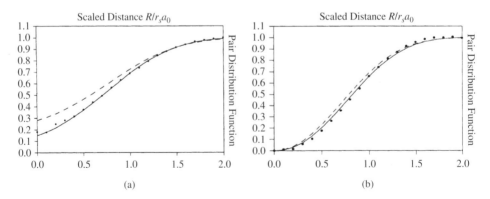

Fig. 4.3 Pair correlation function and its coupling constant average for the uniform electron gas for $r_s = 2$ and $\zeta = 0$ (left panel) and $\zeta = 1$ (right panel). Figures 3 and 6 reprinted with permission from J. P. Perdew and Y. Wang, *Phys. Rev. B* **46**, 12947 (1992). Copyright (1992) by the American Physical Society.

4.3.1 The exchange-correlation hole

We now wish to interpret the exchange-correlation energy $\tilde{E}_{XC}[\rho]$ as the Coulomb interaction between the electronic density and some displaced charge density. This can be done by defining the *exchange-correlation hole* in the following way:

$$\tilde{\rho}_{XC}(\mathbf{r}, \mathbf{r}') = \rho(\mathbf{r}')\left[\tilde{g}(\mathbf{r}, \mathbf{r}') - 1\right], \quad (4.62)$$

so that the exchange-correlation energy is written

$$\tilde{E}_{XC}[\rho] = \frac{1}{2} \int \int \frac{\rho(\mathbf{r})\tilde{\rho}_{XC}(\mathbf{r}, \mathbf{r}')}{|\mathbf{r} - \mathbf{r}'|} d\mathbf{r} \, d\mathbf{r}'. \quad (4.63)$$

The XC hole $\tilde{\rho}_{XC}$ represents a fictitious charge depletion due to exchange and correlation effects, i.e. due to the fact that the presence of an electron at $\mathbf{r}$ reduces the probability of finding a second electron at $\mathbf{r}'$ in the vicinity of $\mathbf{r}$. It corrects for the fact that the Hartree contribution to the energy completely ignores this depletion. Actually, $\tilde{\rho}_{XC}(\mathbf{r}, \mathbf{r}')$ is the exchange-correlation hole averaged over the strength of the interaction, which takes into account kinetic correlations. It is also important to notice that the exchange-correlation hole is a non-local object, where the exchange-correlation charge density at $\mathbf{r}'$ depends on $\mathbf{r}$, i.e. on where the electronic density is evaluated.

The properties of $\tilde{g}(\mathbf{r}, \mathbf{r}')$ and $\tilde{\rho}_{XC}(\mathbf{r}, \mathbf{r}')$ are very interesting and instructive:

(i) $\tilde{g}(\mathbf{r}, \mathbf{r}') = \tilde{g}(\mathbf{r}', \mathbf{r})$, i.e. $\tilde{g}$ is symmetric under exchange of $\mathbf{r}$ and $\mathbf{r}'$.
(ii) $\int \tilde{g}(\mathbf{r}, \mathbf{r}')\rho(\mathbf{r}') d\mathbf{r}' = \int \tilde{g}(\mathbf{r}, \mathbf{r}')\rho(\mathbf{r}) d\mathbf{r} = N - 1$ (normalization condition).
(iii) $\int \tilde{\rho}_{XC}(\mathbf{r}, \mathbf{r}') d\mathbf{r}' = \int \tilde{\rho}_{XC}(\mathbf{r}, \mathbf{r}') d\mathbf{r} = -1$. This means that the exchange-correlation hole contains exactly *one* displaced electron. This sum rule is extremely important, and must be verified by any reasonable approximation to $\tilde{\rho}_{XC}(\mathbf{r}, \mathbf{r}')$.

If we separate the exchange and correlation contributions it is easy to see that the displaced electron arises exclusively from the exchange part. This is a consequence of how the electron–electron interaction has been separated. In the Hartree term we have included the interaction of the electron with itself. This unphysical contribution is exactly canceled by the exchange interaction of the full charge density with the displaced density. However, exchange is more than that. It is a non-local operator whose local component is the negative of the self-interaction.

The correlation hole must then integrate to zero, $\int \tilde{\rho}_C(\mathbf{r}, \mathbf{r}') d\mathbf{r}' = 0$. Therefore, the correlation energy corresponds to the interaction of the charge density with a neutral charge distribution.

4.4 Some observations about Kohn–Sham theory

Kohn–Sham theory was proposed as an alternative to Thomas–Fermi-like approaches, where the kinetic energy was approximated as a local term in the density. While this latter works reasonably well when the electronic density is smooth, as in alkali metals, in general it is quite poor. The reason is that the non-local character of the kinetic term is completely ignored. This non-locality is short-ranged, but the local component is by no means larger than the rest.

The introduction of the non-interacting reference system permits us to account exactly for the most important part of the kinetic energy. The missing part (correlation) is due to the fact that the full many-body wave function is not a single Slater determinant, otherwise Hartree–Fock theory would be exact. In the previous section it has been shown that $T_R[\rho]$ is actually the most important part of $T[\rho]$. The price to pay for having a good description of the kinetic energy is that, instead of solving a single equation for the density in terms of the external potential, an eigenvalue problem has to be solved for the N lowest eigenstates. The latter is computationally much more expensive than inverting a single equation. A thorough analysis of DFT and Kohn–Sham theory has been carried out by Jones and Gunnarsson (1989). Some observations about Kohn–Sham theory are discussed below.

(i) The main difference between the Kohn–Sham and Hartree equations is that the effective potential (here called the reference potential) includes exchange and correlation. Therefore, the computational cost of Kohn–Sham is of the same order as Hartree, but much less than Hartree–Fock, which includes the exact non-local exchange.

(ii) The self-interaction correction term present in the Hartree equations is absent in Kohn–Sham theory. Self-interaction, as in Hartree–Fock theory, should be cancelled by the exchange term. The price paid in the Hartree approach for eliminating the self-interaction is that the effective potential becomes state-dependent. This means that the orbitals do not correspond to the same effective Hamiltonian, and thus a set of N coupled differential equations has to be solved. In Kohn–Sham theory the potential is the same for all states, and thus the problem can be solved in terms of a single eigenvalue equation.

(iii) There is actually nothing wrong with having state-dependent potentials. In fact, Kohn–Sham orbitals are also univocally determined by the electronic density, and are thus functionals of it. Therefore, orbital-dependent potentials are still functionals of the density, and find a place within the realm of DFT. In fact, a self-interaction corrected (SIC), orbital-dependent version of DFT was proposed in the eighties (Perdew and Zunger, 1981), but it was not widely used. This idea was later revived in the context of SIC pseudopotentials, since electronic self-interaction is more important for core than valence electrons (Vogel et al., 1995).

4.4 Some observations about Kohn–Sham theory   71

(iv) *The true many-body wave function of the interacting system is not the Slater determinant of Kohn–Sham orbitals.* It is, however, determined by the density, and thus by the Kohn–Sham orbitals used to construct the density.

(v) The correlation functional has to be modified to account for the missing part in the kinetic energy $T_R[\rho]$, which corresponds to a non-interacting system. The exchange functional remains unchanged.

(vi) At variance with Hartree–Fock (see Section 3.1), in DFT there is no Koopmans' theorem to identify the difference between the ground state energy of an $(N+1)$-electron system and that of an N-electron system, with the Kohn–Sham eigenvalues (Koopmans, 1933). The energy difference is actually obtained by integrating the eigenvalue for a partially occupied state, from no occupancy to full occupancy (Jones and Gunnarsson, 1989).

(vii) However, as in Hartree–Fock theory, Janak's theorem for the ionization potential of molecules is valid (Janak, 1978). If the reference potential vanishes at long distances (as for molecules), the ionization energy is given by the highest occupied Kohn–Sham eigenstate: $I = -\mu = -\varepsilon_N$. In general, the Kohn–Sham eigenvalues are defined as the derivatives of the total energy with respect to the occupation numbers: $\varepsilon_{i,s} = \partial E/\partial n_{i,s}$. The eigenvalue corresponding to the lowest unoccupied Kohn–Sham eigenstate, however, cannot be interpreted as the electron affinity A. The reason is that, as soon as this state becomes occupied, the density changes and so does the Kohn–Sham eigenvalue spectrum.

(viii) Electronic excitations can be calculated by a variety of methods such as Green's function many-body techniques or time-dependent DFT. A particularly simple, yet reasonably accurate way of estimating excitation energies for transitions between states of different symmetry, e.g. $1s \rightarrow 2p$, is provided by the ΔSCF scheme. Within this approach, excitations are calculated as the energy difference between two self-consistent calculations (Jones and Gunnarsson, 1989). Moreover, energy differences can be related to eigenvalues computed at half-occupancy through Slater's transition state method (Slater, 1974), so that excitation energies can be rather successfully estimated as eigenvalue differences, but these are not the eigenvalues corresponding to a single configuration. It has been shown, however, that eigenvalue differences within the same configuration are a well-defined zero-order approximation to excitation energies (Görling, 1996).

(ix) The matter of whether the non-interacting reference system exists or not has been analyzed by several authors (Levy, 1982; Kohn, 1983). This is related to the problem of the v-representability of the density, i.e. which densities of interacting systems correspond to an external potential? Within the Kohn–Sham scheme it is also important to worry about which densities of *non-interacting* systems correspond to an external potential (Kohn, 1983). Kohn concluded that for approximate Kohn–Sham schemes and non-degenerate ground states, the ground state and nearby densities are necessarily v-representable in the non-interacting picture. However, for degenerate ground states it is not difficult to find reasonable densities that are not v-representable for non-interacting systems. By taking a first-row atom like B or C, with an incomplete p shell, and spherically averaging the densities of the degenerate ground states, the resulting density does not correspond to any Slater determinant. In fact, it corresponds to a linear combination of determinants. These cases can still be re-cast within the framework of Kohn Sham theory by virtue of Janak's theorem (Janak, 1978), which allows for fractional occupation of the degenerate single-particle orbitals at the Fermi level (frontier orbitals), maintaining full occupation below and zero occupation above.

(x) There are more subtle cases where the degeneracies are not intrinsic, but accidental (near-degeneracies). This would be the case of the carbon dimer C_2, where the σ_g and π_u states are very close in energy for C–C distances around the equilibrium bond length (Jones and Gunnarsson, 1989). In quantum chemistry these cases of *non-dynamical correlation* are treated within a multi-determinantal CI scheme. The behavior is similar to that of *resonant* systems like those described in Section 3.2.2, where multi-reference methods that include two or more determinants in the variational search, like CASSCF, are required to retrieve the correct behavior. A consistent and promising way of introducing multi-reference approaches within DFT has been proposed by Stoll and Savin (1985), and recently applied to a range of simple cases of closed-shell atoms and diatomic molecules (Leininger *et al.*, 1997). The convergence with the size of the configuration space has been shown by Pollet *et al.* (2002) to be quite reasonable, except for extreme cases.

(xi) Within the DFT scheme, some authors have proposed to overcome the problem of the non v-representability of the non-interacting reference system by carrying out the minimization of the energy functional also with respect to the occupation numbers (Pederson and Jackson, 1991). It has been shown, however, that any choice of occupations different from the conventional one produces densities for which the non-interacting kinetic and exchange-correlation functionals are not differentiable. Therefore, fractional occupations cannot be used within a variational approach (Valiev and Fernando, 1995). This problem may not appear when using approximate functionals and, indeed, it may be possible to minimize the energy with respect to the occupation numbers using the LDA functional. This approach, however, does not stand on solid ground.

In summary, a practical approach to solve the electronic ground state problem by mapping *exactly* the many-body problem onto a self-consistent one-electron problem has been presented. Therefore, given an external potential, we are in a position to find the electronic density, the ground state energy, and any desired ground state property (e.g. equilibrium geometry, vibrational frequencies and normal modes, elastic moduli, dielectric constant, transport properties, etc.). The only remaining task is to devise reliable and practical approximations to the exchange-correlation functional. This is the theme of the next chapter.

References

Dirac, P. A. M. (1930). Note on exchange phenomena in the Thomas atom. *Proc. Cambridge Phil. Soc.* **26**, 376–385.

Fermi, E. (1928). A statistical method for the determination of some properties of atoms. II. Application to the periodic system of the elements. *Z. Phys.* **48**, 73–79.

Görling, A. (1996). Density-functional theory for excited states. *Phys. Rev. A* **54**, 3912–3915.

Hedin, L. (1965). New method for calculating the one-particle Green's function with application to the electron-gas problem. *Phys. Rev.* **139**, A796–823.

Hohenberg, P. and Kohn, W. (1964). Inhomogeneous electron gas. *Phys. Rev.* **136**, B864–867.

Janak, J. F. (1978). Proof that $\partial E/\partial n_i = \epsilon_i$ in density-functional theory. *Phys. Rev. B* **18**, 7165–7168.

Jones, R. O. and Gunnarsson, O. (1989). The density functional formalism, its applications and prospects. *Rev. Mod. Phys.* **61**, 689–746.

Kittel, C. (1996). *Introduction to Solid State Physics*, 7th edn. New York, Wiley.

Kohn, W. (1983). v-representability and density functional theory. *Phys. Rev. Lett.* **51**, 1596–1598.

Kohn, W. and Sham, L. (1965). Self-consistent equations including exchange and correlation effects. *Phys. Rev.* **14**, A1133–1138.

Koopmans, T. A. (1933). Über die Zuordnung von Wellenfunktionen und Eigenwerten zu den einzelnen Elektronen eines Atoms. *Physica* **1**, 104–113.

Langreth, D. C. and Perdew, J. P. (1977). Exchange-correlation energy of a metallic surface: wave-vector analysis. *Phys. Rev. B* **15**, 2884–2901.

Leininger, T., Stoll, H., Werner, H.-J., and Savin, A. (1997). Combining long-range configuration interaction with short-range density functionals. *Chem. Phys. Lett.* **275**, 151–160.

Levy, M. (1982). Electron densities in search of Hamiltonians. *Phys. Rev. A* **26**, 1200–1208.

March, N. H. (1992). *Electron Density Theory of Atoms and Molecules*. London, Academic Press.

Oda, T., Pasquarello, A., and Car, R. (1998). Fully unconstrained approach to noncollinear magnetism: application to small Fe clusters. *Phys. Rev. Lett.* **80**, 3622–3625.

Pederson, M. and Jackson, K. A. (1991). Pseudoenergies for simulations on metallic systems. *Phys. Rev. B* **43**, 7312–7315.

Perdew, J. P. and Wang, Y. (1992). Pair-distribution function and its coupling-constant average for the spin-polarized electron gas. *Phys. Rev. B* **46**, 12947–12954.

Perdew, J. P. and Zunger, A. (1981). Self-interaction correction to density-functional approximations for many-electron systems. *Phys. Rev. B* **23**, 5048–5079.

Petersilka, M., Gossmann, U. J., and Gross, E. K. U. (1996). Excitation energies from time-dependent density-functional theory. *Phys. Rev. Lett.* **76**, 1212–1215.

Pollet, R., Savin, A., Leininger, T., and Stoll, H. (2002). Combining multideterminantal wave functions with density functionals to handle near-degeneracy in atoms and molecules. *J. Chem. Phys.* **116**, 1250–1258.

Slater, J. C. (1951). A simplification of the Hartree–Fock method. *Phys. Rev.* **81**, 385–390. (1974). *Quantum Theory of Molecules and Solids*. New York, McGraw-Hill.

Stoll, H. and Savin, A. (1985). Density functionals for correlation energies of atoms and molecules. In *Density Functional Methods in Physics*, R. M. Dreizler and J. da Providencia, eds. New York, Plenum Press.

Theophilou, A. K. (1979). The energy density functional formalism for excited states. *J. Phys. C* **12**, 5419–5430.

Thomas, L. H. (1927). The calculation of atomic fields. *Proc. Cambridge Phil. Soc.* **23**, 542–548.

Valiev, M. M. and Fernando, G. W. (1995). Occupation numbers in density-functional calculations. *Phys. Rev. B* **52**, 10697–10700.

Vogel, D., Krüger, P., and Pollmann, J. (1995). *Ab initio* electronic-structure calculations for II-VI semiconductors using self-interaction-corrected pseudopotentials. *Phys. Rev. B* **52**, R14316–14319.

Von Barth, U. and Hedin, L. (1972). A local exchange-correlation potential for the spin polarized case. I. *J. Phys. C* **5**, 1629–1642.

Wigner, E. P. (1938). Effects of electron interaction on the energy levels of electrons in metals. *Trans. Faraday Soc.* **34**, 678–685.

5

Exchange and correlation in DFT: approximations and their performances

The strategy to attack the many-body electronic problem presented in the previous chapter consisted of dividing the total energy of an electronic system into a number of different contributions, $E[\rho] = T_R + V_{ext} + E_H + E_X + \tilde{E}_C$, each of which can be addressed separately. These are the non-interacting kinetic energy (T_R), the classical electron–electron interaction or Hartree term (E_H), the interaction of the electrons with external fields, in particular that of the atomic nuclei (V_{ext}), the exchange energy (E_X), and the coupling constant averaged correlation term ($\tilde{E}_C$). The second and third terms are known as explicit functionals of the electronic density. The first and fourth terms are known as functionals of the non-interacting orbitals, which are in turn (unknown) functionals of the density.

The last term, i.e. the correlation energy, is the big unknown. Wigner was the first to address this issue in the context of the homogeneous electron gas, by proposing a correlation energy per unit volume given by Expression (4.6). The exchange energy, although well known as a function of the single-particle orbitals, involves the calculation of computationally expensive integrals – see Expression (3.29) in Section 3.1. In addition, up to date there is no approximation available where the correlation energy is treated at a comparable level of accuracy. Therefore, if exchange is treated exactly as a functional of the orbitals, it will not be able to compensate for any errors introduced when approximating the correlation term. The key issue here is that the really meaningful quantity is the sum of the two terms $E_X + \tilde{E}_C$; the division is a matter of convenience. Therefore, it seems sensible to treat both terms to a similar level of approximation.

The idea now is to look for consistent approximations to exchange and correlation where both terms are treated in a similar manner. One of the natural starting points is the homogeneous electron gas, which is a simplified model for metallic systems. This is the simplest system of correlated electrons, and as such has been studied in great detail. The exchange energy is exactly given by Dirac's expression (4.5):

$$\epsilon_X^D[\rho] = -\frac{3}{4}\left(\frac{3}{\pi}\right)^{1/3}\rho^{1/3} = -\frac{3}{4}\left(\frac{9}{4\pi^2}\right)^{1/3}\frac{1}{r_s} = -\frac{0.458}{r_s} \text{ a.u.,} \quad (5.1)$$

where $r_s = (3/4\pi\rho)^{1/3}$ is the mean interelectronic distance expressed in atomic units (1 bohr = 0.529 177 Å).

Excellent approximations for correlation are also available (von Barth and Hedin, 1972; Vosko et al., 1980; Perdew and Zunger, 1981). The most accurate results are based on the quantum Monte Carlo simulations of Ceperley and Alder (1980). This correlation functional is exact within numerical accuracy, and has been parameterized by Perdew and Zunger (1981) for the spin-polarized (P) and spin-unpolarized (U) homogeneous electron gas.

$$\epsilon_C^{PZ}[\rho] = \begin{cases} A \ln r_s + B + C r_s \ln r_s + D r_s, & r_s \leq 1, \\ \gamma / (1 + \beta_1 \sqrt{r_s} + \beta_2 r_s), & r_s > 1. \end{cases} \quad (5.2)$$

For $r_s \leq 1$ the above expression derives from the random phase approximation, and has been calculated by Gell-Mann and Brueckner (1957). This is valid in the limit of very dense electronic systems, and fixes the values of the leading coefficients: $A^U = 0.0311$, $B^U = -0.048$. Use of scaling relations (Misawa, 1965) allows us also to obtain the values for the fully polarized gas: $A^P = 0.01555$, $B^P = -0.0269$. The remaining coefficients have been fitted to the quantum Monte Carlo results of Ceperley and Alder (1980): $C^U = 0.002$, $D^U = -0.0116$, and $C^P = 0.0007$, $D^P = -0.0048$.

At low densities, Perdew and Zunger used a Padé approximant fitted to the results of Ceperley and Alder (1980). The numerical values of the fitted coefficients are $\gamma^U = -0.1423$, $\beta_1^U = 1.0529$, $\beta_2^U = 0.3334$ for the unpolarized gas, and $\gamma^P = -0.0843$, $\beta_1^P = 1.3981$, $\beta_2^P = 0.2611$ for the fully polarized (spin 1/2) electron gas. Interestingly, the second derivative of the above $\epsilon_C[\rho]$ is discontinuous at $r_s = 1$. This may cause some problems when dealing with very dense, high pressure systems like plasmas.

Another possible parameterization is the one proposed by Vosko et al. (1980), where the correlation functional is given by

$$\frac{\epsilon_C^{VWN}[r_s]}{A} = \ln\left(\frac{r_s}{F(\sqrt{r_s})}\right) + \frac{2b}{\sqrt{4c-b^2}}\tan^{-1}\left(\frac{\sqrt{4c-b^2}}{2\sqrt{r_s}+b}\right)$$

$$-\frac{bx_0}{F(x_0)}\left[\ln\left(\frac{\sqrt{r_s}-x_0}{F(\sqrt{r_s})}\right) + \frac{2(b-2x_0)}{\sqrt{4c-b^2}}\tan^{-1}\left(\frac{\sqrt{4c-b^2}}{\sqrt{r_s}+b}\right)\right], \quad (5.3)$$

with $F(x) = x^2 + bx + c$, and where A, b, c, and x_0 are fitting coefficients that differ for the spin-polarized and spin-unpolarized cases.

Using the homogeneous electron gas as a reference may not seem a particularly good idea for molecular systems, as their electronic densities are far from uniform. Maybe this was the reason why DFT took so long to be adopted by the computational chemistry community, because most of the available approximations are derived from the homogeneous electron gas.

In the following two sections we describe and analyze the most widely used approaches to the exchange-correlation problem within DFT, namely the local density and generalized gradient approximations. Then we discuss a number of extensions, always within Kohn–Sham theory, that have been proposed to go beyond the above approximations. An approach to treat exchange exactly within DFT based on the optimized effective potential method (OEP) is presented in Section 5.5. This method introduces orbital-dependent exchange-correlation functionals, and involves a higher level of computational complexity than the usual Kohn–Sham approach. It is included here to illustrate the general principle of the OEP method, as this represents an interesting approach that can become useful in related contexts, including correlation. Some ideas proposed to improve correlation beyond the usual approximations are finally mentioned. Most of these are not yet common practice, but they represent a state-of-the-art and a showcase of what may become standard procedures in the near future.

5.1 The local density approximation

The local density approximation (LDA) has been for a long time the most widely used approximation to the exchange-correlation energy. It was proposed in the seminal paper by Kohn and Sham (1965), but the philosophy was already present in Thomas–Fermi–Dirac theory. The main idea is to consider a general inhomogeneous electronic system as locally homogeneous, and then to use the exchange-correlation hole corresponding to the homogeneous electron gas, which is known to an excellent accuracy. In practice, energy terms local in the density are calculated by integrating over the volume of the system the corresponding energy density calculated at the values that the electronic density assumes at every point $\mathbf{r}$ in the volume, as illustrated in Fig. 5.1.

This idea is expressed in mathematical terms by re-writing the expression for the (non-local) exchange-correlation hole $\tilde{\rho}_{XC}$, which was defined in (4.62), in the following way:

$$\tilde{\rho}_{XC}^{LDA}(\mathbf{r},\mathbf{r}') = \rho(\mathbf{r})\left\{\tilde{g}^h\left[|\mathbf{r}-\mathbf{r}'|, \rho(\mathbf{r})\right] - 1\right\}, \qquad (5.4)$$

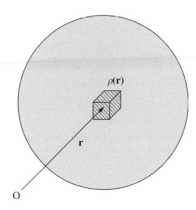

Fig. 5.1 Illustration of the local density approximation (LDA). Local energy terms are obtained by integrating the energy density calculated at the values assumed by the electronic density $\rho(\mathbf{r})$.

with $\tilde{g}^h\left[|\mathbf{r}-\mathbf{r}'|,\rho(\mathbf{r})\right]$ the pair correlation function of the homogeneous gas. This pair correlation function depends only on the distance between $\mathbf{r}$ and $\mathbf{r}'$ (the system is homogeneous), and must be evaluated for the density ρ that locally assumes the value $\rho(\mathbf{r})$. With this definition the exchange-correlation energy can be written as the average of an energy density $\epsilon_{\mathrm{XC}}^{\mathrm{LDA}}[\rho]$:

$$\tilde{E}_{\mathrm{XC}}^{\mathrm{LDA}}[\rho] = \int \rho(\mathbf{r})\, \tilde{\epsilon}_{\mathrm{XC}}^{\mathrm{LDA}}[\rho(\mathbf{r})]\, d\mathbf{r}, \qquad (5.5)$$

weighted with the space-dependent electronic density of the system. The expression for the exchange-correlation energy density in terms of the exchange-correlation hole is

$$\tilde{\epsilon}_{\mathrm{XC}}^{\mathrm{LDA}}[\rho] = \frac{1}{2}\int \frac{\tilde{\rho}_{\mathrm{XC}}^{\mathrm{LDA}}(\mathbf{r},\mathbf{r}')}{|\mathbf{r}-\mathbf{r}'|}\, d\mathbf{r}'. \qquad (5.6)$$

In practice, the exchange-correlation energy within the LDA is calculated via (5.5) using $\tilde{\epsilon}_{\mathrm{XC}}^{\mathrm{LDA}}[\rho] = \epsilon_{\mathrm{X}}^{\mathrm{LDA}}[\rho] + \tilde{\epsilon}_{\mathrm{C}}^{\mathrm{LDA}}[\rho]$, where $\epsilon_{\mathrm{X}}^{\mathrm{LDA}}[\rho]$ is the exchange energy density given by (5.1) and $\tilde{\epsilon}_{\mathrm{C}}^{\mathrm{LDA}}[\rho]$ the correlation energy density given by (5.2) or (5.4).

While the exchange-correlation energy $E_{\mathrm{XC}}[\rho]$ should be a functional of ρ, there is no reason why the energy density should also be so. In fact, in general ϵ_{XC} is not a functional of the density. From its very definition it is clear that it has to be a non-local object, because it reflects the fact that the probability of finding an electron at $\mathbf{r}$ depends on the presence of other electrons in the surroundings, through the exchange-correlation hole. However, in the LDA it becomes a functional of the density because it corresponds to a homogeneous system where ρ is the same everywhere.

5.1 The local density approximation

Looking at Expression (5.4), it may seem that there is an inconsistency in the definition. The exact expression would indicate that we should take $\rho(\mathbf{r}')$ instead of $\rho(\mathbf{r})$ This, however, would make $E_{XC}^{LDA}[\rho]$ a non-local object depending on the densities at $\mathbf{r}$ and $\mathbf{r}'$. In this case, a parameterization in terms of the homogeneous gas is not possible because it is unclear whether to use $\rho(\mathbf{r})$ or $\rho(\mathbf{r}')$ as the uniform density. The uniform gas is characterized by a single density, and not two! The LDA is then equivalent to assuming

$$\tilde{g}(\mathbf{r}, \mathbf{r}') \approx \tilde{g}^h[|\mathbf{r} - \mathbf{r}'|, \rho(\mathbf{r})] \left(\frac{\rho(\mathbf{r})}{\rho(\mathbf{r}')} \right) \tag{5.7}$$

for the pair correlation function (and its coupling constant average $\tilde{g}$). Therefore, the LDA in fact embodies two separate approximations.

(i) The LDA exchange-correlation hole is centered at $\mathbf{r}$, and interacts with the electronic density at $\mathbf{r}$. The true exchange-correlation hole is centered at $\mathbf{r}'$ instead of $\mathbf{r}$.
(ii) The pair correlation function (g) is approximated by that of the homogeneous electron gas of density $\rho(\mathbf{r})$, but corrected by the density ratio $\rho(\mathbf{r})/\rho(\mathbf{r}')$ to compensate for the fact that the LDA exchange-correlation hole is centered at $\mathbf{r}$ instead of $\mathbf{r}'$.

5.1.1 The local spin density approximation

Magnetic and open-shell systems are better described by means of spin density functional theory (SDFT), which was introduced in Section 4.2.5. In SDFT the role of the density is played by the two spin densities $\rho_\uparrow(\mathbf{r})$ and $\rho_\downarrow(\mathbf{r})$, which are related to the total electronic density and the magnetization density by (4.47) and (4.48). The non-interacting kinetic energy (4.50) splits trivially into *spin-up* and *spin-down* contributions, while the external and Hartree potentials depend only on the full density $\rho(\mathbf{r})$. The exchange-correlation functional, however, depends on the individual spin densities. Therefore, the reference potential becomes spin-dependent as in Expression (4.41), where the only difference is in the exchange-correlation potential.

The extension of the LDA to spin-polarized systems is the *local spin density approximation* (LSDA), which basically consists of replacing the exchange-correlation energy density with a spin-polarized expression:

$$E_{XC}^{LSDA}[\rho_\uparrow(\mathbf{r}), \rho_\downarrow(\mathbf{r})] = \int [\rho_\uparrow(\mathbf{r}) + \rho_\downarrow(\mathbf{r})] \epsilon_{XC}^h[\rho_\uparrow(\mathbf{r}), \rho_\downarrow(\mathbf{r})] \, d\mathbf{r}$$

$$= \int \rho(\mathbf{r}) \epsilon_{XC}^h[\rho(\mathbf{r}), \zeta(\mathbf{r})] \, d\mathbf{r}. \tag{5.8}$$

The common practice in LSDA is to interpolate between the fully-polarized (ϵ_{XC}^P) and unpolarized (ϵ_{XC}^U) exchange-correlation energy densities using some interpolation function that depends on the magnetization density ζ:

$$\epsilon_{XC}^h[\rho,\zeta] = f(\zeta)\epsilon_{XC}^U[\rho] + [1-f(\zeta)]\epsilon_{XC}^P[\rho]. \tag{5.9}$$

An appropriate expression for the interpolation function $f(\zeta)$ has been proposed by von Barth and Hedin (1972):

$$f^{\text{vBH}}(\zeta) = \frac{(1+\zeta)^{4/3} + (1-\zeta)^{4/3} - 2}{2^{4/3} - 2}. \tag{5.10}$$

A more realistic formula for correlation, based on the random phase approximation (RPA), was proposed by Vosko et al. (1980).

$$\epsilon_C^{\text{VWN}}[\rho,\zeta] = \epsilon_C^U[\rho] + \left(\frac{f(\zeta)}{f''(0)}\right)[1-\zeta^4]\epsilon_C^A[\rho]$$
$$+ f(\zeta)\zeta^4\left(\epsilon_C^P[\rho] - \epsilon_C^U[\rho]\right), \tag{5.11}$$

where $\epsilon_C^U[\rho]$ and $\epsilon_C^U[\rho]$ are the correlation energy densities for the unpolarized and fully polarized uniform gas, and $\epsilon_C^A[\rho]$ has the same expression as the former, but different fitting coefficients.

5.1.2 Performance of the LDA and LSDA

Jones and Gunnarsson (1989) have discussed thoroughly the LDA and the LSDA, by analyzing their performance for different types of systems, in particular atomic and molecular, but also solids. Many are the successes of these approximations, and these can be traced back to two fundamental properties of the LDA exchange-correlation hole.

(i) It satisfies the sum rule expressing that the exchange-correlation hole contains exactly one displaced electron:

$$\int \tilde{\rho}_{XC}^{\text{LDA}}(\mathbf{r},\mathbf{r}')\,d\mathbf{r}' = \int \rho(\mathbf{r})\tilde{g}^h[|\mathbf{r}-\mathbf{r}'|,\rho(\mathbf{r})]\,d\mathbf{r}' = -1, \tag{5.12}$$

because for each $\mathbf{r}$, $\tilde{g}^h[|\mathbf{r}-\mathbf{r}'|,\rho(\mathbf{r})]$ is the pair correlation function of an existing system, i.e. the homogeneous gas at density $\rho(\mathbf{r})$. Therefore, the expression in the middle is just the integral of the exchange-correlation hole of the homogeneous gas. For this latter, either approximations or numerical results carefully take into account that the integral is -1.

(ii) Even if the exact $\tilde{\rho}_{XC}$ is not spherically symmetrical, what really matters for the exchange-correlation energy (4.63) is the spherical average of the hole:

$$E_{XC}[\rho] = -\frac{1}{2}\int \rho(\mathbf{r})\left(\frac{1}{R(\mathbf{r})}\right)d\mathbf{r}, \tag{5.13}$$

5.1 The local density approximation

with

$$\frac{1}{R(\mathbf{r})} = \int \frac{\tilde{\rho}_{XC}(\mathbf{r},\mathbf{r}')}{|\mathbf{r}-\mathbf{r}'|} d\mathbf{r}' = 4\pi \int_0^\infty s \tilde{\rho}_{XC}^{SA}(\mathbf{r},s) \, ds \quad (5.14)$$

and

$$\tilde{\rho}_{XC}^{SA}(\mathbf{r},s) = \frac{1}{4\pi} \int_\Omega \tilde{\rho}_{XC}(\mathbf{r},\mathbf{r}') \, d\Omega. \quad (5.15)$$

The spherical average $\tilde{\rho}_{XC}^{SA}(\mathbf{r},s)$ is reproduced to a good extent by the LDA, whose $\tilde{\rho}_{XC}$ is already spherical. Figure 5.2 shows the LSDA exchange-correlation hole and its spherical average for the nitrogen atom compared to the exact XC hole, as presented by Jones and Gunnarsson (1989). It is clear that the spherical average is quite well reproduced by the LSDA, while the hole itself is very distant from the exact one. The consequence of this observation is that, while LDA (LSDA) exchange-correlation energies are quite reasonable, the exchange-correlation potential is not that good. This issue has been explored in detail by Filippi, Gonze, and Umrigar, who compared LDA and more elaborate exchange-correlation potentials with exact results (Filippi et al., 1996).

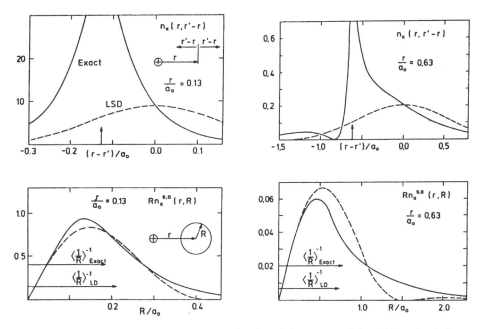

Fig. 5.2 The LSDA exchange hole $\rho_X(\mathbf{r},\mathbf{r}'-\mathbf{r})$ (upper panels) and its spherical average (lower panels) as a function of $|\mathbf{r}'-\mathbf{r}|$ for the nitrogen atom, compared to the exact hole. Left panels correspond to a distance $r = 0.13\,a_0$ and right panels to $r = 0.63\,a_0$ from the nucleus. The exact hole peaks at the nucleus while the LSDA hole peaks at the electron. These differences are averaged out when computing the energy. Figure 7 reprinted with permission from R. O. Jones and O. Gunnarsson, Rev. Mod. Phys. **61**, 689 (1989). Copyright (1989) by the American Physical Society.

Trends within the LDA-LSDA

There are a number of features of the LDA that are rather general and well established. These are the following.

(i) It favors electronic densities that are more homogeneous than the exact ones.
(ii) As a consequence, it tends to overestimate the binding energy of molecules and the cohesive energy of solids. This trend is opposite to Hartree–Fock, which underestimates binding energies.
(iii) Geometries of "well-behaved" systems, i.e. those involving strong bonds (covalent, ionic, or metallic) are remarkably good within the LDA. Bond lengths, bond angles, and vibrational frequencies reproduce experimental values within a few percent. In general the LDA tends to underestimate bond lengths due to the over-binding problem. Elastic constants and phonon frequencies are also well reproduced, although the LDA shows a slight tendency to underestimate them. Dielectric properties like the dielectric constant and piezoelectric coefficients are overestimated by about 10%.
(iv) For weakly bound systems that involve hydrogen bonds or van der Waals closed-shell interactions, bond lengths are too short (over-binding). Dispersion interactions are poorly reproduced.
(v) Chemical trends as a function of the atomic number of atomic quantities, such as ionization potentials and transfer energies between electronic states of different symmetry, are usually correct. The LSDA is superior to the LDA for atomic trends because it is better suited for the description of open electronic shells due to the explicit treatment of the spin. It is also better than Hartree–Fock because it takes into account correlation between opposite spin projections.

Limitations of the LDA-LSDA

The LDA and LSDA are very successful approximations for many systems of interest, especially those where the electronic density is quite uniform such as bulk metals, but also for less uniform systems such as molecules, semiconductors, and ionic crystals. There are, however, a number of features that the LDA is known to fail to reproduce.

(i) Electronic densities of atoms in the core region, where the electrons are quite localized, are poor. The reason is that the LDA fails to cancel the self-interaction, which is important for strongly localized states. The electronic density in the valence (outer) region of atoms is much better reproduced, although it still decays into the vacuum with an incorrect behavior. Hartree–Fock, where the self-interaction is exactly canceled by the exchange term, gives much better results for atoms than the LDA. In atoms, correlation effects are less important and largely overcome by the exchange interaction. Total LSDA atomic energies are systematically higher than HF and experimental values.

(ii) This behavior is common to all finite systems such as molecules or clusters. The exchange-correlation potential decays exponentially instead of exhibiting the correct long-range attractive $-e^2/r$ behavior in the vacuum region (Levy et al., 1984), thus affecting the dissociation limit and ionization energies.

(iii) Similar behavior is observed in low-dimensional systems such as surfaces or wires. For metallic surfaces the exchange-correlation potential decays exponentially into the vacuum region instead of following a power law. This affects the location and shape of the image potential, which does not decay into the vacuum as $-e^2/4z$, with important consequences regarding molecular physisorption (weak adsorption). The energetics and geometries in the chemisorption regime, i.e. when chemical bonds are formed, are quite well described by the LDA.

(iv) The lack of cancelation of the self-interaction has a very serious effect on negatively charged ions. Due to the fact that the potential decays too fast into the vacuum region, in many cases the extra electron does not feel an attractive potential and cannot bind. Therefore, even in cases where negatively charged ions are experimentally known to exist, the LDA finds them unstable.

(v) Molecular dissociation is also affected by the left–right correlation problem, as happens within the Hartree–Fock approach (Handy and Cohen, 2001). This is a different problem. At large separations the two fragments should become independent from each other, and cannot share electrons. In diatomic molecules such as H_2, however, the LDA ground state is always a bonding state made of paired electrons. This state is localized in both fragments even at large separation, while the correct limit is to have one electron in each atom, with any spin projection. As a consequence, the energy in the dissociation limit is too high. The combination of this feature with over-binding makes the LDA overestimate dissociation energies. The LSDA improves this feature over the LDA because it allows for a non-zero spin polarization localized in each one of the fragments. However, it does not cure the problem completely. The same feature in the Hartree–Fock picture is corrected only when correlation is included at a reasonably high level such as MP4 or CISD, but not at the MP2 level (see Fig. 3.1).

(vi) Weak inter-molecular bonds, such as the ubiquitous hydrogen bonds that are crucial in water and biological systems, are poorly reproduced because in the bonding region the density is very small and the binding is dominated by inhomogeneities in the density.

(vii) Van der Waals systems such as closed-shell dimers (Ar_2) are problematic within the LDA. Dispersion interactions can be understood as the coupling of the electric field generated by density fluctuations in one of the fragments with the density in the other separated fragment. This is clearly a dynamical, inherently non-local correlation effect, which is not taken into account by the LDA or LSDA.

(viii) The energy bandgap in semiconductors, i.e. the energy necessary to promote an electron from the valence to the conduction band, can be written as the difference between ionization energy and electron affinity. This, in turn, can be expressed as the energy difference between the eigenvalue at the bottom of the conduction band for $N+\eta$ electrons and that at the top of the valence band for $N-\eta'$ electrons, with $0 < \eta, \eta' < 1$ (Jones and Gunnarsson, 1989). Perdew and Levy (1983) and Sham and Schlüter (1983) showed that this energy difference is made up of two contributions: $E_g = \varepsilon_g + \Delta$. One is the usual Kohn–Sham eigenvalue bandgap for a system with N electrons (ε_g), and the other is a possible additional discontinuity Δ in the exchange-correlation potential. This rigid shift of the XC potential is consistent with a change in density of the order $1/V$, with V the volume of the system. In fact, adding or subtracting a single electron to an extended system modifies the density by a minute amount, proportional to $1/V$, which goes to zero for an infinite system. Kohn (1986) has shown that this discontinuity arises as a consequence of three contributions: the addition of an electron with density $\rho_c(\mathbf{r})$ in the conduction band, the removal of an electron with density $\rho_v(\mathbf{r})$ from the valence band, and a change in the external potential such that the density changes from $\rho_{N-1}(\mathbf{r}) + \rho_v(\mathbf{r})$ to $\rho_{N-1}(\mathbf{r}) + \rho_c(\mathbf{r})$. In the exact theory this last term does not compensate for the other two, and a finite discontinuity appears, whose magnitude strongly depends on the system. However, the LDA and the LSDA expressions for the XC energy are regular, thus giving rise to a vanishing discontinuity (Kohn, 1986). The consequence is that the excitation energy reduces to the Kohn–Sham bandgap, which is typically of the order of 40% smaller than the true gap, and this is also reflected in the underestimation of the dielectric constant. Moreover, some small-gap semiconducting systems such as Ge are found to be metallic within the LDA. The reason for this behavior is that the excitation spectrum of the homogeneous electron gas is gap-less. Therefore, introducing a gap in the excitations of the reference system should help in recovering the discontinuity (Toulouse *et al.*, 2002). The Hartree–Fock approximation has a similar limitation, but the bandgap is severely overestimated because unoccupied states are unbound, although in practice they artificially bind when a finite basis set is used.

(ix) Strongly correlated systems such as transition metal oxides (e.g. NiO or V_2O_3) are characterized by a correlation-induced opening of an energy gap at the Fermi level and satellite structures in the electronic spectrum, which the LDA fails to reproduce.

5.1.3 Possible improvements over the LDA

Once the extent of the approximations involved in the LDA has been understood, better approximations can be constructed. According to the preceding subsection, the main limitations of the LDA are the following:

(i) inhomogeneities in the density are not taken into account;
(ii) the self-interaction present in the Hartree term of the energy is not completely canceled by the LDA exchange-correlation term;
(iii) non-local exchange and correlation effects are not included;
(iv) strong local correlation effects cannot be reproduced because of the treatment of the correlation functional.

The above suggests that there is no unique and obvious way of improving over the LDA. Since the limitations are of different kinds, there have been developments in several different directions. The amount of work done in some of these directions is overwhelming. Here we summarize the main approaches, which are then discussed in more depth in the following sections.

Undoubtedly, due to its computational efficiency and its similarity to the LDA, the most widely exploited approach has been to introduce semi-locally the inhomogeneities of the density, by expanding $E_{XC}[\rho]$ as a series in terms of the density and its gradients. This approach, known as gradient expansion, is easy to implement in practice and computationally more convenient than other directions. It has been very successful in improving over the LDA for some important cases such as hydrogen-bonded systems. This has probably been the catalyzer for the community of computational chemists to adopt DFT in recent years. The self-interaction cannot be fully compensated within this approach, but can be improved over the LDA.

The non-locality of the pair correlation function was addressed in different ways. One of these is the weighted density approximation, where the pair correlation function is weighted with a modified density $\bar{\rho}(\mathbf{r})$ that incorporates information about the density at neighboring points. The non-locality of the exchange interaction was recently cast within the DFT scheme, and proved very successful in correcting the structural and electronic trends of semiconductors. Also, orbital-dependent correlation functionals have been studied. These approaches that treat non-locality in an explicit way are normally computationally much more expensive than gradient corrections.

Another road explored is to employ either standard or advanced many-body tools such as solving Dyson's equation for the electronic Green's function, starting from the LDA solution for the bare Green's function.

In the context of strongly correlated systems, e.g. those exhibiting narrow d or f bands, where the limitation of the LDA is at describing strong on-site correlations of the Hubbard type, these features have been introduced a posteriori within the so-called LDA+U approach, more recently extended to include dynamical mean field correlations (LDA+DMFT).

5.2 Gradient expansions

To address the issue of inhomogeneities in the electronic density, the natural road is to carry out an expansion of the density in terms of the gradient and higher order derivatives. In general, the exchange-correlation energy can be written in the following form:

$$E_{XC}[\rho] = \int \rho(\mathbf{r})\, \varepsilon_{XC}[\rho(\mathbf{r})]\, F_{XC}[\rho(\mathbf{r}), \nabla\rho(\mathbf{r}), \nabla^2\rho(\mathbf{r}), \ldots]\, d\mathbf{r}, \quad (5.16)$$

where the function F_{XC} is an enhancement factor that modifies the LDA expression according to the variation of the density in the vicinity of the considered point. In this sense, gradient corrections constitute a semi-local approach, which will hardly be able to capture non-local effects at longer ranges.

The second order gradient expansion of the *exchange* energy introduces a term proportional to the squared gradient of the density. If this expansion is further carried on to fourth order, as originally done by Gross and Dreizler (1981) and further developed by Perdew (1985), the term also appears proportional to the square of the Laplacian of the density. The Laplacian term was also derived using a different route by Ghosh and Parr (1986), although it was then dropped out when considering the gradient expansion only up to second order. More recently, a general derivation of the exchange gradient expansion up to sixth order, using second order density response theory, was given by Svendsen and von Barth (1996).

The fourth order gradient expansion of the exchange enhancement factor F_X is

$$F_X(p,q) = 1 + \frac{10}{81}p + \frac{146}{2025}q^2 - \frac{73}{405}qp + Dp^2 + \mathcal{O}(\nabla \rho^6), \tag{5.17}$$

where

$$p = \frac{|\nabla \rho|^2}{4(3\pi^2)^{2/3}\rho^{8/3}} \tag{5.18}$$

is the square of the reduced density gradient, and

$$q = \frac{\nabla^2 \rho}{4(3\pi^2)^{2/3}\rho^{5/3}} \tag{5.19}$$

is the reduced Laplacian of the density.

The first two coefficients of the expansion are exactly known. The third one is the result of a difficult many-body calculation, and has only been estimated numerically by Svendsen and von Barth, to an accuracy of the order of 20%. The fourth coefficient, D, has not been explicitly calculated to date, but the best numerical estimate for it is $D = 0$.

5.2.1 The generalized gradient approximation (GGA)

The second order gradient expansion corresponds to an expression of the type

$$E_{XC}[\rho] = \int A_{XC}[\rho]\rho(\mathbf{r})^{4/3} d\mathbf{r} + \int C_{XC}[\rho] \, |\nabla \rho(\mathbf{r})|^2 / \rho(\mathbf{r})^{4/3} d\mathbf{r}, \tag{5.20}$$

which is asymptotically valid for densities that vary slowly in space. The LDA retains only the leading term of Eq. (5.20). It is well known that a straightforward evaluation of this expansion is ill behaved, in the sense that it is not monotonically

5.2 Gradient expansions

convergent, and it exhibits singularities that cancel out only when an infinite number of terms is re-summed. In fact, the first-order correction worsens the results, and the second order correction is plagued with divergences (Ma and Brueckner, 1968). The largest error of this approximation actually arises from the gradient contribution to the correlation term. Provided that the problem of the correlation term can be cured in some way, such as the real space cutoff method of Langreth and Mehl (1981), the biggest problem remains with the exchange energy.

One of the main lessons learnt from the early works of Gross and Dreizler (1981) and Perdew (1985) is that the gradient expansion has to be carried out very carefully in order to retain all the relevant contributions to the desired order. Another important lesson is that these expansions easily violate one or more of the *exact* conditions required for the exchange and correlation holes, such as the normalization condition, the negativity of the exchange density, or the self-interaction cancelation. Perdew showed that imposing these conditions to functionals that originally do not verify them results in a remarkable improvement of the quality of exchange energies (Perdew, 1985). On the basis of this type of reasoning, a number of modified gradient expansions have been proposed along the years, mainly between 1986 and 1996. These have been named generalized gradient approximations (GGAs).

GGAs have been obtained following two different approaches. One of them consists of deriving appropriate expressions by theoretical methods, and then requesting that the coefficients are such that a number of formal conditions are fulfilled, and some exact results in known limits are reproduced. Typical requirements are sum rules, long-range decay, etc. However, this cannot be done by considering directly the bare gradient expansion (5.20). What is needed from the functional is a form that mimics a re-summation to infinite order, and this is the main idea of the GGA. This re-summation is not unique and it is not even carried out explicitly. Functional forms are proposed that verify the required conditions. Clearly, the expressions will depend on the specific approach, thus leading to a forest of different recipes. The distinction between them resides in which formal properties are enforced. Naturally, not all of them can be enforced at the same time. The other approach is to fit the parameters of the functional in order to reproduce a number of experimental results in a molecular database (training set). These can be structural parameters, formation energies, thermochemical data, or others. This approach works very well when applied to molecules similar to those in the training set, but the transferability to other systems such as solids is not guaranteed. Normally GGAs improve over some of the drawbacks of the LDA, although this is not always the case. These aspects will be discussed below, after presenting a few functionals and discussing their properties. A thorough comparison of different GGAs has been done by Filippi *et al.* (1994).

Langreth–Mehl functional

The first GGA proposed in the literature (Langreth and Mehl, 1981) assumes the following form:

$$\varepsilon_X = \varepsilon_X^{LDA} - a\frac{|\nabla\rho(\mathbf{r})|^2}{\rho(\mathbf{r})^{4/3}}\left(\frac{7}{9} + 18f^2\right), \tag{5.21}$$

$$\varepsilon_C = \varepsilon_C^{RPA} + a\frac{|\nabla\rho(\mathbf{r})|^2}{\rho(\mathbf{r})^{4/3}}\left(2e^{-F} + 18f^2\right), \tag{5.22}$$

where $F = b|\nabla\rho(\mathbf{r})|/\rho(\mathbf{r})^{7/6}$, $b = (9\pi)^{1/6}f$, $a = \pi/(16(3\pi^2)^{4/3})$, and $f = 0.15$.

BLYP functional

In 1988, Becke proposed an exchange functional where the parameters were fitted to experimental molecular data (Becke, 1988).

$$\varepsilon_X = \varepsilon_X^{LDA}\left(1 - \frac{\beta}{2^{1/3}A_x}\frac{x^2}{1 + 6\beta x \sinh^{-1}(x)}\right), \tag{5.23}$$

for $x = 2(6\pi^2)^{1/3}s = 2^{1/3}|\nabla\rho(\mathbf{r})|/\rho(\mathbf{r})^{4/3}$, $A_x = (3/4)(3/\pi)^{1/3}$, and $\beta = 0.0042$.

This was complemented by a correlation functional derived also in 1988 by Lee, Yang, and Parr (LYP), thus giving rise to a very widely used combination called the BLYP functional (Lee et al., 1988):

$$\varepsilon_C = -\frac{a}{1 + d\rho^{-1/3}}\left\{\rho + b\rho^{-2/3}\left[C_F\rho^{5/3} - 2t_W + \frac{1}{9}\left(t_W + \frac{1}{2}\nabla^2\rho\right)\right]e^{-c\rho^{-1/3}}\right\}, \tag{5.24}$$

where

$$t_W = \frac{1}{8}\left(\frac{|\nabla\rho|^2}{\rho} - \nabla^2\rho\right), \tag{5.25}$$

$C_F = 3/10(3\pi^2)^{2/3}$, $a = 0.049\,18$, $b = 0.132$, $c = 0.2533$, and $d = 0.349$. This correlation functional is not based on the LDA. It has been derived as an extension to other closed-shell systems of the Colle–Salvetti expression for the electronic correlation in helium (Colle and Salvetti, 1975).

PBE functional and revisions

In 1996, Perdew, Burke, and Ernzerhof (PBE) proposed an exchange and correlation functional that satisfies as many formal properties and limits as possible, sacrificing only those deemed to be energetically less important (Perdew et al., 1996).

The enhancement factor $F_X(\rho, \zeta, s)$ over the local exchange defined in Expression (5.16) depends on the local density ρ, magnetization density ζ (in the spin-dependent case), and the dimensionless density gradient $s = |\nabla \rho(\mathbf{r})|/(2k_F\rho)$. The chosen expression is

$$F_X(s) = 1 + \kappa - \frac{\kappa}{1 + \mu s^2/\kappa}, \qquad (5.26)$$

where $\mu = \beta(\pi^2/3) = 0.21951$ and $\beta = 0.066725$ is related to the second order gradient expansion (Perdew and Wang, 1992a). This form (a) satisfies the uniform scaling condition, (b) recovers the correct uniform electron gas limit because $F_x(0) = 1$, (c) obeys the spin-scaling relationship, (d) recovers the LSDA linear response limit for $s \to 0$ ($F_X(s) \to 1 + \mu s^2$), and (e) satisfies the *local* Lieb–Oxford bound (Lieb and Oxford, 1981), $\varepsilon_X(\mathbf{r}) \geq -1.679 \rho(\mathbf{r})^{4/3}$, i.e. $F_X(s) \leq 1.804$ everywhere, provided that $\kappa \leq 0.804$. PBE choose the largest allowed value, $\kappa = 0.804$. Other authors have proposed the same form, but with values of κ and μ fitted empirically to a database of atomization energies (Becke, 1986; Zhang and Yang, 1998). The values of κ so obtained violate the Lieb–Oxford inequality.

The correlation energy is written in a form similar to an earlier proposal of Perdew and Wang (1992a). It assumes the form

$$E_C^{GGA} = \int \rho(\mathbf{r}) \left[\varepsilon_C^{LDA}(\rho, \zeta) + H[\rho, \zeta, t] \right] d\mathbf{r}, \qquad (5.27)$$

with

$$H[\rho, \zeta, t] = (e^2/a_0)\gamma\phi^3 \ln\left\{ 1 + \frac{\beta}{\gamma}t^2\left[\frac{1 + At^2}{1 + At^2 + A^2t^4}\right]\right\}. \qquad (5.28)$$

Here, $t = |\nabla\rho(\mathbf{r})|/(2\phi k_s \rho)$ is a dimensionless density gradient, with k_s the Thomas–Fermi screening wave number, and $\phi(\zeta) = [(1+\zeta)^{2/3} + (1-\zeta)^{2/3}]/2$ is a spin-scaling factor. The quantity β is the same as for the exchange term $\beta = 0.066725$, and $\gamma = (1 - \ln 2)/\pi^2 = 0.031091$. The function A has the following form:

$$A = \frac{\beta}{\gamma}\left[e^{-\varepsilon_C^{LDA}[\rho]/(\gamma\phi^3 e^2/a_0)} - 1 \right]^{-1}. \qquad (5.29)$$

So defined, the correlation correction term H satisfies the following properties: (a) it tends to the correct second-order gradient expansion in the slowly varying (high density) limit ($t \to 0$), (b) it approaches minus the uniform electron gas correlation $-\varepsilon_C^{LDA}$ for rapidly varying densities ($t \to \infty$), thus making the correlation energy vanish in that limit, as required by the correlation hole sum rule, (c) it cancels the logarithmic singularity of ε_C^{LDA} in the high density limit, thus forcing the correlation energy to scale to a constant under uniform scaling of the density.

This GGA retains the correct features of the LDA (LSDA), and combines them with the inhomogeneity features that are assumed to be energetically the most important ones. It sacrifices a few correct, but less important, features, like the second-order gradient coefficients in the slowly varying limit and the non-uniform scaling of the exchange energy in the rapidly varying density region. The PBE functional is very satisfactory from the theoretical point of view, because it verifies many of the exact conditions for the XC hole, and it does not contain any fitting parameters. In addition, its quality is equivalent to or even better than BLYP (Ernzerhof and Scuseria, 1999). A revised version of PBE, known as RPBE, appears to provide an improved description of adsorption and hydrogen-bonded systems (Hammer *et al.*, 1999).

Trends of the GGAs

The general trends of GGAs in comparison with the LDA and LSDA are the following.

(i) They improve binding energies and also atomic energies.
(ii) They improve bond lengths and angles.
(iii) They improve energetics, geometries, and dynamical properties of water, ice, and water clusters. BLYP and PBE show the best agreement with experiment. In general, they improve the description of hydrogen-bonded systems, although this is not very clear for the case of the F···H bond.
(iv) Semiconductors are marginally better described within the LDA than in GGA, except for the binding energies.
(v) For $4d-5d$ transition metals the improvement of the GGA over LDA is not clear, depending on how well the LDA does in each particular case.
(vi) Lattice constants of noble metals (Ag, Au, Pt) are overestimated in GGAs. Due to a fortuitous error cancelation, the LDA values are very close to experiment. Therefore, any incomplete improvement over the LDA, such as the GGA, can only worsen the results.
(vii) There is some improvement in the gap energy (and consequently in the dielectric constant), but it is not substantial because this feature is related to the description of the screening of the exchange hole when one electron is removed, and this feature is not fully taken into account within the GGA. The XC discontinuity still vanishes, but the Kohn–Sham bandgap increases with respect to the LDA.
(viii) GGAs do not satisfy some known asymptotic behaviors, e.g. for isolated atoms:

- $\mu_{XC}(\mathbf{r}) \sim -e^2/r$ for $r \to \infty$, while $\mu_{XC}^{LDA,GGA}(\mathbf{r})$ vanish exponentially;
- $\mu_{XC}(\mathbf{r}) \to$ const for $r \to 0$, while $\mu_{XC}^{LDA}(\mathbf{r}) \to$ const, but $\mu_{XC}^{GGA}(\mathbf{r}) \to -\infty$.

This latter is a disturbing feature that is normally corrected to avoid numerical instabilities in atomic or pseudopotential calculations.

Therefore, there seems to be a limit in the accuracy that GGAs can reach. The main responsibility for this is the exchange term, whose non-locality is not fully taken into account, but for some systems correlation also plays a major role (strongly-correlated systems). A particularly problematic issue is that GGA functionals still do not compensate satisfactorily for the self-interaction present in the Hartree term.

5.2.2 Meta-generalized gradient approximation (meta-GGA)

The next step beyond the GGA is to consider the fourth order gradient expansion of the exchange and correlation energy (5.17). This was recently achieved by constructing a practical meta-generalized gradient approximation (meta-GGA) that incorporates additional semi-local information via the Laplacian of the density (or the kinetic energy density) (Perdew et al., 1999). The philosophy for constructing the functional is the same as in PBE, namely to retain the good formal properties of the lower level approximation (the PBE-GGA in this case) while adding others.

In the same spirit of PBE, Perdew et al. (PKZB) proposed an exchange enhancement factor that verifies some of the formal relations, and reduces to the gradient expansion (5.17) in the slowly varying limit of the density. The expression is formally identical to that of PBE:

$$F_X^{MGGA}(p, \bar{q}) = 1 + \kappa - \frac{\kappa}{1 + x/\kappa}, \qquad (5.30)$$

but

$$x = \frac{10}{81}p + \frac{146}{2025}\bar{q}^2 - \frac{73}{405}\bar{q}p + \left[D + \frac{1}{\kappa}\left(\frac{10}{81}\right)^2\right]p^2 \qquad (5.31)$$

is a new inhomogeneity parameter that replaces μs^2 in PBE. The variable q in the gradient expansion (5.17), which is basically the reduced Laplacian, is also replaced by a new variable $\bar{q}$ defined as

$$\bar{q} = \frac{3\tau[\rho]}{2(3\pi^2)^{2/3}\rho^{5/3}} - \frac{9}{20} - \frac{p}{12}. \qquad (5.32)$$

This reduces to q in the slowly varying density limit, but remains finite at the position of the nucleus. This is an improvement over the GGA, where the analogous parameter q exhibits the rather unpleasant feature of diverging at the position of the nucleus. In the above expression, $\tau[\rho] = \tau_\uparrow + \tau_\downarrow$ is the kinetic energy density for the non-interacting system, where

$$\tau_\sigma = \frac{1}{2}\sum_\alpha^{occup} |\nabla \psi_{\alpha\sigma}(\mathbf{r})|^2 \qquad (5.33)$$

for $\sigma = \uparrow, \downarrow$. The connection between $\tau[\rho]$ and the density is given by the second-order gradient expansion

$$\tau^{\text{GEA}} = \frac{3}{10}(3\pi^2)^{2/3}\rho^{5/3} + \frac{1}{72}\frac{|\nabla\rho|^2}{\rho} + \frac{1}{6}\nabla^2\rho. \tag{5.34}$$

The formal conditions requested for this functional are: (a) the spin-scaling relation, (b) the uniform density-scaling relation (Levy and Perdew, 1985), and the Lieb–Oxford inequality (Lieb and Oxford, 1981). PKZB proposed a value of $\kappa = 0.804$, exactly as in PBE. A coefficient $D = 0.113$ was obtained by minimizing the absolute error in the atomization energies for a molecular data set. The meta-GGA recovers the exact linear response function up to fourth order in $k/2k_\text{F}$. This is not the case for PBE and other GGAs, which recover only the LSDA linear response, and at the expense of sacrificing the correct second-order gradient expansion.

The correlation part of the meta-GGA retains the correct formal properties of PBE-GGA correlation, such as the slowly varying limit and the finite limit under uniform scaling. In addition, the correlation energy is required to be self-interaction free, i.e. to vanish for a one-electron system. PKZB proposed the following form:

$$E_\text{C}^{\text{MGGA}}[\rho_\uparrow, \rho_\downarrow] = \int d\mathbf{r}\{\rho\varepsilon_\text{C}^{\text{GGA}}(\rho_\uparrow, \rho_\downarrow, \nabla\rho_\uparrow, \nabla\rho_\downarrow)\left[1 + C\left(\frac{\sum_\sigma \tau_\sigma^\text{W}}{\sum_\sigma \tau_\sigma}\right)^2\right]$$

$$-(1+C)\sum_\sigma \left(\frac{\tau_\sigma^\text{W}}{\tau_\sigma}\right)^2 \rho_\sigma \varepsilon_\text{C}^{\text{GGA}}(\rho_\sigma, 0, \nabla\rho_\sigma, 0)\}, \tag{5.35}$$

where $\varepsilon_\text{C}^{\text{GGA}}$ is the PBE-GGA correlation energy density, and τ_σ^W is the kinetic energy density given by von Weizsäcker (1935):

$$\tau_\sigma^\text{W}[\rho] = \frac{1}{8}\frac{|\nabla\rho_\sigma|^2}{\rho_\sigma}, \tag{5.36}$$

which is *exact* for a one-electron density. Therefore, the correlation energy vanishes for any one-electron density, irrespectively of the value of the parameter C. For many-electron systems the self-interaction cancelation is not complete, but the error is shifted to fourth order in the gradient, thus having little effect on systems with slowly varying density. Perdew *et al.* obtained a value of $C = 0.53$ by fitting it to PBE-GGA surface correlation energies for jellium. Atomic correlation energies also agree, but are slightly less accurate. As an example, the correlation energy for He is -0.84 hartree in LSDA, -0.68 hartree in PBE-GGA, and -0.48 hartree in MGGA, which basically coincides with the exact value (Seidl *et al.*, 1999).

An assessment of the general quality of the PKZB meta-GGA in comparison to GGA and hybrid HF-GGA approaches of the B3LYP type (Adamo *et al.*, 2000) showed that the kinetic energy density is a useful additional ingredient. Atomization energies are quite improved in PKZB meta-GGA with respect to PBE-GGA, but unfortunately geometries and frequencies are worsened. In particular, bond lengths are far too long. It has been argued (Adamo *et al.*, 2000) that a possible reason could be that in this functional the long-range part of the exchange hole, which would help localize the exchange hole, thus favoring shorter bond lengths, is missing.

A fix to this problem was proposed by Tao *et al.* (2003) (TPSS), and analyzed by Perdew *et al.* (2004). These authors first recognized that the meta-GGA formalism is unable to reproduce the exact exchange energy density for arbitrary one- and two-electron densities. In place of this, the condition imposed is for the exchange potential to be finite at the position of the nucleus in those two cases. This feature, which is present in the LDA and LSDA, is lost in the GGA. To recover it within the meta-GGA, TPSS proposed a modification of Expression (5.31) such that the exchange enhancement factor F_X is independent of the reduced gradient for values of $s = \sqrt{p} \approx 0.4$, which is generally the case near the nuclei of atoms and molecules, where the electron density is dominated by the $1s$ electronic state. In addition, the parameter D was set to back to zero. The correlation functional was also slightly modified to avoid a spurious enhancement of the correlation energy density in the inter-shell region of atoms. The TPSS meta-GGA proved very accurate for a range of properties and systems, preserving the accuracy of atomization energies while correcting geometries and frequencies, including hydrogen bonds (Staroverov *et al.*, 2003; Perdew *et al.*, 2004).

Similar functionals, including the kinetic energy density, where between 10 and 20 parameters were fitted to chemical data, were also proposed (Becke, 1998; van Voorhis and Scuseria, 1998). The quality of the results is similar to the previous cases. In fact, one of the semiempirical meta-GGA functionals (van Voorhis and Scuseria, 1998) gives very good geometries and frequencies, of quality comparable to TPSS. Also, energy barriers are quite reasonable in meta-GGA, improving over the GGA (Grüning *et al.*, 2004). Other authors explored the idea of using a correlation functional where the reference system is changed from the (gap-less) homogeneous electron gas, where the electronic states are occupied up to the Fermi level, to a uniform electron gas with a gap in the excitation spectrum. This latter is closer to molecular and semiconducting systems (Krieger *et al.*, 1999; Toulouse *et al.*, 2002).

Yet some issues remain unresolved at the meta-GGA level. The most important one remains the incomplete cancelation of the self-interaction by the exchange term. Also static correlation effects in the form of fractional occupations (multi-reference ground states) and van der Waals interactions are beyond the capabilities of meta-GGA, but it has been argued that these can be recovered by including additional terms in the gradient expansion (Perdew and Schmidt, 2001; Perdew et al., 2004).

5.3 Non-locality: the weighted density approximation

The exact pair correlation function $g(\mathbf{r}, \mathbf{r}')$ is a truly non-local object that depends on the electronic density in two different points, $\mathbf{r}$ and $\mathbf{r}'$. According to (4.63), the true XC energy corresponds to the Coulomb interaction between the charge density at $\mathbf{r}$ and the XC hole at $\mathbf{r}'$. The LDA, however, approximates the XC hole as in (5.4). It has already been mentioned that this approximation implies not only using locally the pair correlation function of the homogeneous electron gas, but also modifying the location of the center of the XC hole from $\mathbf{r}'$ to $\mathbf{r}$. This last feature is retained in the various gradient approximations (GGA and meta-GGA), while the expression for the XC energy density is modified to take into account inhomogeneities of the density in a semi-local manner. This is why GGAs and meta-GGAs are very efficient computationally, but struggle to capture non-local XC effects.

A serious limitation in the construction of a good non-local approximation to the pair correlation function of an inhomogeneous system is that the homogeneous $g(\mathbf{r}, \mathbf{r}')$ corresponds to a density that is uniform everywhere. Therefore, even if it is known exactly for the exchange part (4.61), and to an excellent degree for correlation (Perdew and Wang, 1992b), the question arises of which one of the two densities, $\rho(\mathbf{r})$ or $\rho(\mathbf{r}')$, should be used to construct $g(\mathbf{r}, \mathbf{r}')$ in an inhomogeneous environment. One possibility is to calculate the pair correlation function at an *average* density $\bar{\rho}(\mathbf{r})$, which in general is different from $\rho(\mathbf{r})$, and incorporates information about the density in the neighborhood (Alonso and Girifalco, 1978; Gunnarsson and Jones, 1980). Any averaging procedure has to verify a crucial condition: the approximate XC hole must contain exactly one displaced electron. This corresponds to the following normalization condition:

$$\int \rho_{XC}^{WDA}(\mathbf{r}, \mathbf{r}')\,d\mathbf{r}' = \int \rho(\mathbf{r}') \left[g^{WDA}(\mathbf{r}, \mathbf{r}') - 1 \right] d\mathbf{r}' = -1, \quad (5.37)$$

where the pair correlation function is written as

$$g^{WDA}(\mathbf{r}, \mathbf{r}') = 1 + G^{WDA}\left[|\mathbf{r} - \mathbf{r}'|, \bar{\rho}(\mathbf{r}) \right]. \quad (5.38)$$

Approaches of this type receive the name of *weighted density approximation* (WDA) or the related *average density approximation* (ADA). Clearly, there is an enormous freedom in choosing the form of the function G^{WDA}, provided that normalization is verified. In fact, several different approximations have been proposed (Alonso and Girifalco, 1978; Gunnarsson and Jones, 1980; Denton and Ashcroft, 1989; Lutsko and Baus, 1990). The WDA inherits some important properties of the exact exchange, like the correct asymptotic behavior of the XC potential at long distance, and the self-interaction cancelation. Let us remark that there is a formal problem in this approach because in general $\bar{\rho}(\mathbf{r}) \neq \bar{\rho}(\mathbf{r}')$, and then the $\mathbf{r} \to \mathbf{r}'$ symmetry of $g(\mathbf{r}, \mathbf{r}')$ is broken. It has been shown, however, that this has a small effect on calculated properties (García-González et al., 2000).

One of the main limitations of the LDA that can be improved by the WDA is the incorrect location of the center of the XC hole. Apart from earlier work, an exploration in the context of realistic electronic structure calculations was carried out by D. Singh, with not particularly encouraging results (Singh, 1993). More recently, Rushton et al. (2002a,b) studied different forms of the WDA and compared them to variational Monte Carlo results (Nekovee et al., 2001). By generalizing the approach of Gunnarsson and Jones (1980), they proposed a WDA pair correlation of the form

$$G^{\text{WDA}}\left[|\mathbf{r} - \mathbf{r}'|, \bar{\rho}(\mathbf{r})\right] = C(\bar{\rho}) f^{\text{WDA}}\left(\frac{\lambda(\bar{\rho})}{|\mathbf{r} - \mathbf{r}'|}\right). \tag{5.39}$$

Gunnarsson and Jones originally used $f^{\text{WDA}}(x) = 1 - \exp(-x^5)$, while Rushton et al. extended this study to a number of functions of different types, including second and fourth order Gaussians, Lorentzians, and variants of the Gunnarsson–Jones function (Charlesworth, 1996; Rushton et al., 2002b). The functions $\lambda(\bar{\rho})$ and $C(\bar{\rho})$ are determined by requesting that the sum rule (5.37) is verified, and that the XC energy density for a uniform gas is reproduced. This leads to the following expressions:

$$\lambda(x) = -\frac{I_1}{2I_2 \epsilon_{\text{XC}}^{\text{LDA}}(x)}, \tag{5.40}$$

$$C(x) = -\frac{1}{I_2 x \lambda^3(x)}, \tag{5.41}$$

with

$$I_n = 4\pi \int x^n f^{\text{WDA}}(x) \, \mathrm{d}x. \tag{5.42}$$

This approach restores the center of the XC hole to the correct place. It also shows that, for very inhomogeneous systems, the errors introduced by the LDA can be corrected by the introduction of a semi-local term proportional to the Laplacian of the density, as in meta-GGA (Rushton et al., 2002a). In addition, it has been shown that the accuracy in structural properties is connected to the quality of the XC hole in the high density region. The problem of this approach remains in the enormous freedom to choose the pair correlation function, which permits us to fit any desired quantity. While this is unsatisfactory, the promising results indicate that this avenue is worthwhile to continue exploring.

5.4 Hybrid HF-KS approaches

The observation that LDA and GGA trends are opposite to those of Hartree–Fock motivated the development of approximations which combine these two approaches. These involve a DFT correlation with a combination of DFT and Hartree–Fock exchange:

$$E_{XC}^{hyb} = \alpha E_X^{HF} + (1 - \alpha) E_X^{DFT} + E_C^{DFT}, \qquad (5.43)$$

where the coefficient α is either chosen to assume a specific value such as 1/2, or is fitted to some properties of a molecular database.

An example of the latter is the approximation known as B3LYP (Becke, 1993). This reproduces the geometries and binding energies of molecular systems, to the same accuracy of low-level correlated quantum chemistry approaches like second order Møller–Plesset perturbation theory (MP2), and sometimes even at a higher level such as coupled clusters or CI methods. The advantage is a significantly lower computational cost. This idea is appealing and physically sensible, but the approach of fitting the coefficient α to molecular data is not compatible with the *ab initio* approach.

Interestingly, it has been shown that hybrid schemes have a rigorous formal justification within the exact DFT scheme (Görling and Levy, 1997), when this is formulated as a generalized Kohn–Sham approach (Seidl *et al.*, 1996). The reference system is still described by a Slater determinant, but, at variance with the usual Kohn–Sham approach, it does not correspond to the minimization of the expectation value of the kinetic energy $F_{KS} = \langle \Phi | \hat{T} | \Phi \rangle$. Instead, it is obtained by minimizing a different functional under the constraint that the wave function can be written as a Slater determinant. If this functional is chosen to be

$$F_{HF-KS} = \langle \Phi | \hat{T} + \hat{V}_{ee} | \Phi \rangle = F_{KS} + E_H[\Phi] + E_X[\Phi], \qquad (5.44)$$

the scheme resembles the Hartree–Fock method, but it contains an unknown, formally exact correlation term that is absent in standard HF. This approach is known as the Hartree–Fock–Kohn–Sham scheme (HF-KS), and it was first suggested by Baroni and Tuncel (1983). Another possibility is to construct a functional where the exchange term is replaced by a *screened exchange* (sX), with the bare Coulomb interaction replaced with a statically screened interaction of the Yukawa type. This scheme is better suited for approximations, and it has been shown to lead to improved bandgaps in semiconductors (Bylander and Kleinman,

1990; Seidl *et al.*, 1996). The generalized Kohn–Sham equations can be written in the form of the usual KS equations (4.34), supplemented with corrective terms:

$$\left\{\hat{H}_{\text{KS}}[\rho] + \Delta\hat{\mu}_{\text{X}}[\{\varphi_i\}] + \Delta\hat{\mu}_{\text{C}}\right\}\varphi_i(\mathbf{r}) = \varepsilon_i\varphi_i(\mathbf{r}) \quad (5.45)$$

that are likely to be small, but not unimportant.

Hybrid HF-KS schemes like (5.43) also correspond to a precise realization of the generalized Kohn–Sham scheme, where the functional

$$F^\alpha_{\text{HF-KS}} = \langle\Phi|\hat{T} + \alpha\hat{V}_{\text{ee}}|\Phi\rangle = F_{\text{KS}} + \alpha E_{\text{H}}[\Phi] + \alpha E_{\text{X}}[\Phi] \quad (5.46)$$

is minimized with respect to a wave function Φ of the determinantal form, and the rest of the energy is treated within the usual approximations to the Kohn–Sham problem, typically GGA (Becke, 1993) and more recently meta-GGA (Krieger *et al.*, 1999; Staroverov *et al.*, 2003). Therefore, any value of α is in principle acceptable for a hybrid HF-KS scheme. The optimal value obtained by fitting to a database, such as in B3LYP, is the one that maximizes the error cancellation between the approximate exchange and correlation functionals (Seidl *et al.*, 1996).

5.5 Exact exchange: the optimized effective potential method

One of the main limitations of DFT is that, even if the exact exchange-correlation energy is a functional of the density, this functional is not known explicitly, and approximations are required. The kinetic energy functional is not explicitly known in terms of the density either. However, the exact expression *in terms of the orbitals* is known for non-interacting electrons. This is the basis for Kohn–Sham theory, where the problem of interacting electrons in an external potential is mapped onto a problem of non-interacting electrons in an effective *local* potential, provided that they share the same electronic density. This indicates that expressing energy contributions in terms of the orbitals is perfectly allowed within DFT, being the orbitals (unknown) functionals of the density. This opens the door to the exact treatment of exchange, which is also a well-known *non-local* functional of the orbitals, given by the Fock expression. The problem is that this non-local functional does not find a natural place within the Kohn–Sham scheme, which requires a local effective potential.

A way forward is to reformulate Kohn–Sham theory by retaining the exact expression for the exchange functional. The resulting energy expression, which depends explicitly on the one-electron orbitals, is then minimized under the constraint that these orbitals are the solutions of a *non-interacting problem in a local effective potential*, which is the same for all the electrons. Inspired by the work of Slater (1951), in 1953, and as an alternative to HF theory, Sharp and Horton suggested that the electronic many-body problem could also be formulated as a search for a *local potential*. This should make the energy obtained by solving the one-electron equations stationary with respect to small variations of this potential (Sharp and Horton, 1953). This was called the *optimized effective potential method* (OEP). This approach was re-derived by Talman and Shadwick (1976), always within the Hartree–Fock context, and the formal equivalence between the OEP and DFT was demonstrated later (Sahni et al., 1982; Perdew and Norman, 1982).

Mathematically, the OEP method involves two equations. First, the condition that the energy is stationary with respect to the effective potential

$$\frac{\delta E_{\text{KS}}[\{\varphi_{j\tau}\}]}{\delta v_\sigma^{\text{OEP}}} = 0. \quad (5.47)$$

Second, the orbitals $\varphi_{i\sigma}(\mathbf{r})$ are the solutions of the one-electron equation

$$\left\{-\frac{\hbar^2}{2m}\nabla^2 + v_\sigma^{\text{OEP}}[\{\varphi_{j\tau}\}](\mathbf{r})\right\}\varphi_{i\sigma}(\mathbf{r}) = \varepsilon_{i\sigma}\varphi_{i\sigma}(\mathbf{r}). \quad (5.48)$$

As usual, the Kohn–Sham energy is written

$$E_{\text{KS}}[\{\varphi_{j\tau}\}] = T_{\text{R}}[\{\varphi_{j\tau}\}] + \int \rho(\mathbf{r})v_{\text{ext}}(\mathbf{r})d\mathbf{r} + \frac{1}{2}\int\int\frac{\rho(\mathbf{r})\rho(\mathbf{r}')}{|\mathbf{r}-\mathbf{r}'|}d\mathbf{r}d\mathbf{r}'$$
$$+ E_{\text{XC}}[\{\varphi_{j\tau}\}]. \quad (5.49)$$

In the above we have included a spin index (τ) to be consistent with the spin-dependence of the exact-exchange functional. The last expression is simply the definition of E_{XC}, and T_{R} is the non-interacting kinetic energy.

The OEP can be derived from these equations, but here we present an alternative formulation originally proposed by Görling and Levy (1994) (see also Görling, 1996). This is basically an extension of Kohn–Sham formulation, but taking into account that the XC potential is an orbital-dependent functional. The local effective potential is

$$v_\sigma^{\text{OEP}}[\{\varphi_{j\tau}\}] = v_{\text{ext}}(\mathbf{r}) + \int\frac{\rho(\mathbf{r}')}{|\mathbf{r}-\mathbf{r}'|}d\mathbf{r}' + \mu_{\text{XC}\sigma}^{\text{OEP}}[\{\varphi_{j\tau}\}]. \quad (5.50)$$

To calculate $\mu_{\text{XC}\sigma}^{\text{OEP}}$ we have to introduce the intermediate step of varying the XC energy with respect to the orbitals. This is done by applying the chain rule for functional derivatives:

$$\mu_{\text{XC}\sigma}^{\text{OEP}}(\mathbf{r}) = \frac{\delta E_{\text{XC}}[\{\varphi_{j\tau}\}]}{\delta\rho_\sigma(\mathbf{r})} = \sum_\nu\sum_{i=1}^{N_\nu}\int\left(\frac{\delta E_{\text{XC}}}{\delta\varphi_{i\nu}(\mathbf{r}')}\right)\left(\frac{\delta\varphi_{i\nu}(\mathbf{r}')}{\delta\rho_\sigma(\mathbf{r})}\right)d\mathbf{r}' + \text{c.c.}, \quad (5.51)$$

where c.c. indicates the complex conjugate of the first term. The orbitals, in turn, are connected only *implicitly* with the density through the local potential.

5.5 Exact exchange: the OEP method

Therefore, we have to introduce another intermediate step of derivation with respect to v_μ^{OEP}:

$$\mu_{\text{XC}\sigma}^{\text{OEP}}(\mathbf{r}) = \sum_{\nu\mu} \sum_{i=1}^{N_\nu} \int\int \left(\frac{\delta E_{\text{XC}}}{\delta \varphi_{i\nu}(\mathbf{r}')}\right)\left(\frac{\delta \varphi_{i\nu}(\mathbf{r}')}{\delta v_\mu^{\text{OEP}}(\mathbf{r}'')}\right)\left(\frac{\delta v_\mu^{\text{OEP}}(\mathbf{r}'')}{\delta \rho_\sigma(\mathbf{r})}\right) d\mathbf{r}' d\mathbf{r}''. \quad (5.52)$$

The second factor in the product is the variation of the non-interacting orbitals with respect to the potential, and can be calculated using first-order perturbation theory:

$$\frac{\delta \varphi_{i\nu}(\mathbf{r}')}{\delta v_\mu^{\text{OEP}}(\mathbf{r}'')} = \delta_{\nu\mu} \sum_{k\neq i}^{\infty} \left[\frac{\varphi_{k\mu}^*(\mathbf{r}')\varphi_{k\mu}(\mathbf{r}'')}{\varepsilon_{i\mu} - \varepsilon_{k\mu}}\right] \varphi_{i\mu}(\mathbf{r}'') = G_{i\mu}^{\text{R}}(\mathbf{r}', \mathbf{r}'')\varphi_{i\mu}(\mathbf{r}''), \quad (5.53)$$

where $G_{i\mu}^{\text{R}}(\mathbf{r}', \mathbf{r}'')$ is the Green's function of the non-interacting system, given by:

$$G_{i\mu}^{\text{R}}(\mathbf{r}', \mathbf{r}'') = \sum_{k\neq i}^{\infty} \frac{\varphi_{k\mu}(\mathbf{r}')\varphi_{k\mu}(\mathbf{r}'')}{\varepsilon_{i\mu} - \varepsilon_{k\mu}}. \quad (5.54)$$

The third factor is the variation of the potential with respect to the density. This is just the inverse of the linear response function of the non-interacting system, which is related to the Green's function by

$$\chi_\sigma(\mathbf{r}, \mathbf{r}'') = \delta_{\sigma\mu} \frac{\delta \rho_\sigma(\mathbf{r})}{\delta v_\mu^{\text{OEP}}(\mathbf{r}'')} = \delta_{\sigma\mu} \sum_{i=1}^{N_\sigma} G_{i\sigma}^{\text{R}}(\mathbf{r}'', \mathbf{r})\varphi_{i\sigma}(\mathbf{r}'')\varphi_{i\sigma}^*(\mathbf{r}). \quad (5.55)$$

According to (5.54), $G_{i\sigma}^{\text{R}}$ is orthogonal to $\varphi_{i\sigma}$, and then $\int \chi_\sigma(\mathbf{r}'', \mathbf{r})d\mathbf{r}'' = 0$. Therefore, the linear response function is not invertible. In a plane wave representation, this means that the $\mathbf{G} = \mathbf{0}$ component is zero and it should be excluded from the basis set (Städele *et al.*, 1997, 1999). This is simple to do in plane waves, but somewhat more complicated when dealing with atom-centered basis sets (Görling, 1999; Ivanov *et al.*, 1999).

If the restricted $\tilde{\chi}_\sigma(\mathbf{r}'', \mathbf{r})$ (no $\mathbf{G} = \mathbf{0}$ component) is considered, then the expression for the local XC potential corresponding to orbital-dependent functionals assumes the form:

$$\mu_{\text{XC}\sigma}^{\text{OEP}}(\mathbf{r}) = \sum_{i=1}^{N_\sigma} \int\int \left[\frac{\delta E_{\text{XC}}[\rho]}{\delta \varphi_{i\sigma}(\mathbf{r}')} G_{i\sigma}^{\text{R}}(\mathbf{r}', \mathbf{r}'') \varphi_{i\sigma}(\mathbf{r}'') + \text{c.c.}\right] \tilde{\chi}_\sigma^{-1}(\mathbf{r}'', \mathbf{r}) d\mathbf{r}' d\mathbf{r}''. \quad (5.56)$$

To avoid the inversion of the linear response function, we can multiply the LHS of (5.52) by $\chi_\sigma(\mathbf{r}', \mathbf{r})$, integrate with respect to $\mathbf{r}'$, and replace Expression (5.55). In this case, we obtain the *OEP integral equation*:

$$\sum_{i=1}^{N_\sigma} \int \varphi_{i\sigma}^*(\mathbf{r}) \left[\mu_{\text{XC}\sigma}^{\text{OEP}}(\mathbf{r}) - u_{\text{XC}i\sigma}^{\text{OEP}}(\mathbf{r})\right] G_{i\sigma}^{\text{R}}(\mathbf{r}', \mathbf{r}) \varphi_{i\sigma}(\mathbf{r})d\mathbf{r}' + \text{c.c.} = 0, \quad (5.57)$$

where

$$u_{\text{XC}i\sigma}^{\text{OEP}}(\mathbf{r}) = \frac{1}{\varphi_{i\sigma}^*(\mathbf{r})} \frac{\delta E_{\text{XC}}[\{\varphi_{j\tau}\}]}{\delta \varphi_{i\sigma}(\mathbf{r})}. \tag{5.58}$$

If the XC energy functional depends only on the density, then $u_{\text{XC}i\sigma}(\mathbf{r}) = \mu_{\text{XC}\sigma}[\rho]$ coincides with the usual XC potential in Kohn–Sham theory. By choosing $\mu_{\text{XC}\sigma}(\mathbf{r}) = \mu_{\text{XC}\sigma}[\rho]$ the OEP equation is automatically satisfied and the usual Kohn–Sham scheme is recovered. If this is not the case, then the OEP integral equation (5.57) has to be solved for the XC potential, which is then used to construct the local effective potential $v_\sigma^{\text{OEP}}(\mathbf{r})$.

Within this scheme, exchange can be treated exactly using the *orbital-dependent* functional given by the Fock expression:

$$E_{\text{X}}[\{\varphi_{j\tau}\}] = -\frac{1}{2} \sum_\mu \sum_{i,j=1}^{N_\mu} \int\int \frac{\varphi_{j\mu}^*(\mathbf{r})\varphi_{i\mu}^*(\mathbf{r}')\varphi_{i\mu}(\mathbf{r})\varphi_{j\mu}(\mathbf{r}')}{|\mathbf{r}-\mathbf{r}'|} d\mathbf{r}\, d\mathbf{r}', \tag{5.59}$$

which is then used to calculate $u_{\text{X}i\sigma}^{\text{OEP}}(\mathbf{r})$ via (5.58):

$$u_{\text{X}i\sigma}^{\text{OEP}}(\mathbf{r}) = -\frac{1}{\varphi_{i\sigma}^*(\mathbf{r})} \sum_{j=1}^{N_\sigma} \varphi_{j\sigma}^*(\mathbf{r}) \int \frac{\varphi_{i\sigma}^*(\mathbf{r}')\varphi_{j\sigma}(\mathbf{r}')}{|\mathbf{r}-\mathbf{r}'|} d\mathbf{r}'. \tag{5.60}$$

Besides exact exchange, another possible orbital-dependent XC functional is the *self-interaction corrected* (SIC) approach of Perdew and Zunger (1981):

$$E^{\text{SIC}}[\{\varphi_{j\tau}\}] = E_{\text{XC}}^{\text{LSDA}}[\rho_\uparrow,\rho_\downarrow] - \sum_\mu \sum_{i=1}^{N_\mu} \left(E_{\text{XC}}^{\text{LSDA}}[\rho_{i\mu}] + E_{\text{H}}[\rho_{i\mu}]\right), \tag{5.61}$$

where $\rho_{i\mu}(\mathbf{r}) = |\varphi_{i\mu}(\mathbf{r})|^2$, E_{H} is the Hartree energy functional, and $E_{\text{XC}}^{\text{LSDA}}$ is the LSDA exchange-correlation functional. When the SIC approach is formulated following the OEP guidelines, the above expression leads to the potential (Garza et al., 2000):

$$u_{i\sigma}^{\text{SIC}}(\mathbf{r}) = \mu_{\text{XC}}^{\text{LSDA}}[\rho_{i\sigma}] - \int \frac{\rho_{i\sigma}(\mathbf{r}')}{|\mathbf{r}-\mathbf{r}'|} d\mathbf{r}'. \tag{5.62}$$

SIC functionals are thus non-invariant against unitary transformations of the orbitals. This feature is somewhat unsatisfactory because the potential depends on the choice of orbitals, which is not unique. For solids, if extended Bloch orbitals are chosen, the correction is zero. However, if a transformation to localized Wannier orbitals is performed, the SIC is finite.

The OEP equations have to be solved self-consistently, exactly as in the Kohn–Sham scheme, and can be implemented in their exact form or be re-cast in approximate, more easily solvable forms. They have been usually implemented in the exchange-only form (Krieger et al., 1992; Li et al., 1993; van Gisbergen *et al.*, 1999; Ivanov *et al.*, 1999) or augmented with the LDA or GGA correlation functionals (Bylander and Kleinman, 1995; Städele *et al.*, 1999), but there have been other approaches that included orbital-dependent correlation functionals (Grabo *et al.*, 1998; Grabowski *et al.*, 2002).

5.5.1 The Krieger–Li–Iafrate approximation

The OEP formulation, although known for a long time, was not used in practical applications until the early nineties, except for the early work of Talman and Shadwick. The main reason was that solving the full integral equation numerically was an extremely demanding task, which could only be achieved for spherically symmetric systems like atoms (Li *et al.*, 1992; Krieger *et al.*, 1992; Li *et al.*, 1993; Engel and Vosko, 1993) and solids within the atomic sphere approximation (Kotani, 1994).

The following alternative, but still exact, expression for the OEP integral was proposed by Krieger *et al.* (1992)

$$\mu_{XC\sigma}^{OEP}(\mathbf{r}) = \frac{1}{2\rho_\sigma(\mathbf{r})} \sum_{i=1}^{N_\sigma} |\varphi_{i\sigma}(\mathbf{r})|^2 \left[v_{XCi\sigma}(\mathbf{r}) + \left(\bar{\mu}_{XCi\sigma}^{OEP} - \bar{u}_{XCi\sigma}\right)\right] + \text{c.c.}, \quad (5.63)$$

where

$$v_{XCi\sigma}(\mathbf{r}) = u_{XCi\sigma}(\mathbf{r}) - \frac{1}{|\varphi_{i\sigma}(\mathbf{r})|^2} \nabla \cdot [\psi_{i\sigma}^*(\mathbf{r}) \nabla \varphi_{i\sigma}(\mathbf{r})], \quad (5.64)$$

$\psi_{i\sigma}(\mathbf{r})$ are the solutions of the inhomogeneous Kohn–Sham-like equation:

$$\left(\hat{H}_\sigma^{OEP} - \varepsilon_{i\sigma}\right) \psi_{i\sigma}(\mathbf{r}) = -\left[\mu_{XCi\sigma}^{OEP}(\mathbf{r}) - u_{XCi\sigma}(\mathbf{r}) - \left(\bar{\mu}_{XCi\sigma}^{OEP} - \bar{u}_{XCi\sigma}\right)\right] \varphi_{i\sigma}(\mathbf{r}), \quad (5.65)$$

and the bars indicate averages over $|\varphi_{i\sigma}(\mathbf{r})|^2$.

This formulation is strictly equivalent to (5.57), but it admits a reasonably well-controlled mean-field approximation, which is obtained by neglecting the second term in Eq. (5.64), i.e. using $v_{XCi\sigma}(\mathbf{r}) = u_{XCi\sigma}(\mathbf{r})$. Even if this is not generally true, it can be shown that the average of the neglected term is zero (Grabo *et al.*, 1998). This is the Krieger–Li–Iafrate (KLI) approximation, where the averaged XC potential is obtained as the solution of a set of coupled linear equations. These equations are much simpler to solve than the original OEP equation, and have been used in a number of different contexts, mainly in atomic and molecular systems (Grabo *et al.*, 1998). The approach based on solving the full integral equation is usually called exact exchange (EXX). Both EXX and KLI are based on the OEP philosophy.

An alternative road not based in the OEP has been proposed by della Salla and Görling (2001). This consists of constructing a local exchange potential from the non-local Fock expression by solving a self-consistent equation, and has been called *localized Hartree–Fock* (LHF). It has been shown that KLI is an approximation to LHF.

5.5.2 Properties of exact exchange in the OEP approach

The formal properties of both the EXX approach and the KLI approximation have been considered in detail by Grabo *et al.* (1998) (see also Kohanoff and Gidopoulos, 2003). The most important ones concerning the exact exchange (no correlation) functional are the following.

(i) Due to the exact cancelation of the self-interaction, the EXX and KLI exchange potentials decay (correctly) as $-e^2/r$ at long distances in vacuum regions. This is not the case of the LDA and GGA. This corrects some features such as the dissociation limit in molecules and the image potential due to metallic surfaces.

(ii) The principle of integer preference (Perdew *et al.*, 1982), which states that the exact XC potential considered as a continuous function of the number of electrons is discontinuous at integer values – so that integer numbers of electrons are preferred – is verified both in EXX and in KLI, but not in LDA or GGA. A consequence of the lack of integer preference is the well-known underestimation of the bandgap in bulk semiconductors.

(iii) At variance with HF theory, which at first glance may seem equivalent, the long-range decay of the exchange potential into vacuum regions goes as $-e^2/r$ for *all* states, irrespective of whether they are occupied or empty. In HF, the potential corresponding to empty orbitals decays exponentially and thus cannot support empty bound states. In the LDA and GGA the exchange potential decays exponentially for all states, occupied and empty, so that only a few bound empty states are possible. Moreover, many negatively charged ions are not bound. The EXX-OEP solves these problems, and has been shown to support a whole Rydberg molecular series (Görling, 1999). Similar behavior is observed in the SIC-OEP scheme (Garza *et al.*, 2000), where the HOMO-LUMO gaps appear to be good estimates of excitation energies.

(iv) In exchange-only calculations, i.e. when neglecting the correlation term, the spin-unrestricted Hartree–Fock (SUHF) approach gives the variationally lowest ground state energy. It has been shown that the EXX approach gives energies that are only marginally larger than E_{SUHF} (Görling and Ernzerhof, 1995). This fact might appear as an inconsistency, because both the SUHF and the EXX approaches are exact. However, the nature of the HF (non-local X potential) and DFT (local X potential) approaches is different, in the sense that the partition between exchange and correlation energies is different in the two schemes. Therefore, this small difference is related to the fact that correlation has been neglected, and should disappear when the exact correlation is considered.

5.6 Orbital-dependent correlation functionals

While the exchange contribution to the energy and potential is well known, and has been usually approximated in DFT because of its computational cost, the correlation contribution in the general case of an inhomogeneous electronic system is still unknown in a closed form. A few simple cases, like the homogeneous electron gas and some atomic systems (especially He), have been studied numerically very accurately, so that nowadays there are a number of benchmarks to compare the quality of different approximations to correlation.

In order for an approximation to exchange and correlation to be reliable, it is required that both terms be treated consistently. This is one of the main achievements of the LDA, where both energy functionals are approximated in the same limit of a locally homogeneous system. Therefore, even if separately each term is not particularly accurate, the sum of the two terms is rather precise, at least from the energetic point of view. The same can be said about properly constructed GGAs, where exchange and correlation are treated consistently (Perdew *et al.*, 1996). In this case, the reference system is an electron gas whose density is slowly varying in space.

Proper GGAs are, by construction, better approximations than the LDA. They should include the LDA as a limiting case when the density gradient terms in the functional are neglected, although some popular GGAs like the BLYP functional do not satisfy this limit. Therefore, if for some system a proper GGA performs worse than the LDA when compared to experimental data, this means that the good performance of the LDA was actually fortuitous and based mainly on cancelation of errors. This happens, e.g. for noble metals, where the LDA lattice constant is quite close to the experimental value, while PBE produces an expansion of a few percent.

This is the main reason for the seemingly poor performance of exact exchange combined with local correlation functionals, i.e. EXX-LDA and EXX-GGA. This is more dramatic in atoms and molecules than in solids, because there, local correlation functionals based on the homogeneous electron gas are particularly inappropriate. Even the Colle–Salvetti pair correlation function, where the radius of the correlation hole is parameterized to fit the correlation energy of the He atom (Tao *et al.*, 2001), strongly departs from the correct long-range dependence. In fact, for the uniform electron gas it should decay as r^{-4}, but the assumption of a strong Gaussian damping for the pair Jastrow correlation factor prevents such a decay.

This, together with other studies, indicates that the main limitation of such a *hybrid* (exact exchange – local correlation) approach is that the long-range tail of exchange, which is treated exactly, is not properly compensated by a similar long-range tail of opposite sign in the correlation term. Therefore, there is a clear need for improvement at the level of correlation functionals. A possible route to include this limit is to make a connection with the random phase approximation (RPA), which is known to treat long-range correlations properly (Langreth and Perdew, 1977). The short-range behavior of the RPA correlation is, however, rather poor (Singwi et al., 1968). Therefore, a different approach is needed in that region. One possibility is to connect it with the standard GGA (Kurth and Perdew, 1999) – or new variants of the GGA (Yan et al., 2000) – at short distances, where GGAs are rather accurate. Nevertheless, other partitionings between short and long range are also possible (Leininger et al., 1997; Kohn et al., 1998).

5.6.1 The Colle–Salvetti functional

A simple orbital-dependent correlation functional was proposed by Colle and Salvetti (1975) starting from a correlated Jastrow-type many-electron wave function, and performing a series of approximations. The expression of the correlation energy, as given by Lee et al. (1988), is the following:

$$E_C^{CS}(\{\varphi_{j\sigma}\}) = -ab \int d\mathbf{r}\, \gamma(\mathbf{r})\xi(\mathbf{r}) \left[\sum_\sigma \rho_\sigma(\mathbf{r}) \sum_i |\nabla\varphi_{i\sigma}(\mathbf{r})|^2 - \frac{1}{4}|\nabla\rho_\sigma(\mathbf{r})|^2 \right]$$

$$- ab \int d\mathbf{r}\, \gamma(\mathbf{r})\xi(\mathbf{r}) \left[-\frac{1}{4}\sum_\sigma \rho_\sigma(\mathbf{r})\nabla^2\rho_\sigma(\mathbf{r}) + \frac{1}{4}\rho(\mathbf{r})\nabla^2\rho(\mathbf{r}) \right]$$

$$- a \int d\mathbf{r}\, \gamma(\mathbf{r}) \frac{\rho(\mathbf{r})}{\eta(\mathbf{r})}, \qquad (5.66)$$

where

$$\gamma(\mathbf{r}) = 4\frac{\rho_\uparrow(\mathbf{r})\rho_\downarrow(\mathbf{r})}{\rho^2(\mathbf{r})}, \qquad (5.67)$$

$$\eta(\mathbf{r}) = 1 + d\rho(\mathbf{r})^{-1/3}, \qquad (5.68)$$

$$\xi(\mathbf{r}) = \frac{\rho^{-5/3}(\mathbf{r})e^{-c\rho(\mathbf{r})^{-1/3}}}{\eta(\mathbf{r})}, \qquad (5.69)$$

with $a = 0.04918$, $b = 0.132$, $c = 0.2533$, and $d = 0.349$.

The original CS functional is orbital-dependent and can be used in an OEP scheme in conjunction with the EXX or KLI exchange functional. This has been done by Grabo et al. (1998) (KLICS), and the results compared to GGA, correlated quantum chemical calculations, and exact results, for atomic and molecular systems. The spirit of the Colle–Salvetti functional is the same as GGA's in the sense that it is constructed in terms of the local density and the gradients of the orbitals and the density. Therefore, it is an approximation which is valid in the case of slowly varying orbital densities, and cannot constitute a solution for the correlation problem in the general case. In fact, it has been shown that, although short-range correlations are well described by CS, long-range correlations are completely missing, and only a small fraction of the correlation energy is recovered (Tao *et al.*, 2001). In addition, the correlation potential for the He atom has the wrong sign.

5.6.2 Many-body perturbation theory: second order and re-summations

The search for accurate orbital-dependent correlation functionals can be put on a sound basis by making a connection with quantum many-body theory. In quantum chemistry, perturbative approaches such as Møller–Plesset have been used for a long time, and are customarily carried out to second and fourth order (MP2 and MP4). In DFT, the analogous density functional perturbation theory (DFPT) has been developed and discussed in detail by Görling and Levy (1993) (GL theory). We sketch below the salient features of GL theory and discuss some applications.

Görling and Levy considered the many-body Hamiltonian

$$\hat{\mathcal{H}}_\lambda = \hat{T} + \hat{V}_\lambda + \lambda \hat{U}_{ee}, \qquad (5.70)$$

where $0 \leq \lambda \leq 1$ is a coupling constant representing the strength of the electron–electron interaction. $\hat{V}_\lambda$ is constrained to be the local external potential that ensures that the ground state density ρ is independent of λ, and equal to the density of the fully interacting system. The total energy for interaction strength λ is written

$$E_\lambda[\rho] = E^{(0)}[\rho] + \lambda E^{(1)}[\rho] + E_C^\lambda[\rho], \qquad (5.71)$$

where $E^{(0)}[\rho]$ is the energy associated with the non-interacting Hamiltonian, $\lambda E^{(1)}[\rho] = E_X[\rho]$ is the exact exchange energy given by the Fock expression, and $E_C^\lambda[\rho]$ is formally given by

$$E_C^\lambda[\rho] = \left[\langle \Phi_\lambda | \hat{T} + \lambda \hat{U}_{ee} | \Phi_\lambda \rangle - \langle \Phi_0 | \hat{T} + \lambda \hat{U}_{ee} | \Phi_0 \rangle \right], \qquad (5.72)$$

where Φ_λ and Φ_0 minimize $\langle \hat{T} + \lambda \hat{U}_{ee} \rangle$ and $\langle \hat{T} \rangle$, respectively.

To obtain the correlation energy for a given coupling strength λ, use is made of exact scaling relations derived by Levy and Perdew (1985)

$$E_C^\lambda[\rho] = \lambda^2 E_C[\rho_{1/\lambda}], \qquad (5.73)$$

where $E_C[\rho_{1/\lambda}]$ is the correlation energy at full strength ($\lambda = 1$), but at a uniformly scaled density $\rho_{1/\lambda}(x, y, z) = \lambda^{-3}$, and is given by

$$E_C[\rho_{1/\lambda}] = \frac{1}{\lambda^2} \left[\langle \Phi_\lambda | \hat{T} + \lambda \hat{U}_{ee} | \Phi_\lambda \rangle - \langle \Phi_0 | \hat{T} + \lambda \hat{U}_{ee} | \Phi_0 \rangle \right]$$

$$= \frac{1}{\lambda^2} \left[E_\lambda - E_0 - \langle \Phi_0 | \hat{\mathcal{H}}_\lambda - \hat{\mathcal{H}}_0 | \Phi_0 \rangle \right], \quad (5.74)$$

where the λ-interacting Hamiltonian $\hat{\mathcal{H}}_\lambda$ is partitioned in the following way:

$$\hat{\mathcal{H}}_\lambda = \hat{\mathcal{H}}_0 + \lambda \left\{ \hat{U}_{ee} + \sum_{i=1}^{N} \left(-u(\mathbf{r}_i) - v_X(\mathbf{r}_i) - \lambda \frac{\delta E_C[\rho_\lambda]}{\delta \rho_\lambda(\mathbf{r}_i)} \right) \right\}. \quad (5.75)$$

$\hat{\mathcal{H}}_0$ is the non-interacting reference Hamiltonian, and the term in braces is the perturbation. Here, $u(\mathbf{r})$ is the Hartree potential and $v_X(\mathbf{r})$ is the local exchange potential in the OEP sense.

By grouping all the terms in the same power of λ, the following expansion is obtained:

$$E_C[\rho_{1/\lambda}] = \sum_{n=2}^{\infty} \lambda^{n-2} E_C^{(n)}[\rho], \quad (5.76)$$

where the terms $E_C^{(n)}[\rho]$ are calculated perturbatively in the non-interacting reference system, for which the orbitals are just Kohn–Sham's. The total correlation energy is obtained via coupling constant integration (Langreth and Perdew, 1977):

$$E_C[\rho] = \int_0^1 E_C[\rho_{1/\lambda}] d\lambda = \sum_{n=2}^{\infty} \frac{1}{n-1} E_C^{(n)}[\rho]. \quad (5.77)$$

The general expression for $E_C^{(n)}$ has been given by Görling and Levy (1993). While, in general, the perturbative terms in the above expansion are complicated and computationally very expensive, the second order term assumes the familiar form known from usual second order perturbation theory, only applied to Kohn–Sham states:

$$E_C^{(2)}[\rho] = -\sum_{k \neq 0}^{\infty} \frac{|\langle \Phi_0^0 | \hat{U}_{ee} - \hat{V}_H - \hat{V}_X | \Phi_k^0 \rangle|^2}{E_k^0 - E_0^0}, \quad (5.78)$$

with Φ_k^0 the kth (determinantal) excited state of the unperturbed Hamiltonian, and E_k^0 the corresponding energy. In terms of Kohn–Sham single-particle orbitals, the correlation energy is written

$$E_C^{GL2}(\{\varphi_j\}) = -\frac{1}{4} \sum_{ij\alpha\beta} \frac{|\langle \varphi_i \varphi_j | \hat{v}_2 | \varphi_\alpha \varphi_\beta \rangle|^2}{\varepsilon_\alpha + \varepsilon_\beta - \varepsilon_i - \varepsilon_j} - \sum_{i\alpha} \frac{|\langle \varphi_i | \hat{v}_X - \hat{f} | \varphi_\alpha \rangle|^2}{\varepsilon_\alpha - \varepsilon_i}, \quad (5.79)$$

5.6 Orbital-dependent correlation functionals

where $\hat{f}$ is the Fock-like, non-local exchange operator, but formed with the Kohn–Sham orbitals $\{\varphi_j\}$, which are Kohn–Sham single-particle orbitals with associated eigenvalues ε_j. The indices i and j correspond to occupied single-particle orbitals, while α and β indicate empty orbitals.

As in MP theory, the computational cost of higher-order terms in the series is prohibitively expensive, so that normally it would not be possible to afford more than second order DFPT. This is called second-order Görling–Levy theory (GL2). As such, GL2 theory has been used by Ernzerhof to calculate the energetics of atomization of molecular systems (Ernzerhof, 1996), and compared to the more traditional DFT and quantum chemical approaches such as LSDA, GGA, HF, and MP2. One important conclusion of this work is that, if the perturbative series for the correlation is simply cut at the GL2 level, then the resulting atomization energies are particularly bad, even worse than the LSDA values. Ernzerhof suggested that the one important limitation of this type of approximation is that some known exact limits, e.g. the limit of very strongly interacting systems ($\lambda \to \infty$), are not fulfilled. He then proposed an empirical re-summation of the series in the spirit of the GGA, such that these exact limits are verified. The comparison of these results with experimental atomization energies is very favorable, even improving over MP2 results in some difficult cases such as the F_2 and O_3 molecules.

A related approach was proposed under the name of OEP-MBPT(2) (Grabowski et al., 2002), and shown to provide correlation potentials that are qualitatively correct, at variance with other methods, including KLICS. The results, however, depend on which excitations are included in the perturbative analysis, and suggest that perturbative re-summations as in coupled clusters methods are required to achieve quantitative accuracy. The connection between OEP-MBPT(2) and GL2 theories was studied by Ivanov et al. (2003).

The various applications of the *bare* GL2 theory show a general feature of perturbative expansions: unless the perturbation is really very weak, the simplest approach of cutting the expansion at some low order is not quite successful. There are two possible reasons for that: first, that the higher order terms are not significantly smaller than the second order one; second, that for some range of values of the coupling parameter λ, the perturbative series could be divergent. This feature was already noticed by Görling and Levy (1993), who clearly stated that their expansion was based upon the assumption that the correlation energy can be expanded in a Taylor series for *all* values of the coupling constant $0 \leq \lambda \leq 1$.

This hypothesis was tested in Seidl et al. (2000b) on the basis of the atomization energies calculated in the GL2 approximation (Ernzerhof, 1996). Surprisingly, they found that, except for a few notable cases like H_2 and CH_4, the radius of convergence of the perturbative series was always $\lambda_c < 1$. In some cases it was very small ($\lambda_c = 0.06$ for B_2). Numerically, as a function of λ the correlation energy in (5.76) behaves well until λ_c, where it shows a tendency to diverge (it really diverges only if the infinite series is considered). This is reflected in the correlation energy as an oscillatory behavior as a function of the number of terms in the Taylor expansion.

Seidl *et al.* generalized the re-summation ideas proposed by Ernzerhof, according to the GGA spirit. They did this by asking that the correlation functional verifies the limit of very strong interaction ($\lambda \to \infty$) (Seidl *et al.*, 2000a), which is the region of interaction strengths where the perturbative expansion is likely to fail. Then, an interpolation formula was proposed between this limit at $\lambda \to \infty$ and the small λ limit, $E_{XC} \to E_X + E_C^{GL2}$. As a function of the number of terms in the expansion, the series oscillates around the correlation energy given by the interpolation formula (Seidl *et al.*, 2000b). Results presented for atomization energies of molecules, as in the case of Ernzerhof, compare extremely well with experiment. In the limit of weak interactions this approach is deficient and a re-summation scheme such as the RPA is required to obtain a finite correlation energy (Yan *et al.*, 2000).

5.7 Van der Waals (dispersion) interactions

The issue of van der Waals or dispersion interactions is a difficult benchmark in many-body theory that, ultimately, any correlation functional should address. Since the early work of Zaremba and Kohn (1976), up to recent proposals (Rapcewicz and Ashcroft, 1991; Andersson *et al.*, 1996; Dobson and Dinte, 1996; Kohn *et al.*, 1998), it has attracted a lot of attention. The origin of the van der Waals interaction between two non-chemically bonded fragments is the coupling of the electric field generated by fluctuations in the electronic density of one fragment with the density of the other fragment. This is a dynamical correlation effect that the usual local and semi-local functionals such as LDA and GGA cannot capture, and is not related to the exchange, so that EXX alone does not help either. At long distances the van der Waals interaction should approach the classical dipole–dipole interaction, which decays as $E_{vdW} = -C_6/R^6$. Typical van der Waals systems are dimers of closed-shell atoms like He_2, although dispersion interactions constitute a small but important component to the interatomic forces in molecular fluids and biological systems. In fact, the Lennard-Jones 12-6 potential (see Section 10.5) is widely used to represent non-bonding interactions in these contexts.

Most theoretical works concentrate precisely on reproducing this long-range behavior by separating the electron–electron interaction into a short-range and a long-range part. The long-range part is responsible for the van der Waals interaction, and can be typically represented by an effective interaction of the form

$$U_{ee}^{lr}(\mathbf{r}, \mathbf{r}') = \frac{A(\mathbf{r}, \mathbf{r}')}{|\mathbf{r} - \mathbf{r}'|^6}, \qquad (5.80)$$

where $A(\mathbf{r}, \mathbf{r}')$ depends on the choice of an effective density for the exchange-correlation linear response kernel, like $\rho_{\text{eff}} = \sqrt{\rho(\mathbf{r})\rho(\mathbf{r}')}$ (Rapcewicz and Ashcroft, 1991) or $\rho_{\text{eff}} = \{\sqrt{\rho(\mathbf{r})\rho(\mathbf{r}')}[\sqrt{\rho(\mathbf{r})} + \sqrt{\rho(\mathbf{r}')}]\}^{2/3}$ (Andersson *et al.*, 1996). With this, the long-range contribution to the XC energy from two well-separated fragments in volumes V_1 and V_2 is given by

$$E_{\text{XC}}^{lr} = -\frac{3}{\pi} \int_0^\infty du \int_{V_1} d\mathbf{r}_1 \int_{V_2} d\mathbf{r}_2 \frac{\chi_1^z(iu)\chi_2^z(iu)}{|\mathbf{r}_1 - \mathbf{r}_2|^6}, \qquad (5.81)$$

where $\chi_i^z(\omega)$ is the density response of a uniform electron gas of density $\rho(\mathbf{r}_i)$ to a perturbation in the direction of the bond. A more general approach that calculates the density response using real-time propagation has been proposed by Kohn *et al.* (1998). This finally leads to the well-known long-range behavior of the van der Waals interaction between two fragments separated by a distance R:

$$E_{\text{vdW}}(R) = -\left(\frac{3}{\pi} \text{Im} \int_0^\infty \chi_1^z(\omega)\chi_2^z(\omega) d\omega\right) \frac{1}{R^6}. \qquad (5.82)$$

Approaches of this type are able to reproduce this behavior with good accuracy, and also provide a way to introduce the dispersion interaction even if the fragments are not well-separated. In fact, perhaps more important than the long-range tail, the goal of a correlation functional is to reproduce the correct behavior of the whole potential energy surface, especially binding energies and bond lengths.

This aspect has been addressed by Engel *et al.* for the case of He$_2$ and Ne$_2$, using a different approach based on the previously introduced KLI-GL2 approach (KLI exchange combined with second order many-body perturbation theory) (Engel *et al.*, 2000). These authors compared this approach to LDA, HF, KLI (X-only), and MP2 calculations, and also to exact results. As expected, the LDA severely over-binds, while HF and X-only KLI do not bind the dimers at all. The two correlated approaches, MP2 and KLI-GL2, bind the dimer reasonably well. Compared to exact results for Ne$_2$ ($D_e = 3.6$ meV) (Aziz and Slaman, 1991), MP2 tends to under-bind ($D_e = 2.3$ meV) while KLI-GL2 over-binds ($D_e = 8.3$ meV). This is reflected at the level of geometry, where the MP2 bond length is somewhat long (6.06 bohr) against 5.48 bohr in KLI-GL2 and 5.84 bohr in the exact calculation. This indicates that correlation calculated perturbatively to second order is not sufficient, and higher orders are necessary to obtain a quantitative agreement.

5.8 Green's function approach: the GW approximation

A road that has been intensely explored during the past two decades is that of casting the electronic many-body problem in terms of the one particle's Green's function of the system $G(\mathbf{r}, \mathbf{r}'; t)$ (Fetter and Walecka, 1971). The Green's function represents the probability of having an electron at $\mathbf{r}$ at time t given that there was an electron at $\mathbf{r}'$ at time 0. In this sense it constitutes a generalization of the static pair correlation function $g(\mathbf{r}, \mathbf{r}')$ to the time domain. Once the Green's function of the many-electron system is known, then the total energy, electronic density, momentum distribution, density of states, electronic excitations, and any other property of interest can be calculated. In the energy domain, the Green's function is the solution of the Schrödinger-like equation

$$\left\{-\frac{\hbar^2}{2m}\nabla^2 + v_{\text{ext}}(\mathbf{r}) - E\right\} G(\mathbf{r}, \mathbf{r}'; E)$$
$$+ \int \Sigma(\mathbf{r}, \mathbf{r}''; E) G(\mathbf{r}'', \mathbf{r}'; E) d\mathbf{r}'' = -\delta(\mathbf{r} - \mathbf{r}'), \tag{5.83}$$

where $\Sigma(\mathbf{r}, \mathbf{r}''; E)$ is the *self-energy* operator, which in general is a non-local, non-hermitian, energy-dependent operator. A local component such as the Hartree potential $v_H(\mathbf{r})\delta(\mathbf{r} - \mathbf{r}'')$ or $\{v_H(\mathbf{r}) + \mu_{XC}[\rho](\mathbf{r})\}\delta(\mathbf{r} - \mathbf{r}'')$ can be separated out from Σ, and displaced to the first term. In that case, the remaining part of the self-energy describes non-local and dynamical exchange and correlation effects.

If the exact Green's function $G_0(\mathbf{r}, \mathbf{r}'; E)$ for a related reference system is known, then the required Green's function $G(\mathbf{r}, \mathbf{r}'; E)$ can be calculated via many-body perturbation theory through Dyson's equation:

$$G(\mathbf{r}, \mathbf{r}'; E) = G_0(\mathbf{r}, \mathbf{r}'; E) + \int\int G_0(\mathbf{r}, \mathbf{r}_1; E) \Delta\Sigma(\mathbf{r}_1, \mathbf{r}_2; E) G(\mathbf{r}_2, \mathbf{r}'; E) d\mathbf{r}_1 d\mathbf{r}_2, \tag{5.84}$$

where the perturbation is given by $\Delta\Sigma(\mathbf{r}_1, \mathbf{r}_2; E) = \Sigma(\mathbf{r}_1, \mathbf{r}_2; E) - v_2(\mathbf{r}_1, \mathbf{r}_2)$, with $v_2(\mathbf{r}_1, \mathbf{r}_2)$ the interaction potential of the reference system, e.g. $v_2 = 0$ for a reference system of non-interacting electrons.

Green's functions are exactly known for non-interacting electrons and, in general, for single-particle approximations such as Kohn–Sham and Hartree–Fock. These assume the form

$$G_0(\mathbf{r}, \mathbf{r}'; E) = \sum_n \frac{\varphi_n(\mathbf{r}) \varphi_n^*(\mathbf{r}')}{E - \varepsilon_n}, \tag{5.85}$$

where $\varphi_n(\mathbf{r})$ are the solutions of the single-particle equation and ε_n the corresponding eigenvalues. Normally, the reference system is chosen to be either Hartree–Fock or Kohn–Sham within the LDA, although the EXX has also been used recently (Aulbur *et al.*, 2000). If the LDA is chosen as the reference system, then $\Delta\Sigma = \Sigma - v_\mathrm{H} - \mu_\mathrm{XC}$, where $\mu_\mathrm{XC}(\mathbf{r}, \mathbf{r}'; E) = \mu_\mathrm{XC}^\mathrm{LDA}[\rho(\mathbf{r})]\delta(\mathbf{r}-\mathbf{r}')$ is the usual energy-independent LDA exchange-correlation potential and v_H is the (also local) Hartree potential.

The problem now is to calculate the self-energy, which contains all the exchange and correlation effects, static and dynamic, neglected in the reference system. In principle the self-energy should be obtained by solving the many-body problem at a higher level that involves two-particle Green's functions (Fetter and Walecka, 1971). This, however, is not a practicable solution. A sensible way forward is Hedin's *GW approximation* (GWA) (Hedin, 1965), which consists of retaining the first term in an expansion of the self-energy in terms of the dynamically screened Coulomb interaction W:

$$\Sigma_\mathrm{GW}(\mathbf{r}, \mathbf{r}'; t) = iG(\mathbf{r}, \mathbf{r}'; t)W(\mathbf{r}, \mathbf{r}'; t), \tag{5.86}$$

or its equivalent expression in terms of the energy:

$$\Sigma_\mathrm{GW}(\mathbf{r}, \mathbf{r}'; E) = i\int G(\mathbf{r}, \mathbf{r}'; E+E')W(\mathbf{r}, \mathbf{r}'; E)\mathrm{d}E'. \tag{5.87}$$

The screened interaction is related to the bare Coulomb interaction through

$$W(\mathbf{r}, \mathbf{r}'; E) = v_2(\mathbf{r}, \mathbf{r}') + \int\int W(\mathbf{r}, \mathbf{r}_1; E)P(\mathbf{r}_1, \mathbf{r}_2; E)v_2(\mathbf{r}_2, \mathbf{r}')\mathrm{d}\mathbf{r}_1\mathrm{d}\mathbf{r}_2, \tag{5.88}$$

with

$$P(\mathbf{r}, \mathbf{r}'; E) = -iG(\mathbf{r}, \mathbf{r}'; E)G(\mathbf{r}', \mathbf{r}; E), \tag{5.89}$$

the polarization function. The four equations (5.84), (5.87), (5.88), and (5.89) form a closed set that, in principle, can be solved self-consistently without further approximations. At this point it is important to remember that each one of these two-point functions depends on six spatial and one energy (or time) variable. Solving them is quite a demanding task from the computational point of view.

Equation (5.88) can also be written as

$$W(\mathbf{r}, \mathbf{r}'; E) = \int\int \varepsilon^{-1}(\mathbf{r}, \mathbf{r}_1; E)v_2(\mathbf{r}_1, \mathbf{r}')\mathrm{d}\mathbf{r}_1, \tag{5.90}$$

with $\varepsilon(\mathbf{r},\mathbf{r}_1;E)$ the dielectric function, which is normally calculated within the RPA. To simplify the problem, early implementations modeled the frequency dependence of the RPA dielectric function using the *plasmon pole approximation* (Hybertsen and Louie, 1985; Godby *et al.*, 1986). The advantage of this is that the energy integration in (5.87) can be carried out analytically. While plasmon pole models are still in use (Aulbur *et al.*, 2000), more recent approaches are based on the full RPA dielectric function (Rojas *et al.*, 1995; Aryasetiawan, 1992; Aryasetiawan and Gunnarsson, 1995).

Most GW schemes also replace the true Green's function G by the non-interacting G_0 in the GWA self-energy expression (5.87). Abandoning this approximation implies that Dyson's equation should be solved self-consistently, where the self-energy is calculated with the G obtained from Dyson's equation. This is a rather delicate issue, because the true G contains contributions that have to be taken together with the corresponding *vertex corrections*. Otherwise, important sum rules are violated and the number of particles is not conserved (Holm and von Barth, 1998). Such a self-consistent scheme that includes vertex corrections has recently been proposed (Schindlmayr and Godby, 1998). The effect of self-consistency has been studied by Holm, who analyzed the behavior of the total, kinetic, and potential energies for three models: the usual G_0W_0 non-self-consistent method, the GW_0 self-consistent approach on G, where the screened interaction is kept at the RPA level, and the fully self-consistent GW, where also the screened interaction is determined self-consistently (Holm, 1999). It was found that only the GW approach reproduces all the energy pieces individually, but the potential energy is also well reproduced by the GW_0 approach. It was shown how this result can be used to obtain an accurate total energy at this level. The relation between the GW approach and exact exchange methods was explored by Grabo *et al.* (1998).

5.9 Strong correlations: LDA+U and LDA+DMFT

Transition-metal oxides and rare-earth metal compounds are characterized by well-localized d and f orbitals. This localization leads to strong on-site correlations, such that if an electron is occupying a state localized in a particular site, placing a second electron in the same site is penalized with an additional energy U. This idea was originally crystallized at the level of an empirical Hamiltonian by Hubbard (1965). The study of the Hubbard model is a field in itself, being the simplest non-trivial model for correlated electrons. Advanced theoretical and computational techniques have been devised to this end, including exact diagonalization for small clusters, and lattice quantum Monte Carlo methods. The Hubbard phenomenological approach has been combined with density-functional calculations by supplementing the LDA or GGA with a Hubbard-type on-site repulsion term (LDA + U) (Anisimov *et al.*, 1991)

$$E_{\text{LDA+U}} = E_{\text{LDA}} - \frac{1}{2}UN(N-1) + \frac{1}{2}U\sum_{i\neq j} f_i f_j, \qquad (5.91)$$

where the f_i are orbital occupancies. This model produces a splitting into lower and upper Hubbard sub-bands, where the eigenvalues are given by

$$\varepsilon_i = \frac{\partial E_{\text{LDA+U}}}{\partial f_i} = \varepsilon_{\text{LDA}} + U\left(\frac{1}{2} - f_i\right), \qquad (5.92)$$

so that the energy separation is given by the Hubbard U. This reproduces qualitatively the behavior of Mott–Hubbard insulators, where strong correlations induce the opening of an energy gap. The determination of the Hubbard parameter U can be done empirically, by fitting it to experimental data, or by estimating it from LDA calculations from the total energies obtained by varying the occupancy of localized d or f orbitals. Within the LDA + U approach, the Hubbard term is treated at the mean-field level. The LDA + U approach has been reviewed by Anisimov *et al.* (1997), and re-derived as an approximation to the GW scheme, where the self-energy is still a non-local operator, but it is taken to be frequency-independent. The LDA + U idea has more recently been extended to include dynamical mean field correlations (LDA-DMFT) (Savrasov *et al.*, 2000), and combined with the GWA (Biermann *et al.*, 2003).

5.10 Summary of exchange-correlation functionals

In this chapter we have presented an exhaustive analysis of the different approaches to the treatment of exchange and correlation in density functional theory. Figure 5.3 is a diagram indicating these methods, which are grouped according to the type of approximation involved. On the right is Kohn–Sham theory with its local and semi-local approximations, LDA, GGA, and meta-GGA, and the non-local WDA. On the left are exact exchange methods, with orbital-dependent correlation functionals like Colle–Salvetti and more elaborate expressions from MBPT.

Above are the very successful hybrid Hartree–Fock–Kohn–Sham schemes like B3LYP, and also the simplified orbital-free (density-only) approaches based on Thomas–Fermi, including Dirac's exchange and von Weizsäcker's correction to the kinetic energy. Finally, below are many-body approaches like the GW approximation, and other approaches designed to describe excitations within DFT, like ensemble DFT and time-dependent DFT. Successes and failures of the various approximations have been analyzed by Kohanoff and Gidopoulos (2003), and also

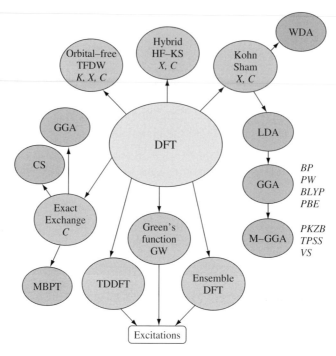

Fig. 5.3 Diagram of the different approximations to the electronic exchange and correlation contribution, proposed within the framework of density functional theory. The letters K, X, and C indicate which parts of the energy are approximated in each approach.

in various validation articles in the literature. The field has been very active for many decades, but there is still room for improvement, especially concerning the efficiency of these methods.

References

Adamo, C., Ernzerhof, M., and Scuseria, G. E. (2000). The meta-GGA functional: thermochemistry with a kinetic energy density dependent exchange-correlation functional. *J. Chem. Phys.* **112**, 2643–2649.

Alonso, J. A. and Girifalco, L. A. (1978). Nonlocal approximation to the exchange potential and kinetic energy of an inhomogeneous electron gas. *Phys. Rev. B* **17**, 3735–3743.

Andersson, Y., Langreth, D. C., and Lundqvist, B. I. (1996). Van der Waals interactions in density-functional theory. *Phys. Rev. Lett.* **76**, 102–105.

Anisimov, V. I., Zaanen, J., and Andersen, O. K. (1991). Band theory and Mott insulators: Hubbard U instead of Stoner I. *Phys. Rev. B* **44**, 943–954.

Anisimov, V. I., Aryasetiawan, F., and Lichtenstein, A. I. (1997). First-principles calculations of the electronic structure and spectra of strongly correlated systems: the LDA+U method. *J. Phys. Condens. Matter* **9**, 767–808.

Aryasetiawan, F. (1992). Self-energy of ferromagnetic nickel in the GW approximation. *Phys. Rev. B* **46**, 13051–13064.

Aryasetiawan, F. and Gunnarsson, O. (1995). Electronic structure of NiO in the GW approximation. *Phys. Rev. Lett.* **74**, 3221–3224.

Aulbur, W. G., Städele, M., and Görling, A. (2000). Exact-exchange-based quasiparticle calculations. *Phys. Rev. B* **62**, 7121–7132.

Aziz, R. A. and Slaman, M. J. (1991). An examination of *ab initio* results for the helium potential energy curve. *J. Chem. Phys.* **94**, 8047–8053.

Baroni, S. and Tuncel, E. (1983). Exact-exchange extension of the local-spin-density approximation in atoms: calculation of total energies and electron affinities. *J. Chem. Phys.* **79**, 6140–6144.

Becke, A. D. (1986). Density functional calculations of molecular bond energies. *J. Chem. Phys.* **84**, 4524–4529.

(1988). Density-functional exchange-energy approximation with correct asymptotic behavior. *Phys. Rev. A* **38**, 3098–3100.

(1993). Density-functional thermochemistry. III. The role of exact exchange. *J. Chem. Phys.* **98**, 5648–5652.

(1998). A new inhomogeneity parameter in density-functional theory. *J. Chem. Phys.* **109**, 2092–2098.

Biermann, S., Aryasetiawan, F., and Georges, A. (2003). First-principles approach to the electronic structure of strongly correlated systems: combining the GW approximation and dynamical mean-field theory. *Phys. Rev. Lett.* **90**, 086402.

Bylander, D. M. and Kleinman, L. (1990). Good semiconductor band gaps with a modified local-density approximation. *Phys. Rev. B* **41**, 7868–7871.

(1995). Energy gaps and cohesive energy of Ge from the optimized effective potential. *Phys. Rev. Lett.* **74**, 3660–3663.

Ceperley, D. M. and Alder, B. J. (1980). Ground state of the electron gas by a stochastic method. *Phys. Rev. Lett.* **45**, 566–569.

Charlesworth, J. P. A. (1996). Weighted-density approximation in metals and semiconductors. *Phys. Rev. B* **53**, 12666–12673.

Colle, R. and Salvetti, D. (1975). Approximate calculation of the correlation energy for the closed shells. *Theor. Chim. Acta* **37**, 329–334.

Della Salla, F. and Görling, A. (2001). Efficient localized Hartree–Fock methods as effective exact-exchange Kohn–Sham methods for molecules. *J. Chem. Phys.* **115**, 5718–5732.

Denton, A. R. and Ashcroft, N. W. (1989). Modified weighted-density-functional theory of nonuniform classical liquids. *Phys. Rev. A* **39**, 4701–4708.

Dobson, J. F. and Dinte, B. P. (1996). Constraint satisfaction in local and gradient susceptibility approximations: application to a van der Waals density functional. *Phys. Rev. Lett.* **76**, 1780–1783.

Engel, E. and Vosko, S. H. (1993). Accurate optimized-potential-model solutions for spherical spin-polarized atoms: evidence for limitations of the exchange-only local spin-density and generalized-gradient approximations. *Phys. Rev. A* **47**, 2800–2811.

Engel, E., Höck, A., and Dreizler, R. M. (2000). Van der Waals bonds in density-functional theory. *Phys. Rev. A* **61**, 032502.

Ernzerhof, M. (1996). Construction of the adiabatic connection. *Chem. Phys. Lett.* **263**, 499–506.

Ernzerhof, M. and Scuseria, G. E. (1999). Assessment of the Perdew–Burke–Ernzerhof exchange-correlation functional. *J. Chem. Phys.* **110**, 5029–5036.

Fetter, A. and Walecka, J. D. (1971). *Quantum Theory of Many-Particle Systems*. New York, McGraw-Hill.

Filippi, C., Umrigar, C. J., and Taut, M. (1994). Comparison of exact and approximate density functionals for an exactly soluble model. *J. Chem. Phys.* **100**, 1290–1295.

Filippi, C., Gonze, X., and Umrigar, C. J. (1996). Generalized gradient approximations to density functional theory: comparison with exact results. In *Recent Developments and Applications of Density Functional Theory*, J. M. Seminario, ed. Amsterdam, Elsevier.

García-González, P., Alvarellos, J. E., Chacón, E., and Tarazona, P. (2000). Image potential and the exchange-correlation weighted density approximation functional. *Phys. Rev. B* **62**, 16063–16068.

Garza, J., Nichols, J. A., and Dixon, D. A. (2000). The optimized effective potential and the self-interaction correction in density functional theory: application to molecules. *J. Chem. Phys.* **112**, 7880–7890.

Gell-Mann, M. and Brueckner, K. A. (1957). Correlation energy of an electron gas at high density. *Phys. Rev.* **106**, 364–368.

Ghosh, S. K. and Parr, R. G. (1986). Phase-space approach to the exchange-energy functional of density-functional theory. *Phys. Rev. A* **34**, 785–791.

Godby, R. W., Schluter, M., and Sham, L. J. (1986). Accurate exchange-correlation potential for silicon and its discontinuity on addition of an electron. *Phys. Rev. Lett.* **56**, 2415–2418.

Görling, A. (1996). Exact treatment of exchange in Kohn–Sham band-structure schemes. *Phys. Rev. B* **53**, 7024–7029.

Görling, A. (1999). New KS method for molecules based on an exchange charge density generating the exact local KS exchange potential. *Phys. Rev. Lett.* **83**, 5459–5462.

Görling, A. and Ernzerhof, M. (1995). Energy differences between Kohn–Sham and Hartree–Fock wave functions yielding the same electron density. *Phys. Rev. A* **51**, 4501–4513.

Görling, A. and Levy, M. (1993). Correlation-energy functional and its high-density limit obtained from a coupling-constant perturbation expansion. *Phys. Rev. B* **47**, 13105–13113.

(1994). Exact Kohn–Sham scheme based on perturbation theory. *Phys. Rev. A* **50**, 196–204.

(1997). Hybrid schemes combining the Hartree–Fock method and density-functional theory: underlying formalism and properties of correlation functionals. *J. Chem. Phys.* **106**, 2675–2680.

Grabo, T., Kreibich, T., Kurth, S., and Gross, E. K. U. (1998). Orbital functionals in density functional theory: the optimized effective potential method. In *Strong Coulomb Correlations in Electronic Structure: Beyond the Local Density Approximation*, V. I. Anisimov, ed. Tokyo, Gordon & Breach.

Grabowski, I., Hirata, S., Ivanov, S., and Bartlett, R. J. (2002). *Ab initio* density functional theory: OEP-MBPT(2). A new orbital-dependent correlation functional. *J. Chem. Phys.* **116**, 4415–4425.

Gross, E. K. U. and Dreizler, R. M. (1981). Gradient expansion of the Coulomb exchange energy. *Z. Phys. A* **302**, 103–106.

Grüning, M., Gritsenko, O. V., and Baerends, E. J. (2004). Improved description of chemical barriers with generalized gradient approximations (GGAs) and meta-GGAs. *J. Phys. Chem. A* **108**, 4459–4469.

Gunnarsson, O. and Jones, R. O. (1980). Density functional calculations for atoms, molecules and clusters. *Phys. Scr.* **21**, 394–401.

Hammer, B., Hansen, L. B., and Nørskov, J. K. (1999). Improved adsorption energetics within density-functional theory using revised Perdew–Burke–Ernzerhof functionals. *Phys. Rev. B* **59**, 7413–7421.

Handy, N. C. and Cohen, A. J. (2001). Left–right correlation energy. *Mol. Phys.* **99**, 403–412.

Hedin, L. (1965). New method for calculating the one-particle Green's function with application to the electron-gas problem. *Phys. Rev.* **139**, A796–823.

Holm, B. (1999). Total energies from GW calculation. *Phys. Rev. Lett.* **83**, 788–791.

Holm, B. and von Barth, U. (1998). Fully self-consistent GW self-energy of the electron gas. *Phys. Rev. B* **57**, 2108–2117.

Hubbard, J. (1965). Electron correlations in narrow energy bands. IV. The atomic representation. *Proc. Roy. Soc. London* **285**, 542–560.

Hybertsen, M. S. and Louie, S. G. (1985). First-principles theory of quasiparticles: calculation of band gaps in semiconductors and insulators. *Phys. Rev. Lett.* **55**, 1418–1421.

Ivanov, S., Hirata, S., and Bartlett, R. J. (1999). Exact exchange treatment for molecules in finite-basis-set Kohn–Sham theory. *Phys. Rev. Lett.* **83**, 5455–5458.

Ivanov, S., Hirata, S., Grabowski, I., and Bartlett, R. J. (2003). Connections between second-order Görling–Levy and many-body perturbation approaches in density functional theory. *J. Chem. Phys.* **118**, 461–470.

Jones, R. O. and Gunnarsson, O. (1989). The density functional formalism, its applications and prospects. *Rev. Mod. Phys.* **61**, 689–746.

Kohanoff, J. and Gidopoulos, N. I. (2003). Density functional theory: basics, new trends and applications. In *Handbook of Molecular Physics and Quantum Chemistry*, S. Wilson, ed., Vol. 2. Chichester, Wiley, Chapter 26, 532–568.

Kohn, W. (1986). Discontinuity of the exchange-correlation potential from a density-functional viewpoint. *Phys. Rev. B* **33**, 4331–4333.

Kohn, W. and Sham, L. (1965). Self-consistent equations including exchange and correlation effects. *Phys. Rev.* **14**, A1133–1138.

Kohn, W., Meir, Y., and Makarov, D. E. (1998). Van der Waals energies in density functional theory. *Phys. Rev. Lett.* **80**, 4153–4156.

Kotani, T. (1994). Exact exchange-potential band-structure calculations by the LMTO-ASA method: MgO and CaO. *Phys. Rev. B* **50**, 14816–14821.

Krieger, J. B., Li, Y., and Iafrate, G. J. (1992). Construction and application of an accurate local spin-polarized Kohn–Sham potential with integer discontinuity: exchange-only theory. *Phys. Rev. A* **45**, 101–126.

Krieger, J. B., Chen, J., Iafrate, G. J., and Savin, A. (1999). Construction of an accurate self-interaction-corrected correlation energy functional based on an electron gas with a gap. In *Electron Correlations and Materials Properties*, A. Gonis and N. Kioussis, eds. New York, Plenum Press.

Kurth, S. and Perdew, J. P. (1999). Density-functional correction of random-phase approximation correlation with results for jellium surface energies. *Phys. Rev. B* **59**, 10461–10468.

Langreth, D. C. and Mehl, M. J. (1981). Easily implementable nonlocal exchange-correlation energy functional. *Phys. Rev. Lett.* **47**, 446–450.

Langreth, D. C. and Perdew, J. P. (1977). Exchange-correlation energy of a metallic surface: wave-vector analysis. *Phys. Rev. B* **15**, 2884–2901.

Lee, C., Yang, W., and Parr, R. G. (1988). Development of the Colle–Salvetti correlation-energy formula into a functional of the electron density. *Phys. Rev. B* **37**, 785–789.

Leininger, T., Stoll, H., Werner, H.-J., and Savin, A. (1997). Combining long-range configuration interaction with short-range density functionals. *Chem. Phys. Lett.* **275**, 151–160.

Levy, M. and Perdew, J. P. (1985). Hellmann–Feynman, virial, and scaling requisites for the exact universal density functionals. Shape of the correlation potential and diamagnetic susceptibility for atoms. *Phys. Rev. A* **32**, 2010–2021.

Levy, M., Perdew, J. P., and Sahni, V. (1984). Exact differential equation for the density and ionization energy of a many-particle system. *Phys. Rev. A* **30**, 2745–2748.

Li, Y., Krieger, J. B., and Iafrate, G. J. (1992). Negative ions as described by Kohn–Sham exchange-only theory. *Chem. Phys. Lett.* **191**, 38–46.

  (1993). Self-consistent calculations of atomic properties using self-interaction-free exchange-only Kohn–Sham potentials. *Phys. Rev. A* **47**, 165–181.

Lieb, E. H. and Oxford, S. (1981). Improved lower bound on the indirect Coulomb energy. *Int. J. Quantum Chem.* **19**, 427–439.

Lutsko, J. F. and Baus, M. (1990). Can the thermodynamic properties of a solid be mapped onto those of a liquid? *Phys. Rev. Lett.* **64**, 761–763.

Ma, S.-K. and Brueckner, K. A. (1968). Correlation energy of an electron gas with a slowly varying high density. *Phys. Rev.* **165**, 18–31.

Misawa, S. (1965). Ferromagnetism of an electron gas. *Phys. Rev.* **140**, A1645–A1648.

Nekovee, M., Foulkes, W. M. C., and Needs, R. J. (2001). Quantum Monte Carlo analysis of exchange and correlation in the strongly inhomogeneous electron gas. *Phys. Rev. Lett.* **87**, 036401.

Perdew, J. P. (1985). Accurate density functional for the energy: real-space cutoff of the gradient expansion for the exchange hole. *Phys. Rev. Lett.* **55**, 1665–1668.

Perdew, J. P. and Levy, M. (1983). Physical content of the exact Kohn–Sham orbital energies: band gaps and derivative discontinuities. *Phys. Rev. Lett.* **51**, 1884–1887.

Perdew, J. P. and Norman, M. R. (1982). Electron removal energies in Kohn–Sham density-functional theory. *Phys. Rev. B* **26**, 5445–5450.

Perdew, J. P. and Schmidt, K. (2001). Jacob's ladder of density functional approximations for the exchange-correlation energy. In *Density Functional Theory and its Application to Materials*. Vol. 577. New York, American Institute of Physics, 1–20.

Perdew, J. P. and Wang, Y. (1992a). Accurate and simple analytic representation of the electron-gas correlation energy. *Phys. Rev. B* **45**, 13244–13249.

  (1992b). Pair-distribution function and its coupling-constant average for the spin-polarized electron gas. *Phys. Rev. B* **46**, 12947–12954.

Perdew, J. P. and Zunger, A. (1981). Self-interaction correction to density-functional approximations for many-electron systems. *Phys. Rev. B* **23**, 5048–5079.

Perdew, J. P., Parr, R. G., Levy, M., and Balduz, J. L. (1982). Density-functional theory for fractional particle number: derivative discontinuities of the energy. *Phys. Rev. Lett.* **49**, 1691–1694.

Perdew, J. P., Burke, K., and Ernzerhof, M. (1996). Generalized gradient approximation made simple. *Phys. Rev. Lett.* **77**, 3865–3868.

Perdew, J. P., Kurth, S., Zupan, A., and Blaha, P. (1999). Accurate density functional with correct formal properties: a step beyond the generalized gradient approximation. *Phys. Rev. Lett.* **82**, 2544–2547.

Perdew, J. P., Tao, J., Staroverov, V. N., and Scuseria, G. E. (2004). Meta-generalized gradient approximation: explanation of a realistic nonempirical density functional. *J. Chem. Phys.* **120**, 6898–6911.

Rapcewicz, K. and Ashcroft, N. W. (1991). Fluctuation attraction in condensed matter: a nonlocal functional approach. *Phys. Rev. B* **44**, 4032–4035.

Rojas, H. N., Godby, R. W., and Needs, R. J. (1995). Space-time method for *ab initio* calculations of self-energies and dielectric response functions of solids. *Phys. Rev. Lett.* **74**, 1827–1830.

Rushton, P. P., Tozer, D. J., and Clark, S. J. (2002a). Description of exchange and correlation in the strongly inhomogeneous electron gas using a nonlocal density functional. *Phys. Rev. B* **65**, 193106.

(2002b). Nonlocal density-functional description of exchange and correlation in silicon. *Phys. Rev. B* **65**, 235203.

Sahni, V., Gruenebaum, J., and Perdew, J. P. (1982). Study of the density-gradient expansion for the exchange energy. *Phys. Rev. B* **26**, 4371–4377.

Savrasov, S., Kotliar, G., and Abrahams, E. (2000). Correlated electrons in d-plutonium within a dynamical mean-field picture. *Nature (London)* **410**, 793–795.

Schindlmayr, A. and Godby, R. W. (1998). Systematic vertex corrections through iterative solution of Hedin's equations beyond the GW approximation. *Phys. Rev. Lett.* **80**, 1702–1705.

Seidl, A., Görling, A., Vogl, P., Majewski, J. A., and Levy, M. (1996). Generalized Kohn–Sham schemes and the band-gap problem. *Phys. Rev. B* **53**, 3764–3774.

Seidl, M., Perdew, J. P., and Levy, M. (1999). Strictly correlated electrons in density-functional theory. *Phys. Rev. A* **59**, 51–54.

Seidl, M., Perdew, J. P., and Kurth, S. (2000a). Density functionals for the strong-interaction limit. *Phys. Rev. A* **62**, 012502.

(2000b). Simulation of all-order density-functional perturbation theory, using the second order and the strong-correlation limit. *Phys. Rev. Lett.* **84**, 5070–5073.

Sham, L. J. and Schlüter, M. (1983). Density-functional theory of the energy gap. *Phys. Rev. Lett.* **51**, 1888–1891.

Sharp, R. T. and Horton, G. K. (1953). A variational approach to the unipotential many-electron problem. *Phys. Rev.* **90**, 317–317.

Singh, D. J. (1993). Weighted-density-approximation ground-state studies of solids. *Phys. Rev. B* **48**, 14099–14103.

Singwi, K. S., Tosi, M. P., Land, R. H., and Sjölander, A. (1968). Electron correlations at metallic densities. *Phys. Rev.* **176**, 589–599.

Slater, J. C. (1951). A simplification of the Hartree–Fock method. *Phys. Rev.* **81**, 385–390.

Städele, M., Majewski, J. A., Vogl, P., and Görling, A. (1997). Exact Kohn–Sham exchange potential in semiconductors. *Phys. Rev. Lett.* **79**, 2089–2092.

Städele, M., Moukara, M., Majewski, J. A., Vogl, P., and Görling, A. (1999). Exact exchange Kohn–Sham formalism applied to semiconductors. *Phys. Rev. B* **59**, 10031–1043.

Staroverov, V. N., Scuseria, G. E., Tao, J., and Perdew, J. P. (2003). Comparative assessment of a new nonempirical density functional: molecules and hydrogen-bonded complexes. *J. Chem. Phys.* **119**, 12129–12137.

Svendsen, P. S. and von Barth, U. (1996). Gradient expansion of the exchange energy from second-order density response theory. *Phys. Rev. B* **54**, 17402–17413.

Talman, J. D. and Shadwick, W. F. (1976). Optimized effective atomic central potential. *Phys. Rev. A* **14**, 36–40.

Tao, J., Gori-Giorgi, P., Perdew, J. P., and McWeeny, R. (2001). Uniform electron gas from the Colle–Salvetti functional: missing long-range correlations. *Phys. Rev. A* **63**, 032513.

Tao, J., Perdew, J. P., Staroverov, V. N., and Scuseria, G. E. (2003). Climbing the density functional ladder: nonempirical meta-generalized gradient approximation designed for molecules and solids. *Phys. Rev. Lett.* **91**, 146401.

Toulouse, J., Savin, A., and Adamo, C. (2002). Validation and assessment of an accurate approach to the correlation problem in density functional theory: the Krieger–Chen–Iafrate–Savin model. *J. Chem. Phys.* **117**, 10465–10473.

Van Gisbergen, S. J. A., Schipper, P. R. T., Gritsenko, O. V., Baerends, E. J., Snijders, J. G., Champagne, B., and Kirtman, B. (1999). Electric field dependence of the exchange-correlation potential in molecular chains. *Phys. Rev. Lett.* **83**, 694–697.

Van Voorhis, T. and Scuseria, G. E. (1998). A novel form for the exchange-correlation energy functional. *J. Chem. Phys.* **109**, 400–410.

Von Barth, U. and Hedin, L. (1972). A local exchange-correlation potential for the spin polarized case. I. *J. Phys. C* **5**, 1629–1642.

Von Weizsäcker, C. F. (1935). Zur Theorie der Kernmassen. *Z. Phys.* **96**, 431–458.

Vosko, S. H., Wilk, L., and Nusair, M. (1980). Accurate spin-dependent electron liquid correlation energies for local spin density calculations: a critical analysis. *Can. J. Phys.* **58**, 1200–1211.

Yan, Z., Perdew, J. P., and Kurth, S. (2000). Density functional for short-range correlation: accuracy of the random-phase approximation for isoelectronic energy changes. *Phys. Rev. B* **61**, 16430–16439.

Zaremba, E. and Kohn, W. (1976). Van der Waals interaction between an atom and a solid surface. *Phys. Rev. B* **13**, 2270–2285.

Zhang, Y. and Yang, W. (1998). Comment on "generalized gradient approximation made simple". *Phys. Rev. Lett.* **80**, 890–891.

Part II

Computational methods

6
Solving the electronic problem in practice

The central mathematical problem in electronic structure theory at the single-particle approximation level is to self-consistently solve a set of N coupled, three-dimensional, partial differential equations. In fact, the two basic methodologies, Hartree–Fock and Kohn–Sham, are the result of a simplification of the $3N$-dimensional many-body problem into tractable, but approximate, schemes. This chapter presents the general aspects involved in the practical, numerical solution of the above problem. The first section concentrates on the two main issues, which are the treatment of the electron–nuclear interaction, and the mathematical representation of the single-particle orbitals. This allows us to introduce the concepts of pseudopotential and basis set. The second section reviews a number of important results from solid state theory, which are relevant to the study of condensed phases with translational invariance, and are central to some of the electronic structure methods described in Chapter 9.

6.1 Kohn–Sham and Hartree–Fock equations

In the Kohn–Sham formulation of density functional theory we have to face the mathematical problem of solving the eigenvalue equation

$$\left\{ -\frac{\hbar^2}{2m} \nabla^2 + v_{\text{ext}}(\mathbf{r}) + \int \frac{\rho(\mathbf{r}')}{|\mathbf{r} - \mathbf{r}'|} d\mathbf{r}' + \mu_{\text{XC}}[\rho] \right\} \varphi_i(\mathbf{r}) = \varepsilon_i \varphi_i(\mathbf{r}), \qquad (6.1)$$

where the electronic density ρ is given by

$$\rho(\mathbf{r}) = \sum_{i=1}^{N} f_i |\varphi_i(\mathbf{r})|^2, \qquad (6.2)$$

N is the number of electrons, and f_i are the occupation numbers corresponding to the one-electron eigenstates. In the case of spin-unpolarized insulators or closed-shell molecules, $f_i = 2$ for the $N/2$ lowest eigenstates and $f_i = 0$ otherwise. For spin-polarized systems or open-shell molecules the exchange-correlation potential, and the external potential when there are external magnetic fields, depends on the spin projection. The same happens when spin–orbit coupling is taken into account. In these cases there are two sets of occupation numbers, one for each spin component. The occupation numbers are $f_{i\sigma} = 1$ for the N lowest eigenstates, but considering both sets of eigenvalues, $\varepsilon_{i\uparrow}$ and $\varepsilon_{i\downarrow}$, together. It may well happen that the number of occupied states is unbalanced, i.e. there are more occupied states with one spin projection than with the other. In that case the system is spin-polarized. It can also happen that the spins are compensated, and the system is globally spin-unpolarized. Nevertheless, it can be *locally* spin-polarized in the sense that the magnetization density $\zeta(\mathbf{r})$ is not uniformly zero, but exhibits spatial variations.

The external potential $v_{\text{ext}}(\mathbf{r})$ represents the interaction between the electrons and the nuclei, and is expressed in the following way:

$$v_{\text{ext}}(\mathbf{r}) = -e^2 \sum_{I=1}^{P} \frac{Z_I}{|\mathbf{r} - \mathbf{R}_I|}; \qquad (6.3)$$

and the exchange-correlation potential μ_{XC} is given by any of the expressions discussed in Chapter 5.

The case of the Hartree–Fock equations is analogous to that of the Kohn–Sham equations, with the difference that the exchange-correlation potential is replaced by the non-local exchange operator $\hat{\mathcal{K}}_j$ defined by (3.36). The direct Coulomb term is exactly the same in both approaches, but here it is normally expressed in a way that is formally similar to that of the exchange term. In the canonical orbital representation, the Hartree–Fock equations read:

$$\left\{ -\frac{\hbar^2}{2m} \nabla^2 + v_{\text{ext}}(\mathbf{r}) + \sum_{j=1}^{N} \left(\hat{\mathcal{J}}_j - \hat{\mathcal{K}}_j \right) \right\} \varphi_i(\mathbf{r}) = \varepsilon_i \varphi_i(\mathbf{r}). \qquad (6.4)$$

At this stage, the solution of either the Kohn–Sham or Hartree–Fock equations requires two important choices:

(i) how to treat the electron–nuclear interaction;
(ii) finding a mathematical way to represent the single-particle orbitals.

6.1.1 The electron–nuclear interaction

The electron–nuclear interaction is given by the bare Coulomb potential as expressed in (6.3). However, a distinction has to be made between *two classes of electrons*: those that participate actively in chemical bonding, named *valence electrons*, and those tightly bound to the nuclei, called *core electrons*, which do not participate in bonding and can be treated as frozen orbitals. Let us remark here that core states are not completely insensitive to the molecular or crystal field. In fact, a useful experimental technique to characterize the environment of an atom is by measuring *core level shifts*. In some cases there is a third class of electrons, called *semi-core electrons*, which do not participate actively in chemical bonding, but are sufficiently close in energy to the valence states to feel the presence of the environment. The semi-core wave functions polarize, and thus cannot be treated as frozen orbitals. Let us remark that, rigorously, there is no such thing as *classes of electrons*. This is a case of language abuse, as all the electrons are exactly the same. The above terminology, however, is widely used to indicate single-particle *electronic states* rather than the electrons themselves.

All-electron methods

This class of methods deals explicitly with all the electrons in the system; core, semi-core, and valence. Due to the different characteristics of these orbitals, they are usually treated in a different way. Core electrons can be taken to be frozen as in the isolated atom situation, but they can also be allowed to polarize in the presence of the environment. One class of methods is based on dividing the space into spherical regions around the atoms and interstitial regions, and requesting that the basis functions be continuous and differentiable across the boundaries. The solutions in the interstitials are *augmented* with partial waves inside every sphere, thus leading to energy-dependent basis functions. This energy dependence arises through the matching conditions on the surface of the spheres. Therefore, the basis functions are flexible, and are determined self-consistently for the energies that solve the Schrödinger equation of the target system. Another class of all-electron methods is based on fixed, energy-independent orbitals. These are easier to handle because there is no need to find the roots of any secular equation.

Within the first approach, core states can be fairly well represented by solutions of the atomic problem, subject to appropriate boundary conditions on the surface of the spheres with very small, usually negligible, interstitial contribution. For valence states the interstitial contribution, related to interatomic bonding, is much more important, but it is smoothly varying in space. These basis functions are augmented with atomic-like orbitals within spherical regions around each atom, which may not be closely related to the orbitals of the isolated atom any more.

In fixed-orbital basis set methods, core electrons are typically represented by a minimal number of basis functions constructed to reproduce accurately the corresponding atomic orbital. This helps reduce the computational burden associated with these almost inert electronic states. Valence electrons are described through more complete sets of basis functions such as two or three sets of orbitals per angular momentum component. This provides the necessary flexibility to describe chemical bonding, where the electronic orbitals are very different from atomic orbitals. Semi-core electrons are treated similarly to valence electrons, but with less variational freedom at the basis set level, to reduce the computational cost.

In augmented partial-wave methods, the logarithmic derivatives of the radial part of the wave functions at the surface of the augmentation spheres are required to be continuous. The matching conditions depend on the eigenvalue associated with that wave function. In principle, these can be fulfilled by re-calculating the logarithmic derivatives at the correct eigenvalue every self-consistent iteration. Another possibility, which is faster and more stable, is to linearize the matching conditions around a reference energy (or a few reference energies) that is (are) representative of the eigenvalue that the wave function assumes in that particular environment. These are called *linear methods*.

Pseudopotential methods

Since core electrons do not participate actively in chemical bonding, it is possible to eliminate the corresponding degrees of freedom by replacing the atomic nuclei with a still point-like, but effective, nucleus of charge $Z_V = Z - Z_{core}$, with Z_{core} the charge associated with the core electrons. This effective nucleus, or *ionic core*, represents the nucleus together with its core electrons. In this way the number of electrons treated explicitly is much smaller, thus reducing significantly the number of required electronic states and the size of the basis set. However, the interaction between the valence electrons and the ionic cores (often called, plainly and confusingly, *the ions*) is not the bare Coulomb interaction anymore; it now includes that part of the electron–electron interaction related to the screening of the nuclear charge by the core electrons. The bare Coulomb interaction must then be replaced by a screened Coulomb potential. This potential should take into account the orthogonalization of the valence orbitals with respect to core orbitals of the same angular momentum, thus producing valence wave functions with the required number of nodes. In principle, this can be achieved as in augmentation all-electron methods, i.e. by matching the valence orbital to an appropriate atomic wave function in the core region.

An alternative is to assume that a good description of the valence wave functions inside the core region is, in most cases, unnecessary, because one is usually concerned with bonding properties. In that case there is no lack of crucial information if the inner solution (inside the core radius) is replaced with a smooth, nodeless pseudo-wave function. This pseudo-wave function is not an orbital corresponding to the original atomic problem. Being nodeless, it now corresponds to the lowest-lying state of an effective, pseudo-atomic problem where the true potential has been replaced by a pseudopotential, or effective core potential.

In principle, it should be possible to construct a pseudopotential that is common to all the angular momenta in the atom. In this case, it is said to be a *local* pseudopotential. The solutions of this pseudo-atomic problem coincide with the solutions of the true atomic problem outside the core region, but are different inside. The democratic treatment of all the angular momenta is a desirable feature, because then the external potential remains local. However, the extent of the core region depends on the angular momentum. In particular, when electrons of the same angular momentum are not present in the core, the valence wave function peaks very strongly close to the nucleus due to the lack of orthogonalization. Consider, for instance, a transition metal like Pt. The $6p$ orbitals are quite external, peaking at around 3.9 Å from the nucleus. The $6s$ orbitals peak at around 2.4 Å, and the main peak of the $5d$ state is located at 1.3 Å. So, which one of the three should be taken as the core radius? If the largest is chosen, then the most contracted orbital will be poorly represented. If the smallest is chosen, then this radius is smaller than the location of the last node of the most external orbital at 1.4 Å. Therefore, the $6p$ orbital cannot be made nodeless for this local pseudopotential, and the lowest-lying p-state should be the $5p$. This is a possible solution, to choose a core radius that is acceptable for all angular momenta, even if for one or more of them this implies that we pseudize for a semi-core electronic state. Another solution to this problem is to abandon the idea of a single pseudopotential for all the angular momenta in favor of angular momentum-dependent potentials, which then become non-local in the sense that the external potential is not multiplicative any longer. Once this step has been taken, then for the same price the pseudopotentials can be tuned to minimize the number of basis functions required to represent the pseudo-wave functions. In unbiased basis sets such as plane waves, this is an extremely important issue because the cost of the calculation can become prohibitive if too many basis functions are required.

6.1.2 Classes of basis sets

The representation of the Kohn–Sham orbitals requires the choice of a basis set for the Hilbert space. Many possibilities have been explored since the early times of quantum mechanics. These are based on the general characteristics of the electronic problem, and on the particular features of the studied problem. They can be classified into four main groups.

(i) *Extended basis sets*: the basis functions are delocalized, either floating (independent of the nuclear positions) or centered at the nuclear positions. The important issue is that they cover all space. They are useful for condensed phases such as solids or liquids. They tend to be inefficient for molecular systems.

(ii) *Localized basis sets*: the basis functions are localized, mainly centered at the atomic positions, but they may also be centered in bonds (e.g. at the mid-point between two covalently bonded atoms) or at the positions of "ghost" atoms. They are mostly used for molecular systems, but can also be used for periodic systems. Localized basis sets can also be floating, i.e. centered at fixed positions instead of atomic positions.

(iii) *Mixed basis sets*: the basis set includes both extended and localized basis functions. They are designed to take the best of both worlds, although they carry with them also the technical difficulties of both (e.g. over-completeness).

(iv) *Augmented basis sets*: where an extended or atom-centered basis set is augmented with atomic-like wave functions in spherical regions around the nuclei (the muffin-tin). Methods based on this type of basis set tend to be very accurate. Since the basis functions are flexible, then fewer of them are needed to achieve basis set convergence, and thus the level of accuracy of the calculation can be pushed to the limits. However, these are more complicated from the technical point of view.

6.2 Condensed phases: Bloch's theorem and periodic boundary conditions

Condensed phases such as solids, liquids, amorphous, and systems of lower dimensionality such as surfaces or wires, are macroscopic objects constituted by a huge number of atoms, of the order of Avogadro's number (6×10^{23}). In fact, they can effectively be treated as infinite. In crystalline solids, a small number of atoms (a *basis*) is replicated periodically *ad infinitum* along one, two, or three directions in space.

There are infinitely many ways of characterizing a crystalline solid, depending on the number of atoms in the basis, i.e. the number of atoms that are periodically repeated. However, there is a minimal choice that contains the whole symmetry of the system. This is called the *unit or Wigner–Seitz* cell and, together with the lattice vectors that indicate the size of the unit cell and the directions of replication, contains all the necessary information to reproduce the infinite crystalline structure. The vectors that serve to reconstruct the infinite solid from the unit cell are called *unit or primitive* vectors, and in general are not orthogonal. The set of points in space corresponding to integer combinations of the primitive vectors receives the name of *Bravais lattice*, of which there are only 32 in three dimensions. The combination of the translational symmetry embodied in the Bravais lattice with the point group symmetry of the basis gives rise to 230 space groups, which are sufficient to classify all known crystalline solids.

Sometimes it is convenient to describe the solid in terms of a cell containing more atoms than the unit cell (the *conventional* cell) in order to simplify the description of the symmetry properties, e.g. so that the lattice vectors are orthogonal. For instance, a body centered cubic (*bcc*) unit cell can also be described as a simple cubic cell containing two atoms in the basis, and a face centered cubic (*fcc*) cell is equivalent to a simple cubic cell with a 4-atom basis. We call $\{\mathbf{a}_i\}$, with $i = 1, 2, 3$, the unit vectors. The volume of the unit cell is given by $\Omega = \mathbf{a}_1 \cdot (\mathbf{a}_2 \times \mathbf{a}_3)$. The Wigner–Seitz cell is enclosed by the intersection of the planes drawn perpendicularly to all the unit vectors exactly at their mid points, and has the smallest possible volume. Hence, the conventional simple cubic cell corresponding to a *bcc* lattice is twice as large as the unit cell, and that of an *fcc* is four times larger.

It is out of the question to study an infinite number of electrons in a computer. Fortunately, this can be avoided thanks to Bloch's theorem (Bloch, 1928), which connects the properties of the electrons in a periodic infinite system with those of the electrons in the unit cell.

Theorem (Bloch)

The wave function of an electron in an external periodic potential $v(\mathbf{r}) = v(\mathbf{r} + \mathbf{a}_i)$ can be written as the product of a function with the same periodicity of the potential, and a purely imaginary phase factor arising from the translational symmetry, i.e.

$$\psi_\mathbf{k}(\mathbf{r}) = e^{i\mathbf{k}\cdot\mathbf{r}} u_\mathbf{k}(\mathbf{r}), \tag{6.5}$$

with $u_\mathbf{k}(\mathbf{r}) = u_\mathbf{k}(\mathbf{r} + \mathbf{a}_i)$.

It is easy to see that the wave function at a location displaced by the unit vector, $\mathbf{r} + \mathbf{a}_i$, assumes the form

$$\psi_\mathbf{k}(\mathbf{r} + \mathbf{a}_i) = e^{i\mathbf{k}\cdot\mathbf{a}_i} \psi_\mathbf{k}(\mathbf{r}), \tag{6.6}$$

so that the probability density $|\psi_\mathbf{k}(\mathbf{r})|^2$ is exactly the same because the purely imaginary phase factor cancels out. The proof of Bloch's theorem can be found in any book on solid state theory (see e.g. Kittel, 1996).

6.2.1 The Brillouin zone: bands and gaps

By looking at Expression (6.6), it can be seen that there is a particular class of vectors $\mathbf{k}$ such that the phase factor $e^{i\mathbf{k}\cdot\mathbf{a}_i} = 1$, and thus the wave function is in phase in all the periodic replicas of the unit cell. The set of the three smallest independent such vectors is sufficient to determine all the *reciprocal lattice vectors*, in the same way as with the primitive lattice vectors in real space. The primitive vectors in reciprocal space are defined by the relation $\mathbf{a}_i \cdot \mathbf{b}_j = 2\pi\delta_{ij}$. It is easy to show that this leads to the form

$$\mathbf{b}_1 = 2\pi\frac{\mathbf{a}_2 \times \mathbf{a}_3}{\Omega}; \quad \mathbf{b}_2 = 2\pi\frac{\mathbf{a}_3 \times \mathbf{a}_1}{\Omega}; \quad \mathbf{b}_3 = 2\pi\frac{\mathbf{a}_1 \times \mathbf{a}_2}{\Omega}, \tag{6.7}$$

with Ω the volume of the unit cell. The primitive reciprocal lattice vectors define a cell in reciprocal space, of volume $\Omega_R = \mathbf{b}_1 \cdot (\mathbf{b}_2 \times \mathbf{b}_3) = (2\pi)^3/\Omega$, which receives the name of *first Brillouin zone*, and sometimes simply the Brillouin zone (BZ), for short. Similarly to real space, larger Brillouin zones, i.e. a second or a third BZ, can be constructed by considering the second or third shell of reciprocal lattice vectors.

To illustrate this point, we consider a system of non-interacting, free electrons in a one-dimensional periodic box of length a. The electronic wave functions are the solutions of the Schrödinger equation

$$-\frac{\hbar^2}{2m}\frac{d^2\psi_k(x)}{dx^2} = \varepsilon(k)\psi_k(x). \tag{6.8}$$

It is easy to see that the wave functions $\psi_k(x) = e^{ikx}$ are solutions of the above equation if the corresponding eigenvalues are

$$\varepsilon(k) = \frac{\hbar^2 k^2}{2m}. \tag{6.9}$$

The relation between energy and wave vector is called the *dispersion relation*, and the allowed values of $\varepsilon(k)$ – in this case all positive values – form the *energy bands*. In Fig. 6.1 the above dispersion relation is shown. The values of k are not confined to the first BZ defined by $|k| \leq \pi/a$. However, it is customary, for the sake of visualization, to transport the wave vectors back to the first BZ by subtracting a reciprocal lattice vector (reduced zone diagram). For example, a wave vector k' in the second BZ *refolds* onto the vector $k = k' - 2\pi/a$, and a wave vector k'' in the third BZ onto the vector $k = k'' - 4\pi/a$, as indicated in Fig. 6.1. The refolded bands correspond then to momentum values $k_n = k + G_n$, with k in the first BZ, $G_n = 2n\pi/a$, and n an integer. Generalizing to three-dimensional lattices, any vector outside the first BZ can be written as $\mathbf{k}' = \mathbf{k} + \mathbf{G}$, where $\mathbf{k}$ is a vector contained in the first BZ and $\mathbf{G} = n_1\mathbf{b}_1 + n_2\mathbf{b}_2 + n_3\mathbf{b}_3$ (with n_i integers) is a reciprocal lattice vector satisfying $\mathbf{a}_i \cdot \mathbf{G} = 2\pi n_i$. In more elaborate one-electron approaches such as Kohn–Sham or Hartree–Fock, the periodic wave function still obeys a Schrödinger-like equation of the same form. However, the external periodic potential is replaced with an effective potential $v_{\text{eff}}(\mathbf{r})$, which also contains electron–electron interaction terms such as the Hartree and exchange-correlation contributions (see Section 6.1).

The main effect of the periodic potential, as opposed to the free electron model, is to modify the dispersion relations opening *energy gaps* at some particular points in the BZ, such that some regions of the eigenvalue spectrum are not accessible to the electrons. The gap opening is due to Bragg reflection between the boundaries

6.2 Condensed phases: Bloch's theorem

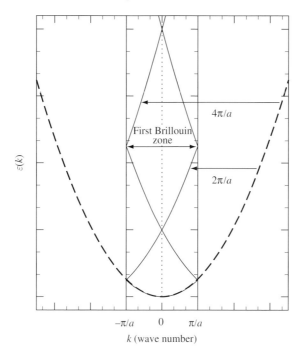

Fig. 6.1 Schematic dispersion relation for a one-dimensional system of non-interacting, free electrons.

of the Brillouin zone. In a one-dimensional example, the wave functions at vectors k and $-k$, $\psi_{\pm k}(x) = e^{\pm ikx} u_{\pm k}(x)$, are degenerate in energy $\varepsilon(k) = \varepsilon(-k)$; the potential has the effect of lifting this degeneracy. This can be seen by considering the potential as a perturbation to the free electron Hamiltonian. According to first-order perturbation theory on degenerate energy levels, the corrected levels are obtained by diagonalizing the matrix

$$H_k = \begin{pmatrix} \varepsilon(k) & U \\ U & \varepsilon(k) \end{pmatrix}, \qquad (6.10)$$

where

$$U = \langle \psi_{-k} | \hat{V}_{\text{eff}} | \psi_k \rangle = \frac{1}{N} \int_{-Na}^{Na} e^{2ikx} v_{\text{eff}}(x) \mathrm{d}x \qquad (6.11)$$

and N is the number of replicated cells. N should be let to go to ∞ at the end of the calculation.

Since $v_{\text{eff}}(x)$ and e^{2ikx} are both periodic functions, the above integral is zero unless the two periods are commensurate. This means that the off-diagonal matrix elements U will be different from zero only if $2k = 2n\pi/a$, i.e. if $2k$ coincides with a lattice vector. Therefore, the degeneracy will be lifted only at the points $k = n\pi/a$, namely at the boundaries of the BZ. For example, if $v_{\text{eff}}(x) = V_0 \cos(2\pi x/a)$, then $U = V_0$, and diagonalizing the above matrix we obtain

$$\varepsilon_\pm(\pi/a) = \varepsilon(\pi/a) \pm V_0/2. \tag{6.12}$$

In the reduced zone diagram these energy gaps will always occur at $k = \pm\pi/a$ or $k = 0$, as shown in Fig. 6.2. The energies between $\varepsilon_+(n\pi/a)$ and $\varepsilon_-(n\pi/a)$ are not accessible to the electrons, i.e. there is no wave vector k that gives rise to a state with energy within the gap. Figure 6.2 shows the dispersion relation for electrons in a simple one-dimensional model of a monoatomic lattice, represented by a periodic potential of the form $V(x) = V_0 \cos(2\pi x/a)$. A more detailed treatment of these aspects of solid state theory can be found in specific books on the subject (see, e.g., Harrison, 1980; Kittel, 1996).

Bloch's theorem indicates that it is not necessary to determine the electronic wave functions everywhere in space. It is sufficient to know the solution in the unit cell. The wave function in the neighboring cell is exactly the same except for a phase factor $e^{i\mathbf{k}\cdot\mathbf{a}}$. We can now exploit Bloch's theorem to write down the

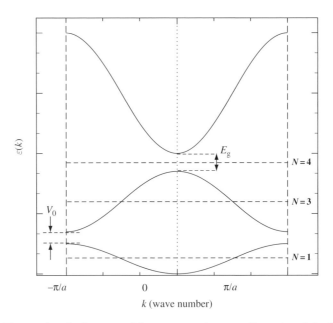

Fig. 6.2 Energy bands for a model monoatomic, one-dimensional lattice. The external potential is $V(x) = V_0 \cos(2\pi x/a)$. Notice the splitting of the first two bands (V_0) at zone boundary, and the splitting of second and third bands at zone center (E_g). Each band can accommodate two electrons (up and down spins). Semiconductors have completely filled bands, while in metals the Fermi level falls in the middle of a band. Horizontal dashed lines indicate the Fermi level of systems with one, three, and four valence electrons per atom. The first two are metallic.

general form of the electronic wave functions in a periodic potential by using the fact that a periodic function can always be represented by a Fourier series:

$$\psi_{\mathbf{k}}(\mathbf{r}) = e^{i\mathbf{k}\cdot\mathbf{r}} \sum_{\mathbf{G}} C_{\mathbf{k}+\mathbf{G}} e^{i\mathbf{G}\cdot\mathbf{r}}, \quad (6.13)$$

where $\mathbf{G} = n_1\mathbf{b}_1 + n_2\mathbf{b}_2 + n_3\mathbf{b}_3$ are *all* the reciprocal lattice vectors.

For an infinite system, any vector $\mathbf{k}$ should be allowed. However, vectors outside the first Brillouin zone can be re-written in the form $\mathbf{k}' = \mathbf{k} + \mathbf{G}_0$, with $\mathbf{k}$ in the first BZ. Therefore, the wave function at $\mathbf{k}'$ can be expressed as

$$\psi_{\mathbf{k}'}(\mathbf{r}) = e^{i\mathbf{k}'\cdot\mathbf{r}} \sum_{\mathbf{G}} C_{\mathbf{k}'+\mathbf{G}} e^{i\mathbf{G}\cdot\mathbf{r}} = e^{i\mathbf{k}\cdot\mathbf{r}} \sum_{\mathbf{G}} C_{\mathbf{k}+\mathbf{G}+\mathbf{G}_0} e^{i(\mathbf{G}+\mathbf{G}_0)\cdot\mathbf{r}}$$

$$= e^{i\mathbf{k}\cdot\mathbf{r}} \sum_{\mathbf{G}'} C_{\mathbf{k}+\mathbf{G}'} e^{i\mathbf{G}'\cdot\mathbf{r}} = \psi_{\mathbf{k}}(\mathbf{r}). \quad (6.14)$$

In conclusion, the calculation of the wave function for *all* the electrons in the infinite solid is mapped – via Bloch's theorem – onto the calculation of the wave function for a finite number of electrons in the unit cell, at an – in principle – infinite number of $\mathbf{k}$ vectors in the first BZ. Obviously, we do not want to solve the electronic problem for an infinite number of Bloch states. Nearby $\mathbf{k}$ vectors carry very similar information. Therefore, it should be possible to reproduce the required physical properties to the desired numerical accuracy by using the wave functions at a finite number of $\mathbf{k}$-points in the first BZ.

6.2.2 Brillouin zone sampling

Let us consider a system that contains two electrons in the unit cell, subject to periodic boundary conditions (PBC). If the system is not spin-polarized, then only the lowest state $\psi_{\mathbf{k}}(\mathbf{r})$ in the box will be occupied by two electrons with opposite spins. Due to Bloch's theorem, the wave function should have the same periodicity as the unit cell, besides a phase factor related to the wave vector $\mathbf{k}$,

$$\psi_{\mathbf{k}}(\mathbf{r}+\mathbf{a}_i) = e^{i\mathbf{k}\cdot\mathbf{a}_i} \psi_{\mathbf{k}}(\mathbf{r}). \quad (6.15)$$

In that case, the associated electronic density is

$$\rho_{\mathbf{k}}(\mathbf{r}) = 2|\psi_{\mathbf{k}}(\mathbf{r})|^2. \quad (6.16)$$

Since the system is periodic, this density is repeated unchanged in all the replicas of the unit cell, i.e. $\rho_{\mathbf{k}}(\mathbf{r}+\mathbf{L}) = \rho_{\mathbf{k}}(\mathbf{r})$, for any linear combination of lattice vectors $\mathbf{L} = n_1\mathbf{a}_1 + n_2\mathbf{a}_2 + n_3\mathbf{a}_3$ (with n_i integers). The wave function is not periodic in the unit cell, unless $\mathbf{k} = \mathbf{0}$. It is, however, periodic in a larger *supercell*. In fact, if

$$\psi_{\mathbf{k}}(\mathbf{r}+n\mathbf{a}_i) = e^{in\mathbf{k}\cdot\mathbf{a}_i} \psi_{\mathbf{k}}(\mathbf{r}), \quad (6.17)$$

the wave function is periodic if $\mathbf{k} \cdot \mathbf{a}_i = 2\pi m/n$, with m and n integers. For example, if $n = 2$, the *allowed* $\mathbf{k}$-vectors are $\mathbf{k} = \mathbf{0}$ and $\mathbf{k} = \mathbf{b}_i/2$, with $\mathbf{b}_i$ the reciprocal lattice vector defined in terms of the lattice vectors $\{\mathbf{a}_i\}$ according to (6.7). For a simple cubic cell $\mathbf{b}_i = (2\pi/a_i)\,\hat{\mathbf{a}}_i$. The next value, $\mathbf{b}_i$, is equivalent to $\mathbf{0}$, and $-\mathbf{b}_i/2$ is equivalent to $\mathbf{b}_i/2$. For $n = 4$ the allowed values are 0, $\pm\mathbf{b}_i/4$, and $\mathbf{b}_i/2$. The next value, $3\mathbf{b}_i/4$, is equivalent to $-\mathbf{b}_i/4$. In general, the non-equivalent $\mathbf{k}$-vectors compatible with PBC in a supercell made of n replicas of the unit cell along the direction of the lattice vector $\mathbf{a}_i$ are

$$\mathbf{k} = \mathbf{0}, \pm\frac{\mathbf{b}_i}{n}, \pm\frac{2\mathbf{b}_i}{n}, \ldots, \pm\frac{(n/2-1)\mathbf{b}_i}{n}, \frac{\mathbf{b}_i}{2}. \quad (6.18)$$

Using the wave functions corresponding to these n vectors, we can construct an electronic density that verifies PBC in this supercell as

$$\rho(\mathbf{r}) = \frac{2}{n} \sum_{j=-(n/2-1)}^{n/2} |\psi_{j\mathbf{b}_i/n}(\mathbf{r})|^2, \quad (6.19)$$

where the factor $2/n$ takes care of normalization to the number of electrons in the unit cell. By taking n very large we recover the limit of an infinite system. The generalization to an arbitrary number of replicas along the direction of the three lattice vectors is straightforward. The following general expression for the electronic density is obtained:

$$\rho(\mathbf{r}) = \sum_{\mathbf{k} \in BZ} \omega_\mathbf{k} |\psi_\mathbf{k}(\mathbf{r})|^2, \quad (6.20)$$

where the sum runs over all $\mathbf{k}$-vectors in the BZ of the unit cell, and $\omega_\mathbf{k}$ are weight factors that depend on the symmetry of the unit cell. If the expansion is carried out around the Γ-point, then there is a reduction in the number of points because $\mathbf{k}$ and $-\mathbf{k}$ are equivalent due to time reversal symmetry. Both choices, centered and de-centered $\mathbf{k}$-point grids, are possible; which one is better depends on the case, but de-centered grids tend to be more efficient.

In two and three dimensions symmetry can be exploited to reduce the portion of the BZ that has to be sampled in the summation above. This requires the concept of the *irreducible wedge* of the BZ, which is the minimal portion that contains all the necessary information to describe the whole BZ. For an orthorhombic unit cell, it will be an octant, but for a tetragonal cell, the C_4 symmetry axis makes one half of the octant equivalent to the other half, so that the irreducible wedge becomes a triangular-base parallelepiped. For a simple cubic cell, the irreducible wedge of the BZ is further reduced to a triangular pyramid, by adding two other C_4 symmetry axes. $\mathbf{k}$-points at the boundaries of the irreducible wedge carry a smaller weight because they are shared with other wedges. The multiplicity factor is $1/M$, with M the number of irreducible wedges sharing this point. Those inside the wedge count one, and the Γ-point counts $1/M_{tot}$, with M_{tot} the total number of wedges needed to fill the whole BZ. Then, the weights are normalized so that $\sum_{i=1}^{N_{tot}} \omega_i = 1$, with N_{tot} the total number of points included in the sum.

In the limit of infinitely closely spaced **k**-points, which corresponds to an infinite crystal, the summation becomes an integral. In practice, however, the only computationally feasible approach is to approximate the full BZ integral with a summation over a finite set of **k**-points. This approach receives the name of *Brillouin zone sampling*. The number of points required will depend on the size of the box and on the specific features of the system. For instance, metallic systems require a very fine sampling to capture properly the shape of the Fermi surface, while semiconductors can be reasonably represented with a few, carefully selected, **k**-points. Sets of special **k**-points for the different symmetries, whose use accelerates the convergence of the BZ summation with the number of points, have been proposed by Baldereschi (1973) and Chadi and Cohen (1973). The following more general recipe for all symmetries, which is well suited for metallic systems, has been proposed by Monkhorst and Pack (1976). A set of q vectors is written in terms of the primitive reciprocal lattice vectors as $\mathbf{k} = n_1 \mathbf{b}_1 + n_2 \mathbf{b}_2 + n_3 \mathbf{b}_3$, where the integers n_i are given by $n_i = (2r - q - 1)/2q$ and $r = 1, 2, \ldots, q$ (see also Moreno and Soler, 1992). The magnitude of the error introduced by sampling the BZ integral with a finite number of **k**-points can always be reduced by using a denser set of points. Most electronic structure computer codes have an automatic generator for one or more types of special points.

It is interesting to notice that the individual wave functions of wave vector $\mathbf{k} \neq 0$ do not fulfill the PBC on the unit cell because of the phase factor. They do verify PBC, but in a larger supercell of size π/k. By varying the wave vector from 0 to π/a, we scan different boundary conditions for the wave functions in the unit cell, from periodic to antiperiodic. The electronic density, however, is always periodic because the phase factor cancels out when the wave functions are squared.

In summary, for solid state applications, one has to solve a set of coupled Kohn–Sham equations, one for each **k**-point included in the BZ sampling:

$$\left\{ -\frac{\hbar^2}{2m}\nabla^2 + v_{ext}(\mathbf{r}) + \int \frac{\rho(\mathbf{r}')}{|\mathbf{r}-\mathbf{r}'|} d\mathbf{r}' + \mu_{XC}[\rho] \right\} \varphi_i^{(\mathbf{k})}(\mathbf{r}) = \varepsilon_i^{(\mathbf{k})} \varphi_i^{(\mathbf{k})}(\mathbf{r}). \quad (6.21)$$

The coupling arises through the electronic density, which is now expressed as a BZ average,

$$\rho(\mathbf{r}) = \sum_{\mathbf{k} \in BZ} \omega_{\mathbf{k}} \sum_{i=1}^{N_{\mathbf{k}}} f_i^{(\mathbf{k})} |\varphi_i^{(\mathbf{k})}(\mathbf{r})|^2, \quad (6.22)$$

where $N_\mathbf{k}$ is the number of electronic states occupied at each $\mathbf{k}$-point, and $f_i^{(\mathbf{k})}$ is the occupation number of band i at wave vector $\mathbf{k}$. If the system is insulating, then $f_i^{(\mathbf{k})} = 1$ independently of i and $\mathbf{k}$, and $N_\mathbf{k} = N$. If there is spin degeneracy the sum is carried up to $N/2$, and then multiplied by a degeneracy factor $f_i^{(\mathbf{k})} = 2$.

6.2.3 Fermi surface sampling for metallic systems

For metallic systems the occupation numbers $N_\mathbf{k}$ are determined by requiring that the associated eigenvalues $\{\varepsilon_i^{(\mathbf{k})}; i = 1, \ldots, N_\mathbf{k}\}$ be smaller than a certain value, which for infinitely fine BZ sampling coincides with the Fermi energy ϵ_F. The Fermi level, or ionization potential, is then given by the highest occupied eigenvalue, which is the only eigenvalue that has a precise meaning within the Kohn–Sham picture (see Section 4.4).

The Fermi level is self-consistently adjusted to fulfill the normalization condition: $\sum_{\mathbf{k} \in BZ} N_\mathbf{k} \omega_\mathbf{k} = N$. The determination of this value can be delicate in the case of metallic systems with Fermi surfaces of complex shape. Slightly different choices of $\mathbf{k}$-points can lead to bands entering or exiting the sum, according to whether $\varepsilon_i^{(\mathbf{k})}$ is below or above ϵ_F. For a sufficiently dense BZ sampling this should not be a problem, but this may imply an impracticable number of $\mathbf{k}$-points. Therefore, schemes that stabilize this finite sum by mimicking BZ integration in some way are necessary.

There are two issues here. One is to find the portion of the BZ that is occupied. This requires assigning a volume in reciprocal space to each $\mathbf{k}$-point in the sum. Initially the BZ was filled with cubic elements (Gilat and Raubenheimer, 1966), but it was soon realized that tetrahedral elements (Jepsen and Andersen, 1971; Lehmann and Taut, 1972) are superior for representing the curved Fermi surface (for free electrons in a cubic cell, the Fermi surface is a sphere). In the *linear tetrahedron* method the eigenvalues $\varepsilon_i^{(\mathbf{k})}$ of the energy bands that cross the Fermi surface are linearly expanded inside the tetrahedron, and the interpolation coefficients are determined from the values at the corners, which are the only ones that are explicitly calculated (Rath and Freeman, 1975). Due to the linear interpolation, finding the fractional volume of the contribution of the tetrahedra that cross the Fermi surface to the integrals becomes a simple linear geometrical problem.

The other issue is more practical, and is concerned with the convergence of the self-consistent procedure. A coarse sampling of the BZ of a metal implies that, for the $\mathbf{k}$-points close to the Fermi surface, the highest occupied bands can enter or exit the sums from one iterative step to the next, just because of the adjustment of the Fermi energy. This introduces an instability that forces the use of small self-consistency mixing coefficients (see Section 11.2). To mitigate this problem, a simple solution is to *smear* the Fermi surface by introducing a distribution of occupation numbers. Typical choices for the smearing function are a Gaussian (Fu and Ho, 1983) or a Fermi function. In the latter, the distribution is expressed as

$$f_i^{(\mathbf{k})} = \frac{1}{e^{(\varepsilon_i^{(\mathbf{k})} - \varepsilon_F)/k_B T_e} + 1}, \qquad (6.23)$$

where $k_B T_e$ is a *broadening* energy parameter that is adjusted to avoid instabilities in the convergence of the self-consistent procedure. Due to the analogy with the *true* Fermi distribution, this parameter is sometimes called the electronic Fermi temperature, but it has to be kept in mind that this is just a technical issue. Only in the case of systems at really finite electronic temperature does this parameter assume a physical meaning. In that case, the finite temperature extension of Kohn–Sham theory due to Mermin (1965) has to be employed, where the energy functional is replaced by a grand potential functional $\Omega[\rho]$ (Alavi et al., 1995).

In the present context, however, T_e is treated as a convergence parameter. The larger it is, the faster the convergence. A caveat of this approach is that for finite T_e the BZ integrals converge faster, but to incorrect values. Therefore, after self-consistency has been achieved for a relatively large value of T_e, this value has to be reduced until the energy becomes independent of it. A way to circumvent this problem has been proposed by Marzari et al. (1997), who introduced a very elegant method within the framework of *ensemble DFT* as formulated by Mermin (1965).

Both issues can be tackled at once by using more sophisticated methods, such as the one proposed by Methfessel and Paxton (1989). Here, the step function that characterizes the Fermi surface and the δ-function associated with the sampling of the BZ using a finite number of points are approximated by a hierarchy of increasingly accurate smooth approximations based on Hermite polynomials. In this way, accurate values of the integrals can be obtained with a small number of **k**-points.

Very often one is interested in comparing the energies of structures having different symmetries. Care must be taken, in this case, that the BZ of all the structures are sampled to the same accuracy. Since for unit cells of different shape it is not possible to choose exactly the same **k**-points, a usual strategy is to try and maintain the same density of **k**-points.

6.2.4 Refolding

Let us consider now a supercell containing just a few unit cells. The larger BZ of the unit cell can be reproduced by repeating the smaller BZ of the supercell using the reciprocal lattice vectors **G**. Suppose that the supercell was sampled with the Γ-point only. When repeating this BZ periodically, the Γ-point will *refold* onto a set of **k**-points corresponding to the BZ of the unit cell. The Γ-point will still appear, but the images of Γ will correspond to other **k**-points in the BZ of the unit cell, as illustrated in Fig. 6.3. The same happens with points other than Γ.

This means that the choice of a larger supercell is equivalent to considering the unit cell, but with a finer sampling of its BZ. The correspondence is not perfect because the refolded **k**-points cannot be chosen at will, as in the case of true **k**-points. They are univocally determined by the shape of the supercell. In the limit of an infinitely large supercell, its BZ becomes a point (the Γ-point, in fact), and its repetition implies a uniform and infinitely fine sampling of the BZ of the unit cell.

A disadvantage of the supercell approach is that the symmetry of the unit cell, which allows us to consider only the irreducible wedge of the BZ, may not necessarily be exploited. In addition, a larger supercell implies a larger number of electrons and basis functions, with an increased computational cost that does not compensate for the smaller number of **k**-points required for BZ sampling. Therefore, supercells are preferred only when atomic rearrangements beyond the size of the unit cell are expected. Otherwise, the most economical strategy is to represent the system with the minimal cell, and sample the BZ with the minimal number of **k**-points that ensures an appropriate convergence of the BZ sums.

An important issue when choosing a supercell is to try to preserve the symmetry of the underlying unit cell. Otherwise, the refolded images of the **k**-points explicitly included will break the symmetry, inducing spurious geometrical distortions. For example, the two-dimensional hexagonal symmetry is respected when using a larger hexagonal cell made of the original and six replicas, but it can also be

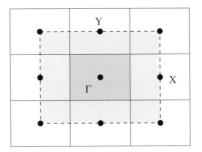

Fig. 6.3 Brillouin zone of a two-dimensional rectangular lattice (dashed line, gray background) including the BZ of a supercell doubled in the two directions (dark rectangle at the center). The reciprocal lattice vectors of the supercell are one half those of the unit cell. When repeating the supercell's BZ, the various **k**-points *refold* onto a denser set of **k**-points in the (larger) BZ of the unit cell.

represented by a rectangular supercell where the unit cell is repeated along the Cartesian axes. This latter breaks the hexagonal symmetry by not including all the members of the *star* of equivalent **k**-points.

6.2.5 Aperiodic systems: molecules, surfaces, and defects

If the periodicity of the system is broken due to the presence of defects, e.g. point defects like interstitials or vacancies, or extended defects such as surfaces, edges, or steps, or if the system is a molecule or a cluster, then, strictly speaking, the replication of a relatively small supercell is not the correct description of the infinite system, which is intrinsically aperiodic.

The usual strategy in these cases is still to adopt the supercell approach with PBC, and make sure that the required physical and chemical properties are converged with respect to the size of the supercell. In the case of molecules or surfaces, the supercell approach consists of introducing a vacuum region, which should be large enough that periodic images corresponding to adjacent replicas of the supercell do not interact significantly. The case of point defects is similar, only the supercell consists of a piece of bulk and no vacuum. In this case the amount of bulk should be large enough that the periodic images of the defect do not interact significantly.

Sometimes convergence may be difficult to achieve, particularly in charged or polar systems that induce long-range electrostatic fields. The case of charged systems is special in that the electrostatic energy of an infinitely periodically replicated charged system diverges (see Section 9.3.1). Therefore, this case must be treated differently by compensating the excess charge in some way. This is valid for charged molecules or clusters, but also for charged defects. Here, the typical approach is to compensate the charge with a uniformly spread background charge of the opposite sign. This option is quite simple to implement, but it may not be particularly realistic. For charged defects neutralization usually occurs via another localized defect in a different region of space. To what extent the uniform background mimics this situation is open to debate.

Neutral finite systems can still introduce difficulties when treated within a supercell approach, if the charge distribution is such that the dipole or higher multi-pole moments are different from zero. The electrostatic interaction between dipoles decays as R^{-3}, while quadrupole–quadrupole interactions decay as R^{-5}. Therefore, in order to minimize spurious effects, the supercell must be chosen large enough that these interactions do not have a significant influence on the properties under scrutiny. Since this is all a consequence of the long-range electrostatics, some alternative approaches, based on a modification of the Poisson equation that determines the electrostatic potential, have been devised. These are designed to eliminate the undesired inter-cell interactions (Hockney and Eastwood, 1981; Makov and Payne, 1995).

Surfaces are extremely important in many scientific areas such as catalysis, gas sensors, and tribology, amongst others. The study of surfaces within the supercell approach is usually carried out in the *slab* geometry, where the semi-infinite bulk is represented by a finite slab with two surfaces. Here, the slab thickness has to be chosen large enough that the two surfaces do not interact with each other. Due to their small screening length (of the order of 1 Å), metals do not require thick slabs. Usually between five and seven atomic layers are sufficient to decouple the two surfaces; the most important effect is perhaps the geometric relaxation of the layers. Insulators have longer screening lengths and more important relaxation and reconstruction effects. Therefore, they usually require a larger number of layers to avoid the interaction between surfaces. A similar situation occurs for interfaces, where sufficient layers of each material have to be considered on both sides in order to avoid the interaction of the interface with the free surfaces.

In addition, the two surfaces can interact through the vacuum between periodic replicas. The direct interaction due to wave function overlap through the vacuum is not very important because these decay rapidly into the vacuum region (exponentially within the LDA). Therefore, the main effect is again due to charged or polarized slabs. For charged slabs, the question arises of where to place the compensating opposite charge. A standard approach has been to distribute it uniformly in the supercell. At first glance this may seem straightforward, but there are a number of subtleties, such as additional forces arising from the removal of the background charge, that must be carefully considered (Lozovoi *et al.*, 2001). For a slab geometry, the compensating charge can also be taken as a distribution localized in the vacuum region (Fu and Ho, 1989). This approach has been recently analyzed in detail by Lozovoi and Alavi (2003) in the context of a study of the reconstruction of charged surfaces in the presence of an electric field generated by the counter-electrode in the vacuum. This scheme is more akin to what happens in electrochemical cells and in field evaporation situations (Sánchez, 2003; Sánchez *et al.*, 2004).

6.2.6 Disordered systems: liquids and solids at finite temperature

At finite temperatures the point group symmetry of a bulk solid is broken, and the concept of a Brillouin zone becomes less evident. If the temperature is such that the system remains in a well-defined crystal structure, and thermal vibrations of the atoms are confined to the vicinity of their equilibrium positions, then the concepts of discrete translational invariance, unit cell, Brillouin zone, and **k**-points hold. When the system melts, this is not true any longer.

In liquids or amorphous systems, PBC break the homogeneity property. This description, however, is still much better than considering isolated clusters as models for fluids, and it represents a very valid alternative to costly, large supercells. In fact, this is the approach invariably used to simulate liquids when particles interact via classical potentials (Allen and Tildesley, 1987). The issue of the effect of PBC on the electronic orbitals is more subtle. It is clear that if there is no translational invariance at all, then the whole machinery derived from Bloch's theorem breaks down. However, the spurious periodicity introduced by the use of PBC in the nuclear degrees of freedom already implies a fictitious translational invariance and, given that this invariance is present anyway, then the solid state machinery can be restored. The difference with bulk solids is that the periodic replication of the supercell is not physically correct. It is an artifact of the supercell construction, which is useful to accelerate convergence with respect to the system size. Within this perspective, the use of BZ sampling in the case of fluid or amorphous systems is correct because it treats the electronic periodicity at a level consistent with the nuclear periodicity introduced through PBC.

An additional complication arises in the case of polar fluids that exhibit long-range spatial correlations, such as those introduced by the hydrogen-bonding network in water. Simulating such systems with a small supercell can be misleading because these correlations are disrupted by PBC. These cases must be treated with great respect, and a careful study of finite-size effects should be carried out before drawing conclusions from simulations in small supercells. A large body of *ab initio* results for liquid water has been produced in the past decade (Sprik *et al.*, 1996), using 32- and 64-molecule supercells. This allowed the extraction of a number of meaningful conclusions about local properties, such as the mechanism for proton transfer and the local structure and dynamics of water. However, it is known from classical simulations that the structure of the H-bond network is far from convergence at these sizes (Arteca *et al.*, 2000).

References

Alavi, A., Kohanoff, J., Parrinello, M., and Frenkel, D. (1995). *Ab initio* molecular dynamics with excited electrons. *Phys. Rev. Lett.* **73**, 2599–2602.

Allen, M. P. and Tildesley, D. J. (1987). *Computer Simulation of Liquids*. Oxford, Clarendon Press.

Arteca, G. A., Cachau, R. E., and Veluri, K. (2000). Structural complexity of hydrogen-bonded networks. *Chem. Phys. Lett.* **319**, 719–724.

Baldereschi, A. (1973). Mean-value point in the Brillouin zone. *Phys. Rev. B* **7**, 5212–5215.

Bloch, F. (1928). Über die Quantenmechanik der Elektronen in Kristallgittern. *Z. Phys.* **52**, 555–560.

Chadi, D. J. and Cohen, M. L. (1973). Special points in the Brillouin zone. *Phys. Rev. B* **8**, 5747–5753.

Fu, C.-L. and Ho, K.-M. (1983). First-principles calculation of the equilibrium ground-state properties of transition metals: applications to Nb and Mo. *Phys. Rev. B* **28**, 5480–5486.

(1989). External-charge-induced surface reconstruction on Ag(110). *Phys. Rev. Lett.* **63**, 1617–1620.

Gilat, G. and Raubenheimer, L. J. (1966). Accurate numerical method for calculating frequency-distribution functions in solids. *Phys. Rev. B* **144**, 390–395.

Harrison, W. A. (1980). *Electronic Structure and the Properties of Solids*. San Francisco, Freeman.

Hockney, R. W. and Eastwood, J. W. (1981). *Computer Simulation Using Particles*. New York, McGraw-Hill.

Jepsen, O. and Andersen, O. K. (1971). The electronic structure of h.c.p. Ytterbium. *Solid State Comm.* **9**, 1763–1767.

Kittel, C. (1996). *Introduction to Solid State Physics*, 7th edn. New York, Wiley.

Lehmann, G. and Taut, M. (1972). Numerical calculation of the density of states and related properties. *Phys. Status Solidi* **54**, 469–477.

Lozovoi, A. Y. and Alavi, A. (2003). Reconstruction of charged surfaces: general trends and a case study of Pt(110) and Au(110). *Phys. Rev. B* **68**, 245416.

Lozovoi, A. Y., Alavi, A., Kohanoff, J., and Lynden-Bell, R. M. (2001). Ab initio simulation of charged slabs at constant chemical potential. *J. Chem. Phys.* **115**, 1661–1669.

Makov, G. and Payne, M. (1995). Periodic boundary conditions in *ab initio* calculations. *Phys. Rev. B* **51**, 4014–4022.

Marzari, N., Vanderbilt, D., and Payne, M. C. (1997). Ensemble density-functional theory for *ab initio* molecular dynamics of metals and finite-temperature insulators. *Phys. Rev. Lett.* **79**, 1337–1340.

Mermin, N. D. (1965). Thermal properties of the inhomogeneous electron gas. *Phys. Rev.* **137**, A1441–1443.

Methfessel, M. and Paxton, A. T. (1989). High-precision sampling for Brillouin-zone integration in metals. *Phys. Rev. B* **40**, 3616–3621.

Monkhorst, H. J. and Pack, J. D. (1976). Special points for Brillouin-zone integrations. *Phys. Rev. B* **13**, 5188–5192.

Moreno, J. and Soler, J. M. (1992). Optimal meshes for integrals in real- and reciprocal-space unit cells. *Phys. Rev. B* **45**, 13891–13898.

Rath, J. and Freeman, A. J. (1975). Generalized magnetic susceptibilities in metals: application of the analytic tetrahedron linear energy method to Sc. *Phys. Rev. B* **11**, 2109–2117.

Sánchez, C. G. (2003). Molecular reorientation of water adsorbed on charged Ag(111) surfaces. *Surface Science* **527**, 1–11.

Sánchez, C. G., Lozovoi, A. Y., and Alavi, A. (2004). Field evaporation from first principles. *Mol. Phys.* **102**, 1045–1055.

Sprik, M., Hutter, J., and Parrinello, M. (1996). *Ab initio* molecular dynamics simulation of liquid water: comparison of three gradient-corrected density functionals. *J. Chem. Phys.* **105**, 1142–1152.

7

Atomic pseudopotentials

The wave functions for free electrons in a periodic crystal can be expanded in plane waves (PWs). If the potential due to the atoms is neglected, then PWs are the exact solution. If the potential is reasonably smooth, then it can be treated as a perturbation, thus leading to the so-called nearly-free electron model (see Section 6.2). The potential originated in the atomic nuclei, however, is far from smooth. In the simplest case of hydrogen the potential is $-1/r$, which diverges at the origin. The $1s$ wave function does not diverge, but it exhibits a cusp at the origin, and decays exponentially with distance. For heavier atoms the wave functions associated with core states are even steeper. Therefore, a PW expansion of the wave functions in a real crystal is a rather hopeless task, because the number of PW components required to represent such steep wave functions is huge. However, it would be desirable to retain the simplicity of the PW approach. Slater (1937) suggested a possible solution to this problem, where the PW expansion was augmented with the solutions of the atomic problem in spherical regions around the atoms, and the potential was assumed to be spherically symmetric inside the spheres, and zero outside (APW method).

In order to overcome this shape approximation of the potential, Herring (1940) proposed an alternative method consisting of constructing the valence wave functions as a linear combination of PW and core wave functions. By choosing appropriately the coefficients of the expansion, this wave function turns out to be orthogonal to the core states. Hence the name of *orthogonalized plane wave* (OPW) method. Since the troublesome region is taken care of by the core orbitals, the part that must be represented by the PWs is rather smooth, and a smaller number of PW components is required to reproduce the valence states. One can go a step beyond the OPW approach, and eliminate the core states altogether by replacing their action with an effective potential, or *pseudopotential*. This pseudopotential, however, cannot be anything. It has to be constructed carefully in order to reproduce accurately the bonding properties of the true potential.

This chapter describes the theory and practical aspects involved in the design of efficient and accurate pseudopotentials.

As mentioned in Section 6.1.1, the electronic states of an atom can be classified into: (1) *core states*, which are highly localized and not involved in chemical bonding, (2) *valence states*, which are extended and responsible for chemical bonding, and (3) *semi-core states*, which are localized and polarizable, but generally do not contribute directly to chemical bonding. The most common pseudopotential approach consists of not allowing the relaxation of core states according to the environment (*frozen core approximation*), although some polarizable core approaches have been proposed. In general, this is a very good approximation that reproduces total atomic energies within 0.01 eV. Semi-core states are often treated as part of the frozen core, but when their contribution is important they have to be included in the valence.

The valence states, due to orthogonalization with respect to the core states of the same symmetry, show a marked oscillatory behavior with a number of nodes equal to $n-l-1$, n being the principal quantum number and l the angular momentum. Nodeless wave functions ($l = n-1$) are not oscillatory but, due to the lack of orthogonalization, they create strongly bound states that are markedly peaked close to the nucleus. This is the case of the $1s$ state in H, the $2p$ states in C, N, O, and F, and the $3d$ states in transition metals. When the basis set chosen is that of plane waves, the computation of Hamiltonian matrix elements requires the Fourier decomposition of the wave functions. Features like the above are very stringent for PW, because sharp peaks require a very large number of PWs to achieve convergence in the expansion, and this translates into a vast amount of computer resources (the dimension of the matrix to diagonalize becomes very large).

Based on the observations that: (1) core states are not fundamental for the description of chemical bonding, and (2) a good description of the valence wave functions inside the core region is not strictly necessary, there is no lack of crucial information if the inner solution (inside some cutoff radius) is replaced with a smooth, nodeless pseudo-wave function, which is not a solution to the original atomic problem. Being nodeless, it now corresponds to the lowest-lying state of an effective, pseudo-atomic problem where the true potential has been replaced by a *pseudopotential* (PS). Also, in the context of quantum chemical calculations for heavy atoms at the Hartree–Fock level, the core electrons can be replaced by a pseudopotential, which is normally referred to as *effective core potential* (ECP).

7.1 Pseudopotential theory

Inspired in the OPW method, the origin of the modern pseudopotential approach can be traced back to the well-known paper of Philips and Kleinman (1959), who showed that one can construct a smooth valence wave function $\tilde{\varphi}_v$ that is not orthogonalized to the core states φ_c, by combining the core and the true valence wave functions φ_v in the following way:

$$|\tilde{\varphi}_v\rangle = |\varphi_v\rangle + \sum_c \alpha_{cv}|\varphi_c\rangle, \tag{7.1}$$

where $\alpha_{cv} = \langle\varphi_c|\tilde{\varphi}_v\rangle \neq 0$. This pseudo-wave function satisfies the modified Schrödinger equation:

$$\left[\hat{\mathcal{H}} + \sum_c (\varepsilon_v - \varepsilon_c)|\varphi_c\rangle\langle\varphi_c|\right]|\tilde{\varphi}_v\rangle = \varepsilon_v|\tilde{\varphi}_v\rangle, \tag{7.2}$$

where $\hat{\mathcal{H}} = \hat{T} + \hat{V}$, $\hat{V} = (Z_c/r)\hat{I}$ is the bare nuclear potential, and $\hat{I}$ is the identity operator. This shows that it is possible to construct a pseudo-Hamiltonian

$$\hat{\mathcal{H}}_{PS} = \hat{\mathcal{H}} + \sum_c (\varepsilon_v - \varepsilon_c)|\varphi_c\rangle\langle\varphi_c| \tag{7.3}$$

with the same eigenvalues of the original Hamiltonian but a smoother, nodeless wave function. The associated potential

$$\hat{V}_{PS} = \frac{Z_c}{r}\hat{I} + \sum_c (\varepsilon_v - \varepsilon_c)|\varphi_c\rangle\langle\varphi_c| \tag{7.4}$$

was called a *pseudopotential*.

It is clear that this pseudopotential acts differently on wave functions of different angular momentum. The most general form of a pseudopotential of this kind is, then, the following:

$$\hat{V}_{PS}(\mathbf{r}) = \sum_{l=0}^{\infty} \sum_{m=-l}^{l} v_{PS}^l(r)|lm\rangle\langle lm| = \sum_{l=0}^{\infty} v_{PS}^l(r)\hat{P}_l, \tag{7.5}$$

where $\langle\mathbf{r}|lm\rangle = Y_{lm}(\theta, \phi)$ are spherical harmonics, $v_{PS}^l(r)$ is the pseudopotential corresponding to the angular component l, and the operator

$$\hat{P}_l = \sum_{m=-l}^{l} |lm\rangle\langle lm| \tag{7.6}$$

is a projection operator onto the lth angular momentum subspace. The meaning of Expression (7.5) is the following: when $\hat{V}_{PS}$ acts on the electronic wave function, the projection operators $\hat{P}_l$ select the different angular components of the wave function, which are then multiplied by the corresponding pseudopotential $v_{PS}^l(r)$. Next, the contributions of all the angular components are added up to form the total pseudopotential contribution to the Hamiltonian matrix elements that enter Schrödinger's equation. Pseudopotentials of this kind are usually called *non-local* because they act differently on the various angular components of the wave function as a consequence of the exchange with the core. In practice, $v_{PS}^l(r)$ is a local operator in the radial coordinate. Therefore, a better name for this type of expression is a *semi-local*, or *angular-dependent*, pseudopotential. If all the angular components of the pseudopotential are taken to be the same, then the pseudopotential is said to be *local*. In principle, local versions can be constructed that verify the required properties for all angular momenta, but they tend to be quite *hard* (many PW components required), and are difficult to construct. That is why it is easier and computationally more effective to use non-local pseudopotentials.

Normally, only a few (l_{max}) low-angular momenta core states are occupied. Therefore, for values of $l > l_{max}$ the ionic core is seen in the same way by all the l components of the wave function, and the summation in (7.5) can be re-cast in the following form:

$$\hat{V}_{PS} = \sum_{l=0}^{\infty} v_{PS}^{loc}(r)\hat{P}_l + \sum_{l=0}^{l_{max}} \left[v_{PS}^l(r) - v_{PS}^{loc}(r) \right] \hat{P}_l$$

$$= v_{PS}^{loc}(r)\,\hat{I} + \sum_{l=0}^{l_{max}} \Delta v_{PS}^l(r)\hat{P}_l, \tag{7.7}$$

where $\Delta v_{PS}^l(r)$ are short-ranged functions confined to the core region, and $v_{PS}^{loc}(r)$ is an average local potential that represents the screened Coulomb interaction.

The choice of the local component is rather arbitrary. It basically should be able to represent reasonably well all the angular components that are not corrected by a non-local component. According to the preceding discussion, it seems that an appropriate choice is to take $v_{PS}^{loc}(r) = v_{PS}^{l_{max}+1}(r)$, i.e. the pseudopotential of the first angular momentum that is not present in the core, e.g. $v_{PS}^p(r)$ for first-row elements, or $v_{PS}^d(r)$ for Si. In methods where the calculation of the non-local contributions to the Hamiltonian is costly, such as PW, there is a computational advantage in making this choice, because then fewer non-local pseudopotential correction terms have to be calculated.

While this works fine as a rule of thumb, reality can be more subtle, because the energetic ordering of atomic levels does not always follow the angular momentum. For instance, the valence shell of fourth-row transition metal atoms such as Ti contains 4s and 3d states, but the 4p states are unoccupied in the neutral atom. The core contains only s and p states. Therefore, the previous rule would suggest to take the d pseudopotential as the local component, and compute a non-local correction for s and p components. This is certainly possible, but the p pseudopotential has to be carefully constructed because it corresponds to an empty atomic state, which could even be unbound. In general these are generated by populating the unoccupied p states in an electronic configuration that is expected to be close to the one realized in the condensed phase. For energetic reasons, these states are the first to hybridize in a crystalline or molecular environment. The first states not including a non-local correction are the f orbitals. So, choosing the p component of the

7.1 Pseudopotential theory

pseudopotential as local may be better in the sense that it could represent more accurately the unoccupied states of higher angular momentum, i.e. the $4f$ states and beyond. Let us remark that there is no profound reason, apart from computational considerations, to choose the local pseudopotential corresponding to a specific angular momentum. In fact, some codes like SIESTA define a local component independently of the angular momenta present in the core (Soler et al., 2002).

The ultimate test is to check the results produced by the pseudopotential in the required environment. Normally, for a solid, the basic properties that are analyzed are lattice parameters and bulk modulus. For a molecule, it is customary to check geometry and vibrational frequencies. There are two levels of comparison: against experimental results, and against all-electron calculations. If the goal is to test the quality of the pseudopotential, then the most fair comparison is that of the second type, because then the effect of pseudization is not masked by the approximation to exchange and correlation (e.g. the LDA). A second issue is whether the LDA or GGA are sufficient to reproduce the above properties, but this can only be addressed after the pseudopotential has been validated again all-electron calculations. Interestingly, it may well happen that the effect of the functional and the pseudopotential compensate each other, and the results are in good agreement with experiment, at least for some properties. This, however, should be taken with caution because other properties may not be well reproduced. The choice of the local component can have some implications regarding the properties of the solid, but far more important is the choice of $l_{\max}$. For example, in the case of silicon, in going from $l_{\max} = 1$ to $l_{\max} = 2$, the lattice constant decreases by 2% and the bulk modulus increases by 8%. Variations due to different choices of the local component are one order of magnitude smaller, and well below the systematic error due to the choice of functional.

Summarizing, there are two basic steps in pseudopotential theory.

(i) Core electrons are removed from the calculation, and the interaction of the valence electrons with the nucleus plus the core states (including orthogonalization) is replaced by an effective, screened potential. The screened potential depends on the angular momentum of the valence electrons because of the different orthogonality conditions. For instance, in the C atom, the $2s$ valence state has to be orthogonal to the $1s$ core state, but the $2p$ valence state does not feel the orthogonality constraint of the $1s$ state because they have different angular quantum numbers. Therefore, within the core region, these two states feel very different potentials from the ionic core. Of course, at large distances the potential is $-Z_V/r$ independently of the angular momentum, because the ionic core is seen as a point charge of magnitude equal to the valence charge Z_V. For each angular momentum l the pseudopotential should have the valence l-state as the ground state. Table 7.1 shows how the electronic configuration of a typical atom is modified when the core electrons are integrated into the core.

Table 7.1 *Electronic states for a true silicon atom and for a pseudo-silicon atom where the 1s, 2s, and 2p states have been integrated into the ionic core of valence charge $Z_V = 4$.*

	core			valence	
true Si atom	$1s^2$	$2s^2$	$2p^6$	$3s^2$	$3p^2$
pseudo-Si atom		—		$1s^2$	$2p^2$

(ii) The full ionic core–electron interaction (often called *ion–electron* interaction), which includes the orthogonality of the valence wave functions to the core states, is replaced by a softer pseudopotential. The solution of the atomic Schrödinger equation for the pseudopotential is a pseudo-wave function different from the true wave function. The pseudopotential, however, is constructed in such a way that its scattering properties (or phase shifts) are the same as those of the all-electron potential, although the radial pseudo-wave function has no nodes inside the core region.

7.2 Construction of pseudopotentials

Clearly, there is an enormous freedom in how pseudopotentials are constructed. Empirical pseudopotentials determined by fitting experimental energy bands have been very popular in the past (Appelbaum and Hamann, 1973; Topp and Hopfield, 1973), but they lacked a very important property, which is *transferability*, namely that a pseudopotential constructed for some specific environment can be used for the same atomic species but in a different environment. The first non-empirical approach to pseudopotentials was the one devised by Philips and Kleinman (1959) (see preceding section). This approach, however, had a severe problem: the normalized pseudo-wave function had an amplitude different from that of the all-electron wave function. Outside the core the shapes were the same but the wave functions were only proportional to each other through a normalization factor. This was not acceptable because it led to an incorrect valence charge distribution, and thus to deviations in the bonding properties. It is important that outside the core region the true and pseudo-wave functions are the same. Hence, an often adopted solution was to *renormalize* the pseudo-wave function to fulfill the above condition. This limitation, however, is not a problem of pseudopotentials in general. It is a consequence of this type of construction where the pseudo-wave functions are orthogonalized to the core states, but this is not the only possibility.

The construction of a pseudopotential is an inverse problem: given a pseudo-wave function that: (1) beyond some distance decays exactly as the all-electron wave function, and (2) is an eigenstate of a pseudo-Hamiltonian with the same

7.2 Construction of pseudopotentials

eigenvalue as the all-electron wave function, the pseudopotential is obtained by inverting the radial Schrödinger equation for that pseudo-wave function. The radial Schrödinger equation

$$\left\{-\frac{\hbar^2}{2m}\frac{d^2}{dr^2} + \frac{l(l+1)}{2r^2} + v(r)\right\} rR(\varepsilon, r) = \varepsilon r R(\varepsilon, r) \quad (7.8)$$

is a second-order linear differential equation. Once ε has been fixed (not necessarily to an eigenvalue), its solution is uniquely determined by the value of the wave function $R(\varepsilon, r)$ and its derivative $R'(\varepsilon, r)$ at any given point r_0. These two conditions can be equally realized by specifying the value of the (dimensionless) radial logarithmic derivative of the wave function at r_0

$$\cot \eta_l(\varepsilon) \propto \left[\frac{d}{dr} \ln R^l(\varepsilon, r)\right]_{r_0} = \frac{1}{R^l(\varepsilon, r_0)} \left[\frac{dR^l(\varepsilon, r)}{dr}\right]_{r_0}, \quad (7.9)$$

together with a normalization condition, and this can be done for all values of angular momentum l. The quantities $\eta_l(\varepsilon)$ are the *phase shifts* of the partial waves in scattering theory.

Therefore, if the all-electron potential and the pseudopotential are the same outside some radius r_c, which is usually called the *cutoff* or *core radius*, then the all-electron and pseudo-wave functions are proportional if the corresponding logarithmic derivatives are the same

$$\frac{1}{R_{AE}^l(\varepsilon, r_c)} \left[\frac{dR_{AE}^l(\varepsilon, r)}{dr}\right]_{r_c} = \frac{1}{R_{PS}^l(\varepsilon, r_c)} \left[\frac{dR_{PS}^l(\varepsilon, r)}{dr}\right]_{r_c}. \quad (7.10)$$

The proportionality becomes an equality only when the pseudo-wave function is further required to preserve the norm inside the cutoff radius

$$\int_0^{r_c} r^2 [R_{PS}^l(\varepsilon, r)]^2 \, dr = \int_0^{r_c} r^2 [R_{AE}^l(\varepsilon, r)]^2 \, dr. \quad (7.11)$$

This property is called *norm-conservation*, and it was first introduced in the field of pseudopotentials by Hamann, Schlüter, and Chiang (1979). These authors (HSC) led a revolution in the field in the late seventies when they proposed a procedure to construct non-local norm-conserving pseudopotentials fitted to first-principles all-electron atomic calculations, without explicit reference to orthogonalization to core states.

7.2.1 Norm-conserving pseudopotentials

A key result of HSC was to realize that the norm of the wave function also appears in a very important identity related to the Friedel sum rule (Topp and Hopfield, 1973; Shaw and Harrison, 1967):

$$-\frac{1}{2} \left\{ [rR^l(\varepsilon, r)]^2 \frac{d}{d\varepsilon} \frac{d}{dr} \ln R^l(\varepsilon, r) \right\}_{r_c} = \int_0^{r_c} r^2 [R^l(\varepsilon, r)]^2 \, dr. \quad (7.12)$$

Therefore, the norm-conservation condition, i.e. the equality of the RHS above for all-electron and pseudo-wave functions, imposes that, to first-order in the eigenvalue, the logarithmic derivatives of the all-electron and pseudo-wave functions vary in the same way. This implies that a small change in the eigenvalue due to changes in the external potential (the environment) produces only a second-order change in the logarithmic derivative. Therefore, condition (7.10), which by construction is strictly verified only for the value of ε used to obtain the wave function – usually called the *reference* energy – becomes approximately valid in a range of eigenvalues around the reference. In this way, pseudopotentials derived from atomic calculations can be exported to other environments. When an atom is part of a molecule or a solid, its electrons feel the influence of the other atoms (the so-called molecular or crystal field). This implies that the electronic eigenvalues are shifted from their atomic values, but the transferability property ensures that the all-electron and pseudo-wave functions still coincide outside the cutoff radius.

The norm-conservation constraint guarantees that the pseudopotential is useful, not in every energy range, but at least in environments such that the eigenvalues do not depart significantly from the eigenvalues used in its construction. For instance, a pseudopotential for H in the H_2 molecule may not be useful for hydrogen at very high pressures because the energy ranges are completely different, but a pseudopotential for Si constructed having in mind the bulk solid will be useful for the Si surface or for liquid Si under similar external conditions. The straightforward recipe for improving transferability is to reduce the cutoff radius, because in this way the pseudo-wave function becomes closer to the all-electron result. However, the reduction of r_c is limited by the (not strictly necessary) condition of a nodeless pseudo-wave function; the cutoff radius cannot be made smaller than the position of the outermost node of the all-electron wave function.

The conditions proposed by Hamann, Schlüter, and Chiang for the construction of norm-conserving pseudopotentials are the following:

(i) the eigenvalues of the pseudo-wave functions coincide with those of the all-electron wave functions for a chosen electronic configuration of the atom;

(ii) R_{PS} is nodeless, and it is identical to the all-electron wave function outside a suitably chosen cutoff radius r_c:

$$R_{PS}(r) = \begin{cases} \tilde{R}_{PS}(r), & r < r_c, \\ R_{AE}(r), & r \geq r_c; \end{cases} \quad (7.13)$$

(iii) the norm of the true and pseudo-wave functions inside the pseudized region ($r < r_c$) is the same (norm-conservation condition):

$$\int_0^{r_c} \left| r\tilde{R}_{PS}(r) \right|^2 dr = \int_0^{r_c} \left| rR_{AE}(r) \right|^2 dr; \quad (7.14)$$

(iv) The logarithmic derivatives of the all-electron and pseudo-wave function agree for $r \geq r_c$.

7.2 Construction of pseudopotentials

Other conditions can be imposed to enhance the smoothness of the potentials, i.e. to reduce the number of high energy Fourier components required for plane wave calculations. This aspect is not particularly relevant in calculations that use localized basis sets, but it is crucial for PWs because this is the single factor that determines the energy cutoff, and thus the cost of the calculation. Typical cutoff radii are between two and three times the core radius, and are located in the region around the maximum of the valence wave function.

Several schemes have been proposed to generate first-principles pseudopotentials that satisfy the above conditions. These differ mostly in the functional form of the potentials and the smoothness conditions. Due to their simple form suitable for analytic integration in both PW and Gaussian basis set calculations, the most widely used for a long time were the pseudopotentials developed by Bachelet et al. (1982) (BHS). There, the pseudopotentials are fitted to the following form:

$$v_{PS}^l(r) = -\frac{Z_V}{r}\left[C \, \mathrm{erf}\left(\sqrt{\alpha_1^{core}} r\right) + (1-C) \, \mathrm{erf}\left(\sqrt{\alpha_2^{core}} r\right)\right]$$
$$+ \sum_{i=1}^{3}(A_{l,i} + r^2 A_{l,i+3})\exp\left(-\alpha_{l,i} r^2\right). \qquad (7.15)$$

The original parameterizations were later revisited by Gonze et al. (1991). More recently, other norm-conserving recipes that improve over the smoothness of the original norm-conserving pseudopotentials have been proposed by other authors (Kerker, 1980; Troullier and Martins, 1991; Rappe et al., 1990; Lin et al., 1993).

Troullier–Martins pseudopotentials

Today, some of the smoothest norm-conserving pseudopotentials are obtained using the recipe by Troullier and Martins (TM), who thoroughly studied the convergence properties of the PW expansion of the pseudopotential (Troullier and Martins, 1991). They generalized Kerker's scheme (Kerker, 1980) by proposing the following analytic form of the wave function inside the cutoff radius:

$$R_{TM}^l(r) = r^l \exp(p(r)), \qquad (7.16)$$

with $p(r) = c_0 + \sum_{i=2}^{n} c_i r^i$. The r^l behavior for small r is included to avoid a hard-core pseudopotential with a singularity at the origin, which would translate into a large number of PWs, but it is also physically correct. In Kerker's scheme (where $n = 4$), the four coefficients of the polynomial are determined by the conditions: (i) charge conservation inside the cutoff radius, and (ii)–(iv) continuity of the pseudo-wave function and its two first derivatives at the cutoff radius. Troullier and Martins added variational freedom in the search for smoothness

Atomic pseudopotentials

by increasing the order of the polynomial. They realized that the asymptotic, large wave number behavior of the pseudopotential depends on the values of its odd derivatives at the origin. This implies that a larger degree of smoothness is achieved when all odd coefficients in the polynomial are set to zero. Additionally, they found that pseudopotentials that are flat at the origin are also smoother. With these ingredients they provided the following practical recipe.

The polynomial is chosen of sixth order in r^2,

$$p(r) = c_0 + c_2 r^2 + c_4 r^4 + c_6 r^6 + c_8 r^8 + c_{10} r^{10} + c_{12} r^{12}, \quad (7.17)$$

with the coefficients determined by the following seven conditions:

(i) norm conservation of the charge within the cutoff radius r_c:

$$2c_0 + \ln\left\{\int_0^{r_c} r^{2(l+1)} \exp[2p(r) - 2c_0]\,dr\right\} = \ln\left\{\int_0^{r_c} r^2 |R_{AE}^l(r)|^2 dr\right\}; \quad (7.18)$$

(ii) continuity of the pseudo-wave function and its first four derivatives at r_c, which in practice can be written as:

$$p(r_c) = \ln\left[\frac{P(r_c)}{r_c^{l+1}}\right],$$

$$p'(r_c) = \frac{P'(r_c)}{P(r_c)} - \frac{l+1}{r_c},$$

$$p''(r_c) = 2v_{AE}(r_c) - 2\varepsilon_l - \frac{2(l+1)}{r_c}p'(r_c) - [p'(r_c)]^2,$$

$$p'''(r_c) = 2v_{AE}'(r_c) + \frac{2(l+1)}{r_c^2}p'(r_c) - \frac{2(l+1)}{r_c}p''(r_c)$$
$$- 2p'(r_c)p''(r_c),$$

$$p''''(r_c) = 2v_{AE}''(r_c) - \frac{4(l+1)}{r_c^3}p'(r_c) + \frac{4(l+1)}{r_c^2}p''(r_c)$$
$$- \frac{2(l+1)}{r_c}p'''(r_c) - 2[p''(r_c)]^2 - 2p'(r_c)p'''(r_c), \quad (7.19)$$

where $P(r) = rR_{AE}^l(r)$, and $v_{AE}(r)$ is the all-electron atomic screened potential (see below);

(iii) zero curvature of the screened pseudopotential at the origin, $v_{sc,l}''(0) = 0$, which translates into

$$c_2^2 + c_4(2l+5) = 0. \quad (7.20)$$

7.2 Construction of pseudopotentials

The derivatives of the wave functions and screened potential can be obtained from the all-electron wave functions and screened potential using, e.g., seventh-order finite differences (Troullier and Martins, 1991).

RRKJ pseudopotentials

Another alternative proposed by Rappe *et al.* (1990) (RRKJ) is to use a pseudo-wave function of the form:

$$R^l_{\text{RRKJ}}(r) = \sum_{i=1}^{n} \alpha_i j_l(G_i r), \tag{7.21}$$

where $j_l(G_i r)$ are spherical Bessel functions of order l, and the values of the wave vectors G_i are chosen to satisfy the conditions

$$\frac{j'_l(G_i r_c)}{j_l(G_i r_c)} = \frac{R'_{\text{AE}}(r_c)}{R_{\text{AE}}(r_c)} \tag{7.22}$$

at the cutoff radius r_c. Equation (7.22) ensures the continuity of the logarithmic derivative of the pseudo-wave function at the cutoff radius, so as to reproduce the phase shifts from all-electron calculations.

In this construction, the fitting parameters are the n expansion coefficients α_i. These have to be chosen according to some criterion, which, in the case of PW calculations, is the smoothness of the pseudo-wave function, namely that the Fourier transform of the pseudo-wave function should decay as quickly as possible. The expression in terms of spherical Bessel functions is particularly transparent in this respect, because it introduces a natural cutoff in reciprocal space given by the highest wave vector G_n (see Lin *et al.*, 1993, Fig. 1).

Since the dominant energetic contribution for high-**G** PW components is the kinetic term, RRKJ proposed to fix a cutoff value G_c, and to find the expansion coefficients α_i by minimizing the kinetic energy arising from PW components with $|\mathbf{G}| > G_c$, with the three additional constraints of normalization and continuity of the first two derivatives of $R^l_{\text{PS}}(r)$ at the cutoff radius. The value of G_c is then refined iteratively to reduce the convergence error in the kinetic energy down to a desired tolerance. Proposed values of n are of the order of 10.

This approach was later revisited by Lin *et al.* (1993) (LQPH), who realized that better stability and convergence properties can be achieved by setting $G_c = G_n$. This makes use of the natural cutoff of the spherical Bessel functions, and is able to generate pseudo-wave functions with an almost negligible contribution from PWs of momentum above G_c. The prescription suggested by LQPH is to set $n = 4$ and $G_c = G_4$, and then obtain the four coefficients in (7.21) by minimizing the kinetic energy above G_c subjected to the normalization and continuity constraints.

The transferability of the pseudopotential is then controlled exclusively by the value of the cutoff radius r_c.

7.2.2 Pseudopotential generation

The general procedure for obtaining a pseudopotential begins by solving the all-electron radial Schrödinger equation for a chosen atomic configuration, i.e. for a given distribution of electrons in the atomic energy levels. This is called the *reference* configuration:

$$\left\{-\frac{1}{2}\frac{d^2}{dr^2} + \frac{l(l+1)}{2r^2} + v[\rho](r)\right\} rR_{AE}^{nl}(r) = \varepsilon_{nl} rR_{AE}^{nl}(r), \qquad (7.23)$$

where

$$v[\rho](r) = -\frac{Z}{r} + \int \frac{\rho(r')}{|r-r'|} dr' + \mu_{XC}[\rho] \qquad (7.24)$$

and $\rho(r)$ is the sum of the electronic densities for the occupied states. Z is the bare nuclear charge.

Then, the pseudo-wave function constructed according to any of the available prescriptions is used to invert the radial Schrödinger equation for the screened pseudopotential:

$$v_{PS}^{(sc)l}(r) = \varepsilon_l - \frac{l(l+1)}{2r^2} + \frac{1}{2rR_{PS}^l(r)} \frac{d^2}{dr^2}[rR_{PS}^l(r)], \qquad (7.25)$$

where the index n has been dropped to indicate that the pseudization is done for the lowest-lying valence state of each angular momentum, while the core states disappear from the description. The inversion can always be done because of the nodeless condition. The core states enter the pseudopotential generation procedure only through the self-consistent potential $v[\rho](r)$. Higher-lying valence states of the same angular momentum correspond to excited states of the pseudopotential. In fact, in some cases pseudopotentials are generated for ionic configurations such as K^+, where the lowest-lying $l=0$ valence shell is the $3s$, which is completely filled. The $4s$ state, instead, appears as the first excited state of the pseudopotential generated for the $3s$ state. In fact, this is one of the checks for transferability, namely that the relevant all-electron excited states ($4s$ in the above example) are well reproduced by the pseudopotential.

The pseudopotential is finally obtained by subtracting (unscreening) the Hartree and exchange-correlation potentials calculated *only* for the valence electrons (with the valence pseudo-wave functions):

$$v_{PS}^l(r) = v_{PS}^{(sc)l}(r) - \int \frac{\rho_v(r')}{|r-r'|} dr' - \mu_{XC}[\rho_v], \qquad (7.26)$$

with

$$\rho_v(r) = \sum_{l=0}^{l_{max}} \sum_{m=-l}^{l} |rR_{PS}^l(r)|^2, \quad (7.27)$$

where l_{max} is the highest angular momentum present in the isolated atom in the reference configuration. Notice that only one pseudo-wave function per angular momentum component enters the pseudo-valence density.

A very important point that is sometimes overlooked is that the exchange-correlation functional E_{XC} in the DFT all-electron used to construct the pseudopotential has to be the same as in the target calculation. Inconsistencies at this level, which are perfectly avoidable, can lead in some cases to unpredictable results.

7.2.3 Non-linear core corrections (NLCC)

If there is a significant overlap between the core and valence charge densities, the unscreening process in the construction of the pseudopotentials leads to an error because the exchange-correlation potential and energy are not linear functions of the density. This is particularly important in the case of systems with few valence electrons such as alkali atoms (Na, K, Rb, and Cs), and also in transition metals like Ti, Fe, or Zr, where the valence d band overlaps spatially with the core s and p states (Zhu et al., 1992). In those cases, the magnitude of the core density in the overlap region is comparable to the valence density, and the exchange-correlation energy cannot be written as $E_{XC}[\rho_c] + E_{XC}[\rho_v]$, where ρ_v and ρ_c are the valence and core charge densities, respectively. Another situation of this type arises when the valence orbitals of one atom overlap with the core orbitals of another atom, as in the case of II–VI semiconducting compounds (Engel and Needs, 1990). A solution to this problem that avoids including semi-core states explicitly in the valence was proposed by Louie et al. (1982). This approach is generally known as *non-linear core correction* (NLCC), and can be summarized in the following steps:

(i) The above unscreening expression is replaced by:

$$v_{PS}^l(r) = v_{PS}^{(sc)l}(r) - \int \frac{\rho_v(r')}{|r-r'|} dr' - \mu_{XC}[\rho_v + \rho_c]. \quad (7.28)$$

(ii) In the actual electronic structure calculations performed with this pseudopotential, the exchange-correlation contribution is computed for the full electronic charge, $\rho_v + \rho_c$, instead of the usual valence charge. The *frozen* core charge density of the isolated atoms is used for ρ_c.

(iii) Since $\rho_c(r)$ is not a smooth function, it is replaced by:

$$\rho_c(r) = \frac{A \sin(Br)}{r} \quad \text{for } r \leq R_{cc}, \quad (7.29)$$

where the parameters A and B are determined by the continuity condition for ρ_c and its first derivative at the core pseudization radius R_{cc}. This latter is chosen such that the valence charge density is negligible compared to the core one for $r < R_{cc}$.

7.2.4 Relativistic corrections

In the case of heavy atoms, the core electrons in the deepest shells have such high energies that they have to be treated relativistically. Therefore, the kinetic operator in Schrödinger's equation has to be replaced by Dirac's expression, which is invariant against rotations and reflections, so that the solutions have well-defined total angular momentum ($\hat{\mathbf{J}}$) and parity ($\hat{P}$) (Schiff, 1955). The total angular momentum is made of orbital ($\hat{\mathbf{L}}$) and spin ($\hat{\mathbf{S}}$) components. Since the electron spin is always 1/2, the composition of these two angular momenta can only give rise to two values of the quantum number j for each angular momentum l, namely $j = l \pm 1/2$.

For a given value of j and parity, Dirac's equation can be written as a pair of coupled differential equations on the auxiliary functions F and G, which receive the name of *minor* and *major* wave function components (Schiff, 1955; Bachelet et al., 1982).

$$\frac{dF_{nj}(r)}{dr} - \frac{\kappa_j}{r} F_{nj}(r) + \alpha \left[\varepsilon_{nj} - v(r) \right] G_{nj}(r) = 0,$$

$$\frac{dG_{nj}(r)}{dr} + \frac{\kappa_j}{r} G_{nj}(r) - \alpha \left[\frac{2}{\alpha^2} + \varepsilon_{nj} - v(r) \right] F_{nj}(r) = 0, \quad (7.30)$$

with

$$\kappa_j = \begin{cases} l & \text{for } j = l - 1/2, \\ -(l+1) & \text{for } j = l + 1/2; \end{cases} \quad (7.31)$$

and the norm of the wave function is given by $\int (|F|^2 + |G|^2) \, dr$. Dirac's equation includes all relativistic and spin–orbit coupling effects, and yields spin–orbit splitting energies. In Eqs. (7.30), the index n labels the principal quantum number and $\alpha = 1/137.04$ is the fine structure constant.

The minor and major components are strongly mixed only in the core region. On the basis of this observation, Kleinman (1980), and later Bachelet and Schlüter (1982), showed that the Dirac equation for valence electrons can be replaced by a Schrödinger-like equation in the major component

$$\left\{ -\frac{1}{2} \frac{d^2}{dr^2} + \frac{\kappa_j(\kappa_j + 1)}{2r^2} + v(r) \right\} G_{nj}(r) = \varepsilon_{nj} G_{nj}(r). \quad (7.32)$$

This expression, which is correct to order α^2, has been used by BHS as a starting point for generating relativistic pseudopotentials and pseudo-wave functions

(Bachelet *et al.*, 1982). This has been done using the procedure described in Section 7.2.2 where $rR_{AE}^{nl}(r)$ is replaced by $G_{nj}^{AE}(r)$, the major component of the all-electron relativistic wave function obtained by solving Dirac's equation.

If the atomic wave functions are described in terms of the angular l and spin $s = 1/2$ quantum numbers, instead of the total angular quantum number j, then a spin–orbit coupling term appears. In that case the total pseudopotential to be used in relativistic calculations becomes

$$\hat{v}_{PS}(r) = \sum_l v_{PS}^l(r)\,\hat{P}_l = \sum_l \left[v_{PS}^{(SR)l}(r) + v_{PS}^{(SO)l}(r)\,\mathbf{L}\cdot\mathbf{S} \right] \hat{P}_l, \qquad (7.33)$$

where

$$v_{PS}^{(SR)l}(r) = \frac{1}{2l+1}\left[l v_{PS}^{l-1/2}(r) + (l+1) v_{PS}^{l+1/2}(r) \right] \qquad (7.34)$$

is an average pseudopotential that is appropriate for *scalar relativistic* calculations, i.e. whenever spin–orbit effects on the valence electrons are not relevant. The other term,

$$v_{PS}^{(SO)l}(r) = \frac{2}{2l+1}\left[v_{PS}^{l+1/2}(r) - v_{PS}^{l-1/2}(r) \right], \qquad (7.35)$$

is a difference pseudopotential describing the strength of the spin–orbit coupling.

An alternative formulation for the scalar relativistic Schrödinger equation which avoids the major and minor components of the wave function has been given by Koelling and Harmon (1977):

$$\left\{ -\frac{1}{2}\frac{d^2}{dr^2} - \frac{\alpha^2}{4M}\frac{dv(r)}{dr}\left[\frac{d}{dr} - \frac{1}{r}\right] + \left[\frac{l(l+1)}{2r^2} + M(v(r) - \varepsilon_{nl})\right] \right\} rR_{AE}^{nl}(r) = 0, \qquad (7.36)$$

with

$$M = 1 - \frac{1}{2}\alpha^2\left(v(r) - \varepsilon_{nl}\right). \qquad (7.37)$$

Since relativistic corrections are important only in the core region, these can be incorporated directly into the pseudopotential in such a way that a solution of the non-relativistic Schrödinger equation in this potential contains relativistic core effects through the logarithmic derivatives of the wave function at the cutoff radius, exactly as for the usual non-relativistic pseudopotentials (Kleinman, 1980).

In the context of DFT-LDA calculations, the exchange-correlation functional should also be modified to account for relativistic effects. This can be done quite simply, as suggested by Bachelet et al. (1982), by multiplying the exchange energy density and potential by density-dependent correction factors given by MacDonald and Vosko (1979). This correction is important only for quite high densities such as those deep into the atomic core.

7.2.5 Choosing the reference configuration

One important aspect in pseudopotential generation is how to choose the electronic configuration of the isolated atom – the *reference* configuration – so that the pseudopotential remains useful in molecular systems or condensed phases (the target system). In principle, transferability should help in the sense that the effect of the environment should be taken into account by the norm-conservation condition. Even if the choice of the reference configuration is arbitrary, in principle it should be irrelevant, provided that the pseudopotential has good transferability properties. In practice this is most often the case, but it must be carefully checked by comparing the pseudo-wave functions and eigenvalues for electronic configurations different from the one used to fit the pseudopotential. These *test* configurations should be similar to those realized in the target system. In general these are unknown, but they can be guessed quite well by chemical intuition. Another useful check is to compare pseudo-wave functions and eigenvalues for pseudopotentials generated using different reference configurations. These aspects will be illustrated in Section 7.2.7. Transferability is expected to work best for electronic configurations close to the reference one, but it is not necessarily obvious for rather different configurations, such as neutral oxygen and O^- as in oxides, or neutral K and K^+ as in potassium salts.

The obvious first choice for the reference configuration is the ground state configuration of the neutral isolated atom. However, states of angular momenta that are unoccupied in the neutral atom hybridize with the occupied states in the presence of a different environment, becoming partially occupied, and thus important. Therefore, it is necessary to include these angular momenta as non-local components of the pseudopotential. A problem frequently faced when constructing pseudopotentials for DFT calculations is that angular momentum states that are unoccupied in the reference configuration are normally unbound, or at best weakly bound. This is a consequence of the approximate character of the exchange-correlation functional, most precisely of the lack of cancelation of the self-interaction correction, which makes the potential fall too quickly into the vacuum region, thus precluding the formation of bound states at negative energies. The traditional solution to this problem has been to obtain those angular components of the pseudopotential using additional, ionized reference configurations where the

7.2 Construction of pseudopotentials

state corresponding to the desired angular momentum is partially occupied. A table of excited (ionized) configurations to be used in the generation of pseudopotentials has been provided in BHS for basically all the elements in the periodic table (see Bachelet *et al.*, 1982, Table II).

7.2.6 Generalized norm-conserving pseudopotentials

The above procedure, although customary, is not entirely satisfactory because the final pseudopotentials are obtained after unscreening, i.e. after subtracting the electrostatic and exchange-correlation potential of the pseudo-charge density. Using different configurations implies that the various l-components of the pseudopotential are obtained by unscreening the Kohn–Sham potential with different pseudo-charge densities, while in the calculation for a target system there is a unique charge density for all the components. This inconsistency is not unlikely to introduce inaccuracies, especially for systems where non-linear core corrections (NLCC) are important, such as alkali atoms.

To overcome this limitation, Hamann (1989) proposed a procedure to generate pseudopotentials at energies that do not correspond to eigenvalues of the atomic Hamiltonian. The advantage of this approach is that a single configuration can be used to generate all the angular components of the pseudopotential. For angular momenta that sustain bound states, the pseudopotential is obtained as usual, by solving the atomic Kohn–Sham equations at the bound-state energy. For other angular momenta, the energy of the highest occupied valence state (of different l) is chosen.

The usual procedure to solve the atomic problem for bound states consists of matching outward and inward solutions of the radial differential equation for a trial energy, and improving on the energy until there is no discontinuity in the slope of the wave function at the matching point (Herman and Skillman, 1963). The main problem to be overcome when calculating atomic wave functions at arbitrary energies in the valence region is that these are not normalizable. Therefore, in the generalized case, since the inward integration cannot be carried out, Hamann proposed the following procedure.

(i) Choose a cutoff radius r_c^l, which for unbound states is taken to be 2 to 3 times the position of the outermost peak of the highest-lying core state of the same angular momentum. For bound states, this criterion gives values similar to the usual cutoff radii. If there is no core state of the same angular momentum, choose the largest of the core radii of the bound valence states.
(ii) Choose an outer radius $R^l \approx 2.5 r_c^l$, beyond which the pseudo-wave function converges accurately to the all-electron wave function.

(iii) For the reference energy ε_l, integrate the Schrödinger equation outwards to obtain the all-electron wave function R_{AE}^l. Then calculate the norm of R_{AE}^l enclosed within the sphere of radius R^l, and normalize it according to

$$\tilde{R}_{AE}^l(\varepsilon_l, r) = \frac{R_{AE}^l(\varepsilon_l, r)}{\left(4\pi \int_0^{R^l} [rR_{AE}^l(\varepsilon_l, r)]^2 \, dr\right)^{1/2}}. \quad (7.38)$$

(iv) For a trial pseudopotential, integrate the radial pseudo-Schrödinger equation outwards, and adjust the pseudo-energy iteratively until the logarithmic derivative of the pseudo-wave function R_{PS}^l coincides with that of the modified all-electron wave function $\tilde{R}_{AE}^l$ in (7.38) at the matching radius R^l, while R_{PS}^l is normalized within this radius.

(v) Adjust the pseudopotential iteratively until the pseudo-energy coincides with the all-electron energy ε_l.

This procedure retains the transferability properties of standard norm-conserving pseudopotentials, but now at arbitrary energies instead of eigenvalues. This property has been used by other authors to propose pseudopotential schemes with improved transferability, by requesting that the scattering properties are reproduced at two or more energies (Blöchl, 1990; Vanderbilt, 1990).

7.2.7 Examples of norm-conserving pseudopotentials

Norm-conserving Troullier–Martins pseudopotentials for oxygen and potassium, within the PBE-GGA approximation to DFT, have been generated using P. Giannozzi's code (Giannozzi, 2004). We first consider the case of oxygen. The pseudopotential has been generated for the neutral configuration $[1s^2]2s^22p^4$, where the $1s$ orbital is a core state and $2s$ and $2p$ are in the valence.

Figure 7.1(a) shows the all-electron and pseudo-wave functions for the two pseudized states, $2s$ and $2p$. The cutoff radii are 0.84 Å and 0.79 Å, and the eigenvalue energies are -1.7578 eV and -0.6645 eV, respectively. The pseudo-energies are virtually the same. The total energies are different because the pseudo-atom does not contain the $1s$ electrons explicitly. Notice how little pseudization can do for the $2p$ state, which is already nodeless at the all-electron level. The effect is more important for the $2s$ state, where pseudization has eliminated the node, thus making the pseudo-wave function much smoother. Figure 7.1(b) shows the actual $2s$ and $2p$ components of the oxygen pseudopotential, together with the unscreened Coulomb potential $-6/r$. Notice how the pseudopotentials approach the Coulomb potential and merge with it at the cutoff radii. The strange behavior at the origin is due to a pathology of the GGA, which produces a divergence at the nuclear position. In the present case the divergence has been regularized

7.2 Construction of pseudopotentials

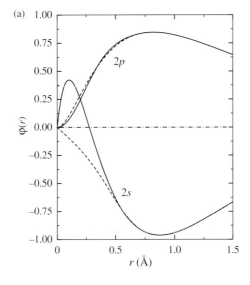

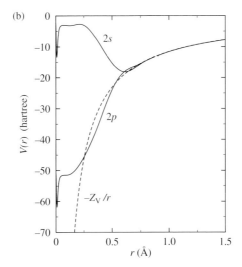

Fig. 7.1 (a) All-electron (solid lines) and pseudo-wave functions (dashed lines) for the O $2s$ (negative values) and $2p$ (positive values) valence states. (b) Pseudopotentials for the $2s$ and $2p$ states of O (solid lines), and the unscreened $-Z_V/r$ Coulomb potential ($Z_V = 6$).

for numerical stability, without causing any appreciable effect on the wave functions.

Figure 7.2 shows a comparison of the $2s$ all-electron and pseudo-wave functions for two different electronic configurations, the neutral $1s^2 2s^2 2p^4$ and the doubly ionized $[1s^2] 2s^1 2p^3$. The pseudo-wave functions for this latter have been obtained

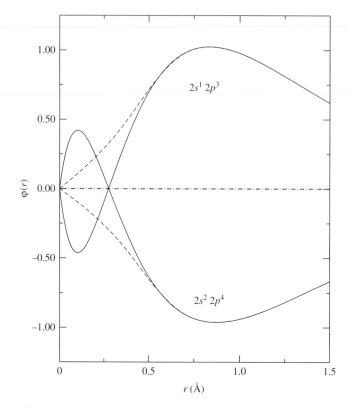

Fig. 7.2 All-electron (solid lines) and pseudo-wave functions (dashed lines) for the O $2s$ valence state in the neutral $2s^22p^4$ configuration (negative values) and $2s^12p^3$ ionized configuration (positive values).

with the pseudopotential generated from the neutral configuration. The eigenvalue of the all-electron $2s$ state is -4.2645 eV, while that of the pseudo-state is almost the same, -4.2650 eV. In fact, there are no observable differences between the two wave functions. Notice that the $2s$ state of the ionized configuration is slightly more compressed than that of the neutral atom, as it should be. Checking the performance of the pseudopotential in a different configuration is a test for transferability, especially if this configuration is representative of the appropriate environment.

We now consider a pseudopotential for the $4s$ state of neutral K. The electronic configuration is $[1s^22s^22p^63s^23p^6]4s^1$. Figure 7.3 shows all the all-electron wave functions of s symmetry, and the pseudo-wave function for the $4s$ state. Notice the spatial superposition of the $3s$ and $4s$ states, which is responsible for the fact that the exchange-correlation energy and potential require non-linear core corrections.

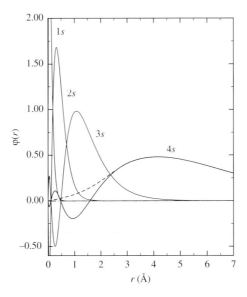

Fig. 7.3 All-electron s wave functions of neutral K (solid lines), and the pseudo-wave function for the $4s$ state (dashed lines).

The $3p$ state is even more spread out than the $3s$. The cutoff radius is 3.88 Å and the radius for pseudizing the core charge is 0.25 Å.

Non-linear core-corrections can be avoided if the $3s$ state is included in the valence, but for this it is necessary to remove the $4s$ state, because we cannot have more than one pseudopotential per angular momentum within the present scheme. The $4s$ state should appear as an excited state of the $3s$ pseudopotential, which must be generated for the K$^+$ ion. Since the $3s$ and $3p$ states have similar energies, we choose the electronic configuration $[1s^22s^22p^6]3s^23p^6$, where both $3s$ and $3p$ states are included into the valence. The core radii are 1.01 Å and 1.11 Å, respectively. The $3s$ and $4s$ all-electron and pseudo-wave functions are shown in Fig. 7.4. Notice that the $3s$ pseudo-wave function is nodeless, but the $4s$ is pseudized only up to the cutoff radius given by the $3s$ pseudopotential. In fact, this is an excited state of the pseudopotential, which reproduces very well the all-electron $4s$ state. For comparison, the all-electron wave function of the $4s$ state of the neutral K atom is also shown together with the pseudo-wave function generated with the K$^+$ and K^0 pseudopotentials. The eigenvalues are virtually the same, even for the $4s$ state, which was empty in the ionic configuration. This means that the ionic pseudopotential is able to reproduce the $4s$ orbital of both neutral and positively charged K, with high accuracy.

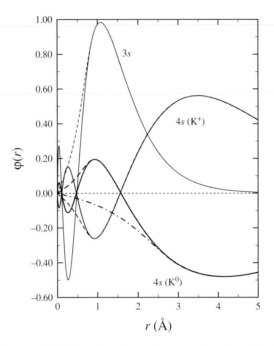

Fig. 7.4 All-electron (solid lines) and pseudo-wave functions (dashed lines) for the $3s$ and $4s$ states of the K^+ ion (positive values); all electron and pseudo-wave functions for the $4s$ state of neutral K (negative values). Pseudo-wave functions generated with the K^+ (dashed line) and K^0 (dot-dashed line) pseudopotentials are shown.

7.3 Separable form of atomic pseudopotentials

In the calculation of the different contributions to the Hamiltonian and the energy of a target system, the evaluation of the non-local pseudopotential part represents one of the most costly operations. Matrix elements of the pseudopotential in some basis $|\phi_\alpha\rangle$ assume the form

$$V_{\text{PS}}^{\alpha\beta} = \langle \phi_\alpha | \hat{V}_{\text{PS}} | \phi_\beta \rangle = \langle \phi_\alpha | v_{\text{PS}}^{\text{loc}}(r)\hat{I} + \sum_{l=0}^{l_{\max}} \Delta v_{\text{PS}}^l(r)\hat{P}_l | \phi_\beta \rangle$$

$$= V_{\text{PS}}^{\text{loc}}(\alpha)\delta_{\alpha\beta} + \sum_{l=0}^{l_{\max}} \Delta V_{\text{PS}}^l(\alpha, \beta), \quad (7.39)$$

with

$$\Delta V_{\text{PS}}^l(\alpha, \beta) = \langle \phi_\alpha | \Delta v_{\text{PS}}^l(r)\hat{P}_l | \phi_\beta \rangle$$

$$= \sum_{m=-l}^{l} \int\int \phi_\alpha^*(\mathbf{r})Y_{lm}(\mathbf{r})\Delta v_{\text{PS}}^l(r)Y_{lm}^*(\mathbf{r}')\phi_\beta(\mathbf{r}')d^3\mathbf{r}d^3\mathbf{r}', \quad (7.40)$$

where, due to the semi-local character of the pseudopotential in the radial coordinate, a factor $\delta(r - r')$ has to be understood so that the double integral runs only over the angular variables.

The most common basis functions are of two types: (1) floating, i.e. PW or any other basis set that is not atom-centered; in that case every basis function contains all the atom-centered angular components, (2) atom-centered, composed by the product of a radial function and a spherical harmonic, such as the case of LCAO, STO, or GTO basis sets. In either case the above integral factorizes into two angular-dependent parts that can be integrated separately, and a radial integral of the form

$$G_{\alpha\beta} = \int r^2 \varphi_\alpha^*(r) \Delta v_{PS}^l(r) \varphi_\beta(r) \, dr, \qquad (7.41)$$

where $\varphi_\alpha(r)$ is the radial part of the basis function, or the spherical Bessel functions $j_l(G_\alpha r)$ in the case of plane waves.

The number of such terms to be calculated scales like $\mathcal{O}(NM^2)$, where N is the number of atoms in the system and M is the number of basis functions. Especially in the case of PWs, the number of basis functions M can be very large indeed, and also scales linearly with the number of atoms. Therefore, this is essentially an $\mathcal{O}(N^3)$ operation, and is the bottleneck for electronic structure calculations with a large number of atoms, particularly when iterative diagonalization methods are used to solve the eigenvalue problem. The other $\mathcal{O}(N^3)$ operation is the orthogonalization step in the diagonalization procedure. The pre-factor, however, is smaller than that required by the computation of the non-local pseudopotential contributions. In the case of localized basis functions, the computational gain is not that dramatic, but there are important advantages at the programming level, because terms of the form (7.41) become cumbersome three-center integrals.

It was this observation, and also the fact that this problem does not arise in fully non-local pseudopotentials such as the original Philips–Kleinman form (7.4), that drove Kleinman and Bylander to propose a different description of the non-local component of the pseudopotential, where the radial semi-local operator is replaced by a fully non-local form that is *separable* in the radial variables (Kleinman and Bylander, 1982), i.e. $\Delta v_{sep}^l(r, r') = \zeta^l(r) \zeta^{l*}(r')$. With this, Expression (7.41) assumes the convenient form

$$\Delta V_{sep}^l(\alpha, \beta) = \sum_{m=-l}^{l} F_{\alpha lm}^* F_{\beta lm}, \qquad (7.42)$$

with

$$F_{\alpha lm} = \int \zeta^{l*}(r) Y_{lm}(\mathbf{r}) \phi_\alpha(\mathbf{r}) \, d^3\mathbf{r}. \qquad (7.43)$$

The general expression for a separable non-local potential of the Kleinman–Bylander form of the above type is the following:

$$\Delta \hat{V}^l_{\text{sep}} = \sum_{m=-l}^{l} \frac{|\zeta^{lm}\rangle \langle \zeta^{lm}|}{\langle \zeta^{lm}|\Phi^{lm}_{\text{PS}}\rangle}, \qquad (7.44)$$

where $\Phi^{lm}_{\text{PS}}(\mathbf{r})$ are the atomic, reference pseudo-wave functions defined in (7.13). This can be seen as a projection operator onto a one-dimensional subspace, where the only relevant aspect is to reproduce the all-electron calculation for the reference configuration. If we consider the action of the separable form of the non-local pseudopotential onto the atomic pseudo-wave function

$$\Delta \hat{V}^l_{\text{sep}} |\Phi^{lm}_{\text{PS}}\rangle = |\zeta^{lm}\rangle. \qquad (7.45)$$

Then, for the separable form $\Delta \hat{V}^l_{\text{sep}}$ to reproduce the all-electron scattering properties and energy derivatives at the reference energy ε_l, it is sufficient to construct the projection function in the following way (Blöchl, 1990; Vanderbilt, 1990):

$$|\zeta^{lm}\rangle = \left(\varepsilon_l - \hat{T} - v^{\text{loc}}_{\text{PS}}(r)\hat{I}\right) |\Phi^{lm}_{\text{PS}}\rangle. \qquad (7.46)$$

In that case, it is easy to see that the reference pseudo-wave function, which coincides with the all-electron wave function beyond the cutoff radius, is an eigenstate of the pseudo-Hamiltonian with (all-electron and pseudo) eigenvalue ε_l:

$$\left(\hat{T} + \hat{V}^{\text{loc}}_{\text{PS}} + \Delta \hat{V}^l_{\text{sep}}\right) |\Phi^{lm}_{\text{PS}}\rangle = \varepsilon_l |\Phi^{lm}_{\text{PS}}\rangle. \qquad (7.47)$$

It is clear that the projection function depends heavily on the choice of the local potential.

7.3.1 Kleinman–Bylander separable form

In 1982, Kleinman and Bylander (KB) developed a fully non-local, separable form of the pseudopotential, $\Delta \hat{V}^l_{\text{sep}}$, by requesting that its action on the reference pseudo-wave functions be the same as that of the original, HSC semi-local form $\Delta \hat{V}^l_{\text{PS}}$. To this end, they proposed (Kleinman and Bylander, 1982)

$$|\zeta^{lm}_{\text{KB}}\rangle = |\Delta \hat{V}^l_{\text{PS}} \Phi^{lm}_{\text{PS}}\rangle. \qquad (7.48)$$

By applying the operator $\Delta \hat{V}^l_{\text{sep}}$ to the reference pseudo-wave function $|\Phi^{lm}_{\text{PS}}\rangle$, it is straightforward to prove the requested property, i.e.

$$\Delta \hat{V}^l_{\text{KB}} |\Phi^{lm}_{\text{PS}}\rangle = \left[\frac{|\Delta \hat{V}^l_{\text{PS}} \Phi^{lm}_{\text{PS}}\rangle \langle \Phi^{lm}_{\text{PS}} \Delta \hat{V}^l_{\text{PS}}|}{\langle \Phi^{lm}_{\text{PS}}|\Delta \hat{V}^l_{\text{PS}}|\Phi^{lm}_{\text{PS}}\rangle}\right] |\Phi^{lm}_{\text{PS}}\rangle = \Delta \hat{V}^l_{\text{PS}} |\Phi^{lm}_{\text{PS}}\rangle. \qquad (7.49)$$

7.3 Separable form of atomic pseudopotentials

The Kleinman–Bylander projector is then written as

$$\Delta \hat{V}_{KB}^l = \sum_{m=-l}^{l} E_{KB}^{lm} |\xi_{KB}^{lm}\rangle \langle \xi_{KB}^{lm}|, \quad (7.50)$$

where

$$|\xi_{KB}^{lm}\rangle = \frac{|\zeta_{KB}^{lm}\rangle}{\langle \zeta_{KB}^{lm} | \zeta_{KB}^{lm} \rangle} \quad (7.51)$$

are normalized projection functions. The strength of the non-locality is determined by the energies E_{KB}^{lm}, which are given by

$$E_{KB}^{lm} = \alpha_{lm} \langle \Phi_{PS}^{lm} | (\Delta \hat{V}_{PS}^l)^2 | \Phi_{PS}^{lm} \rangle \quad (7.52)$$

with

$$\alpha_{lm} = \langle \Phi_{PS}^{lm} | \Delta \hat{V}_{PS}^l | \Phi_{PS}^{lm} \rangle^{-1}. \quad (7.53)$$

7.3.2 Ghost states

An undesirable consequence of the introduction of the separable form of norm-conserving pseudopotentials is the possibility that unphysical states appear at energies below the true ground state eigenvalue of a specific angular momentum. The reason for the appearance of such states is that in the KB Hamiltonian the eigenstates are not ordered by increasing number of nodes, so that solutions with nodes can be lower in energy than nodeless ones. These are called *ghost* states, and can be graphically observed as divergencies in the energy plot of the logarithmic derivative at unphysically low energies.

The question is why and when is the natural ordering of the eigenstates violated. This issue has been studied by Gonze, Käckell, and Scheffler, who showed that the existence of ghosts is related basically to the choice of the local component of the pseudopotential (Gonze et al., 1990). If ε_0^l and ε_1^l are the two lowest eigenvalues of a modified Hamiltonian where the non-local pseudopotential contribution of angular momentum l, i.e. $\Delta \hat{V}_{PS}^l$, has been removed, then:

(i) if $E_{KB}^l < 0$ there is a ghost below the reference energy ε_l if and only if $\varepsilon_l > \varepsilon_0^l$;
(ii) if $E_{KB}^l > 0$ there is a ghost below the reference energy ε_l if and only if $\varepsilon_l > \varepsilon_1^l$.

These problems typically arise when $|E_{KB}^l|$ is very large (Gonze et al., 1990), but this can be avoided by modifying the pseudopotential. One possibility is to slightly modify the value of the cutoff radius for the problematic angular momentum. This program has been successfully carried out for many elements in Gonze et al. (1991). A typical example is the case of the valence *p*-state of the Se atom when the *d* component is taken as the local part of the pseudopotential. By varying the cutoff radius r_c^d, E_{KB}^p goes through a pole when $r_c^d \approx 0.52 r_{max}$, with r_{max} the position of the outermost peak of the all-electron wave function. Values of r_c^d around that are thus bound to be problematic. However, by reducing the cutoff radius to $r_c^d \approx 0.48 r_{max}$, the ghost disappears and the KB construction becomes useful again.

The above conditions are verified when the local part of the pseudopotential $v_{PS}^{loc}(r)$ is much deeper than the total, i.e. local plus non-local, contribution $v_{PS}^{loc}(r) + \Delta v_{PS}^l(r)$ (Blöchl, 1990). In that case, the local potential, apart from the ground, nodeless state, can also sustain states with one or more nodes in the energy region where chemical bonding occurs. Since the KB projector is constructed from the nodeless ground state wave function $|\Phi_{PS}^{lm}\rangle$, then it is likely that states with nodes are orthogonal to the KB projection functions, and thus unaffected by the non-local part of the pseudopotential. Therefore, these eigenstates of the local Hamiltonian will remain eigenstates of the full Hamiltonian, thus resulting in an inversion of the node sequence.

Therefore, an alternative is to modify the definition of the local component of the pseudopotential so as to avoid such deep excited states. This is done in the SIESTA code (Soler *et al.*, 2002), where the local potential is not taken as the potential corresponding to the highest angular momentum, but is a suitably chosen function that ensures that none of the $v_{PS}^l(r)$ is much shallower than $v_{PS}^{loc}(r)$.

Another possibility proposed by Blöchl (1990) is to generalize the KB separable form to include two or more projectors obtained at different eigenvalue energies. These additional terms interact with the ghost state, pushing it up in energy to a region outside the chemical bonding area. Blöchl suggested that a single additional state is likely to be sufficient to this purpose. These multi-reference projectors have been called generalized separable pseudopotentials, and will be discussed below.

7.3.3 Generalized separable forms

In 1990, independently, P. Blöchl and D. Vanderbilt showed that the semi-local form of the pseudopotential is not necessary, and one can simply use a separable form where the projection functions obey Eq. (7.46) (Blöchl, 1990; Vanderbilt, 1990). Moreover, there is no reason to restrict the separable form to a single projector. In fact, it can be shown that a potential of the form

$$\Delta \hat{V}_{gsep}^l = \sum_{i,j} B_{ij}^l \sum_{m=-l}^{l} |\beta_i^{lm}\rangle\langle\beta_j^{lm}| \tag{7.54}$$

has the set of atomic pseudo-wave functions $|\Phi_{PS}^{ilm}\rangle$ as eigenstates with eigenvalues ε_{il},

$$\left(\hat{T} + \hat{V}_{PS}^{loc} + \Delta \hat{V}_{gsep}^l\right)|\Phi_{PS}^{ilm}\rangle = \varepsilon_{il}|\Phi_{PS}^{ilm}\rangle, \tag{7.55}$$

provided that the functions $|\beta_i^{lm}\rangle$ are constructed as appropriate linear combinations of a set of local wave functions $|\zeta_i^{lm}\rangle$, which are obtained through Expression (7.46) from the $|\Phi_{\text{PS}}^{ilm}\rangle$. This can be achieved using Hamann's procedure for obtaining pseudopotentials at arbitrary energies (Hamann, 1989). The expression for the functions $|\beta_i^{lm}\rangle$ is

$$|\beta_i^{lm}\rangle = \sum_j \left(B^{-1}\right)_{ij}^l |\zeta_j^{lm}\rangle, \quad (7.56)$$

with $B_{ij}^l = \langle \phi_{\text{PS}}^{ilm} | \zeta_j^{lm} \rangle$. These matrix elements do not depend on the quantum number m because the azimuthal angular part is the same in the two wave functions, $|\phi_{\text{PS}}^{ilm}\rangle$ and $|\zeta_j^{lm}\rangle$.

An important issue is to ensure that the generalized separable potential is hermitian. It can be shown (Vanderbilt, 1990) that $B_{ij}^l - B_{ji}^{*l} = (\varepsilon_{il} - \varepsilon_{jl}) Q_{ij}^l$, with

$$\begin{aligned}
Q_{ij}^l &= \int_0^{r_c} r^2 \Phi_{\text{AE}}^{ilm*}(r) \Phi_{\text{AE}}^{jlm}(r)\, dr - \int_0^{r_c} r^2 \Phi_{\text{PS}}^{jlm*}(r) \Phi_{\text{PS}}^{jlm}(r)\, dr \\
&= \langle \Phi_{\text{AE}}^{ilm} | \Phi_{\text{AE}}^{jlm} \rangle_{r_c} - \langle \Phi_{\text{PS}}^{ilm} | \Phi_{\text{PS}}^{jlm} \rangle_{r_c},
\end{aligned} \quad (7.57)$$

a generalized norm-conservation indicator. In the last line, the subscript r_c means that the inner product is calculated within a sphere of radius r_c. The diagonal elements refer to the usual norm-conservation condition for the different reference energies, and the off-diagonal elements correspond to wave functions at different reference energies. Therefore, the condition for hermiticity is a generalized norm-conservation criterion, i.e. $Q_{ij}^l = 0$.

In summary, for a given angular momentum, a set of norm-conserving pseudo-wave functions $|\phi_{\text{PS}}^{ilm}\rangle$ (usually between one and three) is generated from the all-electron wave functions at different energy values that span the range of eigenvalues expected in the target system. Using these functions, a generalized non-local separable pseudopotential is constructed according to (7.54). With this construction, the all-electron wave functions and eigenvalues are reproduced at all the reference energies used. Since the sum rule for the energy derivative of the logarithmic derivative is preserved for all the reference energies, the scattering properties of the all-electron potential can be reproduced with arbitrary accuracy. This is still a norm-conserving scheme, but with improved transferability properties with respect to single-reference pseudopotentials.

7.4 Ultrasoft pseudopotentials

The norm-conservation constraint is the main factor responsible for the hardness of some pseudopotentials, especially p states in first-row elements and d states in second-row transition metals, e.g. O $2p$ or Cu $3d$. For these states there is no core state of the same angular momentum to which they have to be orthogonal. Therefore, the all-electron wave function is nodeless and quite compressed compared to the other valence states, thus requiring a large number of PWs to be represented accurately. Pseudization of this wave function does not help much because the pseudo-charge has to match the charge of the all-electron wave function, and this latter is already nodeless.

The norm-conservation constraint is tightly linked to the concept of transferability through the sum rule (7.12). This expression shows that the first order energy variation of the phase shift is proportional to the norm of the wave function in the pseudized region. However, apart from respecting the transferability criterion embodied in (7.12), it is not strictly necessary that the norm of the all-electron and pseudo-wave functions coincide. Therefore, efforts directed towards the reduction of the PW cutoff should focus on relaxing the norm-conservation condition by generalizing the sum rule (7.12). This has been done in 1990 by Vanderbilt, who showed that much smoother, but still highly transferable, pseudopotentials can be obtained in this way (Vanderbilt, 1990). These have received the name of *ultrasoft* (VUS or US) pseudopotentials.

Vanderbilt's construction starts from generalized, multi-reference separable non-local pseudopotentials, by redefining the non-local potential operator as

$$\Delta \hat{V}_{US}^l = \sum_{i,j} D_{ij}^l \sum_{m=-l}^{l} |\beta_i^{lm}\rangle \langle \beta_j^{lm}|, \qquad (7.58)$$

where

$$D_{ij}^l = B_{ij}^l + \varepsilon_{il} Q_{ij}^l, \qquad (7.59)$$

so that

$$\Delta \hat{V}_{US}^l = \sum_{i,j} B_{ij}^l \sum_{m=-l}^{l} |\beta_i^{lm}\rangle \langle \beta_j^{lm}| + \sum_{i,j} \varepsilon_{jl} Q_{ij}^l \sum_{m=-l}^{l} |\beta_i^{lm}\rangle \langle \beta_j^{lm}|. \qquad (7.60)$$

The first term is the same as the generalized separable norm-conserving pseudopotentials, while the second term vanishes only if the generalized norm-conservation condition $Q_{ij}^l = 0$ is enforced. If this is not the case, then the relation between the norm of the all-electron and the pseudo-wave functions is:

$$\langle \Phi_{AE}^{ilm} | \Phi_{AE}^{jlm} \rangle_{r_c} = \langle \Phi_{PS}^{ilm} | \Phi_{PS}^{jlm} \rangle_{r_c} + Q_{ij}^l$$

$$= \langle \Phi_{PS}^{ilm} | \left(\hat{I} + \sum_{i,j} Q_{ij}^l |\beta_i^{lm}\rangle \langle \beta_j^{lm}| \right) | \Phi_{PS}^{jlm} \rangle_{r_c}. \qquad (7.61)$$

7.4 Ultrasoft pseudopotentials

This can be written in a more compact form by defining a *non-local overlap operator*:

$$\hat{S} = \hat{I} + \sum_l \sum_{i,j} Q_{ij}^l \sum_{m=-l}^{l} |\beta_i^{lm}\rangle\langle\beta_j^{lm}|, \qquad (7.62)$$

so that the norm-conservation condition reads

$$\langle\phi_{\text{AE}}^{ilm}|\phi_{\text{AE}}^{jlm}\rangle_{r_c} = \langle\phi_{\text{PS}}^{ilm}|\hat{S}|\phi_{\text{PS}}^{jlm}\rangle_{r_c}. \qquad (7.63)$$

With these definitions, if the Hamiltonian is written as

$$\hat{\mathcal{H}} = \hat{T} + \hat{V}_{\text{PS}}^{\text{loc}} + \sum_{l'} \Delta\hat{V}_{\text{sep}}^{l'} + \sum_{l'} \sum_{i,j} \varepsilon_{il'} Q_{ij}^{l'} \sum_{m'=-l'}^{l'} |\beta_i^{l'm'}\rangle\langle\beta_j^{l'm'}|, \qquad (7.64)$$

the pseudo-wave functions $|\phi_{\text{PS}}^{ilm}\rangle$ are the solutions of a generalized atomic eigenvalue problem

$$\hat{\mathcal{H}}|\Phi_{\text{PS}}^{ilm}\rangle = \varepsilon_{il}|\Phi_{\text{PS}}^{ilm}\rangle + \sum_{l'}\sum_{i,j} \varepsilon_{il'} Q_{ij}^{l'} \sum_{m'=-l'}^{l'} |\beta_i^{l'm'}\rangle\langle\beta_j^{l'm'}|\Phi_{\text{PS}}^{ilm}\rangle$$

$$= \varepsilon_{il}\hat{S}|\Phi_{\text{PS}}^{ilm}\rangle, \qquad (7.65)$$

where only the angular component l of the non-local and overlap operators produces non-zero matrix elements with a state of well-defined angular momentum l such as $|\phi_{\text{PS}}^{ilm}\rangle$. It can be shown that $\hat{Q}$ and $\hat{D}$, and thus $\hat{\mathcal{H}}$ and $\hat{S}$, are hermitian operators (Vanderbilt, 1990).

For this generalized eigenvalue problem the identity (7.12) is modified to

$$-\frac{1}{2}\left\{[rR^l(\varepsilon,r)]^2 \frac{\text{d}}{\text{d}\varepsilon}\frac{\text{d}}{\text{d}r}\ln R^l(\varepsilon,r)\right\}_{r_c} = \langle\phi_{\text{PS}}^{ilm}|\phi_{\text{PS}}^{ilm}\rangle_{r_c} + Q_{ii}^l, \qquad (7.66)$$

and the transferability criterion becomes that the norm of the all-electron wave function should be matched by that of the pseudo-wave function plus the diagonal elements of $\hat{Q}$. The matrix elements Q_{ii}^l represent the amount of charge missing in the pseudo-wave function of angular momentum l, calculated using the reference energy ε_{il}.

Therefore, by relaxing the norm-conservation constraint all the wave functions at different reference energies can be pseudized independently by requesting *only* the matching of the logarithmic derivatives at the cutoff radius. In practice, this means that the cutoff radius can be chosen quite large, well beyond the maximum of the radial wave function. For larger cutoff radii the derivative of the wave function is smaller, and then, since there is no constraint on the norm, sharp peaks in the pseudo-wave function can be avoided, thus resulting in smoother wave functions. This is illustrated in Fig. 7.5, from Vanderbilt's original paper.

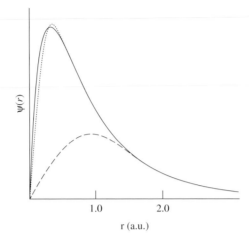

Fig. 7.5 Oxygen 2p radial wave function (solid line) and corresponding norm-conserving (dotted line) and ultrasoft (dashed line) pseudo-wave functions. Figure 1 reprinted with permission from D. Vanderbilt, *Phys. Rev. B* **41**, 7892 (1990). Copyright (1990) by the American Physical Society.

In principle, it looks as if having to solve a generalized eigenvalue equation could increase the computational cost of the calculation. The matrix elements of the overlap operator (7.62), however, have the same form as those of the non-local pseudopotential operator (7.58), and both can be taken together without incurring any significant additional computational expense.

More delicate is the fact that, within this scheme, even if the pseudo- and all-electron wave functions become identical beyond the cutoff radius, the charge enclosed in the pseudized region is different. In a self-consistent electronic structure calculation of the Kohn–Sham type, the potential depends on the charge density. If the valence charge density is defined only through the pseudo-wave function, then the lack of charge in the pseudized region will make the Kohn–Sham potential incorrect. This can be taken into account by writing the valence charge density as

$$\rho_v(\mathbf{r}) = \sum_n |\varphi_n(\mathbf{r})|^2 + \sum_l \sum_{i,j} \rho^l_{ij} Q^l_{ji}(\mathbf{r}), \qquad (7.67)$$

where the first sum runs over the occupied states,

$$\rho^l_{ij} = \sum_n \sum_{m=-l}^{l} \langle \beta^{lm}_i | \varphi_n \rangle \langle \varphi_n | \beta^{lm}_j \rangle \qquad (7.68)$$

and

$$Q^l_{ij}(\mathbf{r}) = \Phi^{ilm*}_{\text{AE}}(\mathbf{r}) \Phi^{jlm}_{\text{AE}}(\mathbf{r}) - \Phi^{ilm*}_{\text{PS}}(\mathbf{r}) \Phi^{jlm}_{\text{PS}}(\mathbf{r}). \qquad (7.69)$$

7.4 Ultrasoft pseudopotentials

Integrating this density we obtain

$$\int \rho_V(\mathbf{r})\,d\mathbf{r} = \sum_n \langle \varphi_n | \varphi_n \rangle + \sum_n \sum_l \sum_{i,j} Q_{ij}^l \sum_{m=-l}^{l} \langle \varphi_n | \beta_j^{lm} \rangle \langle \beta_i^{lm} | \varphi_n \rangle$$

$$= \sum_n \langle \varphi_n | \left(\hat{I} + \sum_l \sum_{i,j} Q_{ij}^l \sum_{m=-l}^{l} |\beta_j^{lm}\rangle \langle \beta_i^{lm}| \right) | \varphi_n \rangle$$

$$= \sum_n \langle \varphi_n | \hat{S} | \varphi_n \rangle. \tag{7.70}$$

If the solutions of the self-consistent generalized eigenvalue problem are normalized according to $\langle \varphi_n | \hat{S} | \varphi_m \rangle = \delta_{nm}$, then the above charge density $\rho_v(\mathbf{r})$ integrates exactly to the number of valence electrons N_v in the system. The functions $Q_{ij}^l(\mathbf{r})$ are a byproduct of the pseudopotential generation procedure, and are calculated once and for all in the same way as pseudopotentials. The quantities ρ_{ij}^l represent a sort of density matrix, which depends on the self-consistent orbitals.

The self-consistent secular equation retains the same generalized eigenvalue form (7.65), where now the local potential includes the usual Hartree and exchange-correlation contributions from the electron–electron interaction, $v_{\text{loc}}[\rho, (\mathbf{r})] = v_{\text{PS}}^{\text{loc}}(\mathbf{r}) + v_H[\rho_v(\mathbf{r})] + \mu_{XC}[\rho_v(\mathbf{r})]$. In addition, due to the non-locality of the overlap operator $\hat{S}$, there is also a contribution of these two terms to the non-local part of the potential, whose matrix elements become

$$\tilde{D}_{ij}^l = D_{ij}^l + \int \{v_H[\rho_v(\mathbf{r})] + \mu_{XC}[\rho_v(\mathbf{r})]\} Q_{ij}^l(\mathbf{r})\,d\mathbf{r}. \tag{7.71}$$

The secular equation is obtained by minimization of the energy functional with the modified non-local pseudopotential term, subject to the generalized normalization constraint. Quite interestingly, according to the second term in (7.71) the non-local pseudopotential contribution depends self-consistently on the charge density, adjusting itself to changes in the charge configuration due to the environment. This feature is supposed to improve transferability over norm-conserving, non-polarizable pseudopotentials.

The quantities $Q_{ij}^l(\mathbf{r})$ that represent the missing valence charge density in the pseudized region can be quite steep functions and, in the spirit of NLCC, they have to be pseudized below some small cutoff radius R_{cc}. In solid-state applications all the sums over states in the above expressions must be supplemented with sums over **k**-points in the Brillouin zone.

This scheme has been extremely successful during the past decade, leading to substantial reductions in the energy cutoff for PW calculations that involved first row and transition metal atoms. For example, for the same level of accuracy and convergence, the energy cutoff for oxygen can be reduced from 150 Ry to 40 Ry (Koval et al., 2005). It is important to remark that in a calculation for a system with more than one atomic species, one can combine ultrasoft and norm-conserving pseudopotentials for different species, i.e. in SiO_2, O is more convenient to treat using ultrasoft pseudopotentials, while for Si the simpler and less costly norm-conserving ones are sufficient.

7.5 Some practical aspects of pseudopotentials

Given a system, the first decision is whether an all-electron calculation is required, or a pseudopotential is sufficient to address the desired issues. When results, e.g. energies, require an extreme accuracy, then it is safer to stay within an all-electron framework. If the pseudopotential approach has been selected, there are a number of additional decisions to make.

(i) What type of pseudopotential to use? Norm-conserving or ultrasoft? This choice depends very much on the basis set. If this is plane waves, then ultrasoft pseudopotentials are the most efficient ones by far. However, they are more delicate to construct, they carry additional technical difficulties at the programming level, and they introduce a small computational overhead if more than one projector is used. In general, this is largely overcome by a massive reduction in the PW cutoff, and ultrasoft versions are preferred. For atom-centered basis sets, the softness of the pseudopotential is not a big issue because the basis functions are tailored to reproduce atomic (or pseudo-atomic) orbitals. Therefore, norm-conserving and ultrasoft are basically equivalent and, for the same price, it is convenient to stay with the simpler norm-conserving pseudopotentials.

(ii) If norm-conserving pseudopotentials have been chosen, the next decision is which flavor, whether BHS, Troullier–Martins, RRKJ, or other. Differences between these reside mostly in their smoothness, which is relevant only for PW basis sets. Both TM and RRKJ pseudopotentials are state-of-the-art in this respect. Transferability can be improved by using generalized separable forms with more than one reference energy.

(iii) Ultrasoft and norm-conserving pseudopotentials can be used for different chemical species in the same system, e.g. ultrasoft for O and norm-conserving for Si in a calculation for SiO_2. Also, different flavors of norm-conserving versions can be mixed.

(iv) The next step is to decide on the electronic configuration to be used for the pseudopotential generation. The default is the neutral atom configuration, but it may be necessary to use another configuration if some important valence state is unbound in the neutral atom. It is acceptable to construct pseudopotentials for different angular momenta using different configurations. In those cases, one tries to choose a configuration that is representative of the environment, e.g. an ionic configuration for alkali metals in salts.

(v) For a given configuration, it has to be decided which states are going to be treated as valence, and which as core. In general, the valence states are those occupied in the last electronic shell of the atom. For example, $4s^1$ for K or $2s^2 2p^4$ for O. When some of the highest lying core states are close in energy to the valence states, then these may have to be considered explicitly as valence electrons, e.g. the $3s$ and $3p$ states in K. In that case, since the pseudopotential is usually constructed for a single principal quantum number and angular momentum, an ionic configuration is chosen where the valence state of the same symmetry is empty, e.g. $3s^2 3p^6$ for K$^+$. The $4s$ valence state appears as the first excited state of this pseudopotential.

(vi) Non-linear core corrections are also an option at this stage. It is far cheaper than semi-core pseudopotentials, but only useful if semi-core states are not polarized. Scalar-relativistic corrections are important for atoms below the fifth row. For even heavier atoms, the full Dirac atomic equation, including spin–orbit coupling, should be solved.

(vii) Finally comes the cutoff (or core) radius. A practical rule of thumb is to choose it around the maximum of the atomic wave function. Another recipe is to take it around 2–3 times the position of the outermost maximum of the highest core state of the same symmetry. These two criteria are similar, and lead to radii that depend on the angular momentum. Accuracy can be improved by decreasing the cutoff radius, but at the expenses of hardening the potential. Hydrogen, even if it is exactly described by the bare Coulomb potential, can be conveniently pseudized in order to reduce the number of required PW components. In this case it is important to choose a pseudization radius small enough to avoid distorting the bonding region ($r_c \approx 0.9$ bohr is a good choice that leads to a PW cutoff of 60 Ry). The cutoff in a PW calculation is dictated by the *hardest* pseudopotential. Therefore, there is little point in trying to improve on the smoothness of an Si pseudopotential, which will anyway converge for a cutoff of the order of 20–30 Ry, if it is going to be used together with H or O, which require a much higher cutoff.

(viii) Once the pseudopotential has been constructed, it must be tested. A first check is to look at the logarithmic derivatives in the region around the reference energy, and verify that pseudo- and all-electron results are close in the desired energy range. Another check is to compare the all-electron and pseudo-wave functions for electronic configurations different from the one used to generate the pseudopotential. These are tests for transferability. The accuracy is tested in the target system, by checking the performance in reproducing structural and vibrational properties against all-electron calculations. It is important to preserve the same functional when doing these comparisons. Here, the choices are how many and which non-local components to include, and how to choose the local component. The first of the two is much more important. The usual procedure is to consider as non-local components the pseudopotentials for all the angular momenta that are present in the neutral atom, plus an extra angular momentum to describe polarization properly. The local component can be one of these pseudopotentials, or some other reasonable choice.

(ix) The performance of a pseudopotential has to be analyzed in conjunction with the basis set. In the case of PWs, the energy cutoff has to be increased to convergence, and only then can the quality of the potential be assessed. The choice of a lower PW cutoff in order to reproduce experimental quantities or all-electron calculations is to be avoided. If at convergence the pseudopotential is not good enough, then it has to be generated again by modifying some of the choices described above.

References

Appelbaum, J. A. and Hamann, D. R. (1973). Self-consistent pseudopotential for Si. *Phys. Rev. B* **8**, 1777–1780.

Bachelet, G. B. and Schlüter, M. (1982). Relativistic norm-conserving pseudopotentials. *Phys. Rev. B* **25**, 2103–2108.

Bachelet, G. B., Hamann, D. R., and Schlüter, M. (1982). Pseudopotentials that work: from H to Pu. *Phys. Rev. B* **26**, 4199–4228.

Blöchl, P. E. (1990). Generalized separable potentials for electronic-structure calculations. *Phys. Rev. B* **41**, 5414–5416.

Engel, G. E. and Needs, R. J. (1990). Calculations of the structural properties of cubic zinc sulfide. *Phys. Rev. B* **41**, 7876–7878.

Giannozzi, P. (2004). *Notes on Pseudopotential Generation*. Scuola Normale Superiore di Pisa. www.nest.sns.it/giannozz/software.html.

Gonze, X., Käckell, P., and Scheffler, M. (1990). Ghost states for separable, norm-conserving, *ab initio* pseudopotentials. *Phys. Rev. B* **41**, 12264–12267.

Gonze, X., Stumpf, R., and Scheffler, M. (1991). Analysis of separable potentials. *Phys. Rev. B* **44**, 8503–8513.

Hamann, D. R. (1989). Generalized norm-conserving pseudopotentials. *Phys. Rev. B* **40**, 2980–2987.

Hamann, D. R., Schlüter, M., and Chiang, C. (1979). Norm-conserving pseudopotentials. *Phys. Rev. Lett.* **43**, 1494–1497.

Herman, F. and Skillman, S. (1963). *Atomic Structure Calculations*. New Jersey, Prentice-Hall.

Herring, C. (1940). A new method for calculating wave functions in crystals. *Phys. Rev.* **57**, 1169–1177.

Kerker, G. P. (1980). Nonsingular atomic pseudopotentials for solid state applications. *J. Phys. C* **13**, L189–L194.

Kleinman, L. (1980). Relativistic norm-conserving pseudopotential. *Phys. Rev. B* **21**, 2630–2631.

Kleinman, L. and Bylander, D. M. (1982). Efficacious form for model pseudopotentials. *Phys. Rev. Lett.* **48**, 1425–1428.

Koelling, D. D. and Harmon, B. N. (1977). A technique for relativistic spin-polarized calculations. *J. Phys. C* **10**, 3107–3114.

Koval, S. F., Kohanoff, J., Lasave, J., Colizzi, G., and Migoni, R. L. (2005). Ferroelectricity and isotope effects in hydrogen-bonded crystals: the case of KDP. *Phys. Rev. B* **71**, 184102.

Lin, J. S., Qteish, A., Payne, M. C., and Heine, V. (1993). Optimized and transferable nonlocal separable *ab initio* pseudopotentials. *Phys. Rev. B* **47**, 4174–4180.

Louie, S. B., Froyen, S., and Cohen, M. L. (1982). Nonlinear ionic pseudopotentials in spin-density-functional calculations. *Phys. Rev. B* **26**, 1738–1742.

MacDonald, A. H. and Vosko, S. H. (1979). A relativistic density functional formalism. *J. Phys. C* **12**, 2977–2990.

Philips, J. C. and Kleinman, L. (1959). New method for calculating wave functions in crystals and molecules. *Phys. Rev.* **116**, 287–294.

Rappe, A. M., Rabe, M., Kaxiras, E., and Joannopoulos, J. D. (1990). Optimized pseudopotentials. *Phys. Rev. B* **41**, 1227–1230.

Schiff, L. I. (1955). *Quantum Mechanics*. New York, McGraw-Hill.

Shaw, R. W. and Harrison, J. W. A. (1967). Reformulation of the screened Heine–Abarenkov model potential. *Phys. Rev.* **163**, 604–611.

Slater, J. C. (1937). Wave functions in a periodic potential. *Phys. Rev.* **51**, 846–851.

Soler, J. M., Artacho, E., Gale, J. D., García, A. I., Junquera, J., Ordejón, P., and Sánchez-Portal, D. (2002). The SIESTA method for *ab initio* order-N materials simulation. *J. Phys. Condens. Matter* **14**, 2745–2780.

Topp, W. C. and Hopfield, J. J. (1973). Chemically motivated pseudopotential for sodium. *Phys. Rev. B* **7**, 1295–1303.

Troullier, N. and Martins, J. L. (1991). Efficient pseudopotentials for plane-wave calculations. *Phys. Rev. B* **43**, 1993–2006.

Vanderbilt, D. (1990). Soft self-consistent pseudopotentials in a generalized eigenvalue formalism. *Phys. Rev. B* **41**, 7892–7895.

Zhu, J., Wang, X. W., and Louie, S. G. (1992). First-principles pseudopotential calculations for magnetic iron. *Phys. Rev. B* **45**, 8887–8893.

8
Basis sets

Solving the electronic structure problem in practice, either within DFT or within Hartree–Fock and post-Hartree–Fock approaches, requires us to choose a mathematical representation for the one-electron orbitals. A possibility is certainly to represent them on a three-dimensional grid in real space, and to solve the partial differential equations using finite differences. However, more efficient alternatives that make use of the specific characteristics of the system under study are possible. In this chapter we describe different types of basis sets that have been proposed and adopted for electronic structure calculations.

We start by expanding the one-electron wave functions in a generic basis set described by the orbitals $|\phi_\alpha\rangle$. In a real-space representation these orbitals become $\langle \mathbf{r}|\phi_\alpha\rangle = \phi_\alpha(\mathbf{r})$. The Kohn–Sham or Hartree–Fock orbitals are then written as a linear combination of these basis orbitals:

$$\varphi_j(\mathbf{r}) = \sum_{\alpha=1}^{M} c_{j\alpha} \phi_\alpha(\mathbf{r}), \tag{8.1}$$

where j labels the wave function (or band), the sum runs over all the basis functions up to the dimension (or size) of the basis set M, and $c_{j\alpha}$ are the expansion coefficients of wave function j.

The wave functions are solutions of the time-independent Schrödinger equation. As such, they depend on the energy value, but not every energy is possible. Only those energies for which the solutions are normalizable are allowed. These are the eigenvalues ε_j, which can form a discrete set or a continuum. Therefore, if the correct functional form of the wave function as a function of energy is known, solving the eigenvalue problem is only a matter of finding the appropriate energies that satisfy Schrödinger's equation. This suggests the idea that a carefully designed basis set may require only one, or at least very few, energy-dependent basis functions. The price for using such a basis set is that the equation to solve is a partial differential equation that depends non-linearly on the energy. It must be solved iteratively, as is done for atomic systems (see Section 7.2.6), but in more complicated molecular or condensed phase geometries.

Atomic systems have spherical symmetry, and the wave functions are usually obtained by integrating the radial Schrödinger equation numerically using finite differences, forward from the origin and backward from infinity, and requesting that the two solutions match in amplitude and derivative at a pre-fixed radius. This cannot be achieved for arbitrary energies. Those energies that allow for a smooth matching of inward and outward solutions are precisely the eigenvalues. If for a certain energy value the derivative is discontinuous, then the energy can be corrected iteratively until the continuity conditions are verified within the desired tolerance. If this procedure is to be generalized to a polyatomic system, continuity conditions have to be imposed at the surface of spheres centered on each one of the atoms. There are two additional complications arising from this: first, since the geometry is not spherical any longer, the continuity conditions have to be secured at every point in the surface of the sphere, and they will in general depend on the polar and azimuthal angles. Therefore, this problem requires an expansion in partial waves, i.e. in components of the wave functions for each angular momentum (l, m). This can be done by expanding in spherical harmonics around each atomic site. Second, the continuity conditions have to be verified *simultaneously* on the surface of *all* the spheres, and not just a single one. This program can indeed be implemented, and there have been different methods proposed along these lines. Probably the most finished examples are Slater's APW approach (Slater, 1937), and the LCMTO method (Andersen, 1971).

We shall return to this in Section 8.6, but now we shall consider another possibility, which is to represent the one-electron wave functions in an energy-independent basis set, where the basis functions are chosen according either to simplicity, to flexibility, or to efficiency. In this case the Schrödinger equation becomes a generalized linear eigenvalue problem:

$$\sum_{\beta=1}^{M} \left(H_{\alpha\beta} - \varepsilon_j S_{\alpha\beta} \right) c_{j\beta} = 0, \tag{8.2}$$

or, in condensed matrix notation,

$$\mathbf{Hc} = \Lambda \mathbf{Sc}, \tag{8.3}$$

where Λ is the diagonal matrix of the eigenvalues, matrix $\mathbf{c}$ has the eigenvectors (the expansion coefficients of the wave functions) as columns, and $\mathbf{H}$ and $\mathbf{S}$ are the energy-independent Hamiltonian and *overlap* matrices, respectively.

The Hamiltonian matrix elements between two basis orbitals are given by

$$H_{\alpha\beta} = \langle \phi_\alpha | \hat{\mathcal{H}} | \phi_\beta \rangle = \int \phi_\alpha^*(\mathbf{r}) \, \hat{\mathcal{H}} \, \phi_\beta(\mathbf{r}) \, d\mathbf{r}, \tag{8.4}$$

while the overlap matrix elements

$$S_{\alpha\beta} = \langle \phi_\alpha | \phi_\beta \rangle = \int \phi_\alpha^*(\mathbf{r}) \phi_\beta(\mathbf{r}) \, d\mathbf{r} \tag{8.5}$$

take into account the possible non-orthogonality of the basis orbitals. With this notation, the wave function normalization condition is

$$\int \varphi_j^*(\mathbf{r}) \varphi_j(\mathbf{r}) \, d\mathbf{r} = \sum_{\alpha=1}^{M} \sum_{\beta=1}^{M} c_{j\alpha}^* S_{\alpha\beta} c_{j\beta} = 1. \tag{8.6}$$

In the Kohn–Sham case, the Hamiltonian operator appearing in (8.4) is the one-electron effective Hamiltonian

$$\hat{\mathcal{H}}_{KS} = -\frac{\hbar^2}{2m} \nabla^2 + v_{ext}(\mathbf{r}) + \int \frac{\rho(\mathbf{r}')}{|\mathbf{r} - \mathbf{r}'|} \, d\mathbf{r}' + \mu_{XC}[\rho]. \tag{8.7}$$

The same scheme is valid in the Hartree–Fock approximation within the *canonical orbital representation*. When the Hartree–Fock equations are expressed in a basis set, more precisely in an atom-centered basis set, they too assume the form of a generalized eigenvalue problem,

$$\mathbf{Fc} = \Lambda \mathbf{Sc}, \tag{8.8}$$

where the matrix elements are

$$F_{\alpha\beta} = \langle \phi_\alpha | \hat{\mathcal{F}} | \phi_\beta \rangle = \int \phi_\alpha^*(\mathbf{r}) \hat{\mathcal{F}} \phi_\beta(\mathbf{r}) \, d\mathbf{r}, \tag{8.9}$$

and

$$\hat{\mathcal{F}} = -\frac{\hbar^2}{2m} \nabla^2 + v_{ext}(\mathbf{r}) + \sum_{j=1}^{N} \left(\hat{\mathcal{J}}_j - \hat{\mathcal{K}}_j \right). \tag{8.10}$$

These equations have been derived for a closed-shell system, where all the spins are paired, independently by Roothaan and Hall, and are consequently called the *Roothaan–Hall equations* (Roothaan, 1951; Hall, 1951). For open-shell systems with unpaired spins, a straightforward extension is to consider a *restricted* many-body wave function of the single-determinantal form where the orbitals corresponding to occupied and empty unpaired spins are the same. However, an improved wave function can be obtained by releasing this constraint, thus allowing for all the spin orbitals to vary independently. This is called the *unrestricted* approach, and it was originally introduced by Slater (1930b) and re-cast in the basis set approximation by Pople and Nesbet (1954). The operators $\hat{\mathcal{J}}_j$ and $\hat{\mathcal{K}}_j$ are the direct and exchange two-body Coulomb operators defined in Chapter 3, and the sum runs over all the occupied states. The only formal difference between Kohn–Sham and Roothaan–Hall is that the exchange-correlation potential is replaced by the non-local exchange operator.

In the following section we extend this approach to periodic systems. Next we describe in detail the plane wave (PW) basis set as an example of a floating set of extended basis functions useful for condensed phase calculations. Other floating basis sets are reviewed, but the emphasis of the rest of this chapter is on atom-centered basis sets, in particular Gaussians, which are the way of choice for calculations in molecular systems. After mentioning mixed basis sets, we finally concentrate on basis sets based on augmentation spheres, which are not only important historically, but also a central component in many all-electron approaches.

8.1 Periodic systems

When studying periodic systems such as solids it has to be ensured that Bloch's theorem is verified, in the sense that the combination of basis orbitals that represents a solution of the Schrödinger equation verifies the translational periodicity of the supercell. In that case, the previous expression has to be modified in the following way:

$$\varphi_j^{(\mathbf{k})}(\mathbf{r}) = e^{i\mathbf{k}\cdot\mathbf{r}} \sum_{\alpha=1}^{M} c_{j\alpha}^{(\mathbf{k})} \phi_\alpha(\mathbf{r}) = \sum_{\alpha=1}^{M} c_{j\alpha}^{(\mathbf{k})} \phi_\alpha^{(\mathbf{k})}(\mathbf{r}), \tag{8.11}$$

where **k** indicates the wave vector in the Brillouin zone, and the modified basis functions are

$$\phi_\alpha^{(\mathbf{k})}(\mathbf{r}) = e^{i\mathbf{k}\cdot\mathbf{r}} \phi_\alpha(\mathbf{r}). \tag{8.12}$$

While this representation is adequate for basis functions that already respect periodic boundary conditions, such as plane waves, care must be taken when using other types of basis set. For atom-centered basis sets a more appropriate expression for the wave functions that is compatible with Bloch's theorem is

$$\varphi_j^{(\mathbf{k})}(\mathbf{r}) = \sum_{\alpha=1}^{M} c_{j\alpha}^{(\mathbf{k})} \left(\frac{1}{\sqrt{\Omega}} \sum_{\mathbf{T}} e^{i\mathbf{k}\cdot\mathbf{T}} \phi_\alpha(\mathbf{r}-\mathbf{T}) \right), \tag{8.13}$$

where the terms in parentheses are the PBC-adapted basis functions, with the sum running over the infinite periodic replicas of the simulation box, i.e. $\mathbf{T} = n_1\mathbf{a}_1 + n_2\mathbf{a}_2 + n_3\mathbf{a}_3$, with $\mathbf{a}_i$ the three lattice vectors. It can be shown that, with this definition, Bloch's theorem $\varphi_j^{(\mathbf{k})}(\mathbf{r}+\mathbf{R}) = e^{i\mathbf{k}\cdot\mathbf{R}} \varphi_j^{(\mathbf{k})}(\mathbf{r})$ is verified.

The generalized eigenvalue equation (8.2) is still valid, although now there is an equation for each **k**-point in the Brillouin zone:

$$\sum_{\beta=1}^{M} \left(\mathcal{H}_{\alpha\beta}^{(\mathbf{k})} - \varepsilon_j^{(\mathbf{k})} S_{\alpha\beta}^{(\mathbf{k})} \right) c_{j\beta}^{(\mathbf{k})} = 0. \tag{8.14}$$

For basis functions that respect the periodicity of the crystal, the Hamiltonian and overlap matrix elements between two basis orbitals are given by:

$$\mathcal{H}_{\alpha\beta}^{(\mathbf{k})} = \langle \phi_\alpha^{(\mathbf{k})} | \hat{\mathcal{H}} | \phi_\beta^{(\mathbf{k})} \rangle = \int \phi_\alpha^{*(\mathbf{k})}(\mathbf{r}) \hat{\mathcal{H}} \phi_\beta^{(\mathbf{k})}(\mathbf{r}) \, d\mathbf{r} \tag{8.15}$$

and

$$S_{\alpha\beta} = \langle \phi_\alpha | \phi_\beta \rangle = \int \phi_\alpha^{*(\mathbf{k})}(\mathbf{r}) \phi_\beta^{(\mathbf{k})}(\mathbf{r}) \, d\mathbf{r} = \int \phi_\alpha^*(\mathbf{r}) \phi_\beta(\mathbf{r}) \, d\mathbf{r}, \tag{8.16}$$

respectively. Notice that in this case the overlap matrix elements do not depend on the wave vector $\mathbf{k}$.

For atom-centered basis functions the expressions are:

$$\mathcal{H}_{\alpha\beta}^{(\mathbf{k})} = \sum_{\mathbf{T}} e^{i\mathbf{k}\cdot\mathbf{T}} \int \phi_\alpha^*(\mathbf{r}) \hat{\mathcal{H}} \phi_\beta(\mathbf{r} - \mathbf{T}) \, d\mathbf{r} \tag{8.17}$$

and

$$S_{\alpha\beta}^{(\mathbf{k})} = \sum_{\mathbf{T}} e^{i\mathbf{k}\cdot\mathbf{T}} \int \phi_\alpha^*(\mathbf{r}) \phi_\beta(\mathbf{r} - \mathbf{T}) \, d\mathbf{r}. \tag{8.18}$$

These expressions are valid whether in the Kohn–Sham ($\hat{\mathcal{H}} = \hat{\mathcal{H}}_{KS}$) or in the Hartree–Fock approximation ($\hat{\mathcal{H}} = \hat{\mathcal{F}}$).

8.2 Plane waves

For solids, and in general condensed phases, Bloch's theorem (6.5) prescribes that the wave functions must be composed of a phase factor and a periodic part that verifies $u_\mathbf{k}(\mathbf{r}) = u_\mathbf{k}(\mathbf{r} + \mathbf{a}_j)$, with $\mathbf{a}_j$ any lattice vector. This can be used to introduce naturally (for solid state applications) the basis set of plane waves (PW). Plane waves are solutions of the Schrödinger equation in the presence of a constant external potential, as is approximately verified in interstitial regions in condensed phases. Closer to the atomic nuclei, however, the external potential is far from constant and, hence, the solution of Schrödinger's equation is no longer a single PW; it has to be written as a linear combination of PWs.

In general, any function in real space can be written as the Fourier transform of a function in reciprocal space,

$$u_\mathbf{k}(\mathbf{r}) = \int e^{i\mathbf{g}\cdot\mathbf{r}} \tilde{u}_\mathbf{k}(\mathbf{g}) \, d\mathbf{g}, \tag{8.19}$$

but due to the periodicity of $u_\mathbf{k}(\mathbf{r})$, the only allowed values of $\mathbf{g}$ are those that verify $e^{i\mathbf{g}\cdot\mathbf{a}_j} = 1$, i.e. $\mathbf{g}\cdot\mathbf{a}_j = 2n\pi$ for $j = 1, 2, 3$ – the three lattice vectors. This implies that $\mathbf{g} = n_1 \mathbf{b}_1 + n_2 \mathbf{b}_2 + n_3 \mathbf{b}_3$, where

$$\mathbf{b}_i = 2\pi \frac{\mathbf{a}_j \times \mathbf{a}_k}{\Omega} \tag{8.20}$$

and $\mathbf{n} = (n_1, n_2, n_3)$ is a vector of integer numbers. Therefore, the $\mathbf{g}$ vectors in the Fourier transform (8.19) are restricted precisely to the reciprocal lattice vectors $\mathbf{G}$ defined by Eq. (6.7), and the general expression for the wave function is:

$$\varphi^{(\mathbf{k})}(\mathbf{r}) = \frac{e^{i\mathbf{k}\cdot\mathbf{r}}}{\sqrt{\Omega}} \sum_{\mathbf{G}=0}^{\infty} C_{\mathbf{k}}(\mathbf{G}) e^{i\mathbf{G}\cdot\mathbf{r}}. \tag{8.21}$$

Hence, due to the periodicity, the Fourier transform (8.19) becomes a Fourier series, where the Fourier coefficients are $C_{\mathbf{k}}(\mathbf{G})$. This restriction of the possible values of $\mathbf{g}$ to the reciprocal lattice vectors ensures that PBC are automatically verified. We now define the PW basis functions

$$\phi_{\mathbf{G}}(\mathbf{r}) = \frac{1}{\sqrt{\Omega}} e^{i\mathbf{G}\cdot\mathbf{r}}, \tag{8.22}$$

which are suitably normalized in the supercell

$$\langle \phi_{\mathbf{G}} | \phi_{\mathbf{G}'} \rangle = \frac{1}{\Omega} \int_{\Omega} e^{i(\mathbf{G}-\mathbf{G}')\cdot\mathbf{r}} \, d\mathbf{r} = \frac{1}{\Omega} (\Omega \delta_{\mathbf{G},\mathbf{G}'}) = \delta_{\mathbf{G},\mathbf{G}'}, \tag{8.23}$$

so that PWs corresponding to different wave vectors, $\mathbf{G} \neq \mathbf{G}'$, are orthogonal. With this definition, the wave functions for the different eigenstates j can be written as

$$\varphi_j^{(\mathbf{k})}(\mathbf{r}) = e^{i\mathbf{k}\cdot\mathbf{r}} \sum_{\mathbf{G}=0}^{\infty} C_{j\mathbf{k}}(\mathbf{G}) \phi_{\mathbf{G}}(\mathbf{r}). \tag{8.24}$$

Notice that the phase factor $e^{i\mathbf{k}\cdot\mathbf{r}}$ involves a wave vector $\mathbf{k}$ in the first BZ, while the reciprocal lattice vectors $\mathbf{G}$ entering the PW expansion, except for $\mathbf{G} = \mathbf{0}$, lie always outside the BZ. Wave functions corresponding to different $\mathbf{k}$ vectors obey separate Schrödinger equations. In the case of non-interacting electrons these are completely independent, but in the case of DFT (or any other many-body theory) the equations couple in the sense of self-consistency through the one-electron orbitals and the electronic density. This latter is expressed as a BZ average, as indicated in (6.20) or (6.22).

The phase factor can be incorporated into the definition of the basis functions, which will become dependent on the particular point in the BZ:

$$\phi_{\mathbf{G}}^{\mathbf{k}}(\mathbf{r}) = \frac{1}{\sqrt{\Omega}} e^{i(\mathbf{k}+\mathbf{G})\cdot\mathbf{r}}, \tag{8.25}$$

so that

$$\varphi_j^{(\mathbf{k})}(\mathbf{r}) = \sum_{\mathbf{G}=0}^{\infty} C_{j\mathbf{k}}(\mathbf{G}) \phi_{\mathbf{G}}^{\mathbf{k}}(\mathbf{r}) \tag{8.26}$$

assumes the usual form of a basis set expansion (8.1). Conversely, the PW coefficients assume the following form in terms of the real-space wave function:

$$C_{j\mathbf{k}}(\mathbf{G}) = \int_\Omega \phi_\mathbf{G}^{\mathbf{k}*}(\mathbf{r})\varphi_j^{(\mathbf{k})}(\mathbf{r})\,d\mathbf{r} = \frac{1}{\sqrt{\Omega}}\int_\Omega e^{-i(\mathbf{k}+\mathbf{G})\cdot\mathbf{r}}\varphi_j^{(\mathbf{k})}(\mathbf{r})\,d\mathbf{r}. \qquad (8.27)$$

8.2.1 PW Matrix elements

In a PW basis set, the matrix elements (8.4) and (8.5) are very easy to calculate. Firstly,

$$S_{\mathbf{G},\mathbf{G}'}^{\mathbf{k}} = \delta_{\mathbf{G},\mathbf{G}'} \qquad (8.28)$$

because the PWs are orthogonal, as shown in (8.23). Next, the kinetic and potential contributions to the Hamiltonian matrix elements $H_{\mathbf{G},\mathbf{G}'}^{\mathbf{k}} = T_{\mathbf{G},\mathbf{G}'}^{\mathbf{k}} + V_{\mathbf{G},\mathbf{G}'}^{\mathbf{k}}$ are given by

$$T_{\mathbf{G},\mathbf{G}'}^{\mathbf{k}} = -\frac{\hbar^2}{2m}\langle \phi_\mathbf{G}^{\mathbf{k}}|\nabla^2|\phi_{\mathbf{G}'}^{\mathbf{k}}\rangle = \frac{\hbar^2}{2m}|\mathbf{k}+\mathbf{G}|^2\delta_{\mathbf{G},\mathbf{G}'} \qquad (8.29)$$

and

$$V_{\mathbf{G},\mathbf{G}'} = \langle\phi_\mathbf{G}^{\mathbf{k}}|\hat{V}|\phi_{\mathbf{G}'}^{\mathbf{k}}\rangle = \frac{1}{\Omega}\int V(\mathbf{r})e^{-i(\mathbf{G}-\mathbf{G}')\cdot\mathbf{r}}\,d\mathbf{r} = \tilde{V}(\mathbf{G}-\mathbf{G}'), \qquad (8.30)$$

where $\tilde{V}(\mathbf{G}-\mathbf{G}')$ is the Fourier transform of the potential and, if $\hat{V}$ is a local potential, the matrix elements are independent of the wave vector $\mathbf{k}$ in the BZ. The kinetic term is, quite conveniently, diagonal in reciprocal space.

Therefore, in a PW basis set the time-independent Schrödinger equation (8.2) becomes the eigenvalue equation

$$\sum_{\mathbf{G}'}\left(\frac{\hbar^2}{2m}|\mathbf{k}+\mathbf{G}|^2\delta_{\mathbf{G},\mathbf{G}'} + \tilde{V}(\mathbf{G}-\mathbf{G}')\right)C_{j\mathbf{k}}(\mathbf{G}') = \varepsilon_{j\mathbf{k}}C_{j\mathbf{k}}(\mathbf{G}), \qquad (8.31)$$

which has a particularly simple and appealing form.

8.2.2 PW Energy cutoff

Bloch's theorem implies that the wave function of an electron in a periodic potential can be naturally expanded in a plane-wave basis set. The $\mathbf{G}$ vectors allowed in the PW expansion are the reciprocal lattice vectors and, in principle, an infinite number of such vectors is required to represent the wave functions with infinite accuracy. In practice, however, the Fourier coefficients $C_\mathbf{k}(\mathbf{G})$ of the wave functions decrease with increasing $|\mathbf{k}+\mathbf{G}|$, so that the PW expansion can be effectively truncated at a finite number of terms, i.e. limited to all waves with kinetic energy lower than some energy cutoff E_{cut}:

$$\frac{\hbar^2}{2m}|\mathbf{k}+\mathbf{G}|^2 < E_{cut}. \tag{8.32}$$

For the Γ-point ($\mathbf{k} = \mathbf{0}$), the above expression defines a sphere in the space of wave vectors, of radius G_{cut} given by

$$|\mathbf{G}| < G_{cut} = \sqrt{\frac{2mE_{cut}}{\hbar^2}}, \tag{8.33}$$

or, in atomic units ($\hbar = m = 1$), $G_{cut} = \sqrt{2E_{cut}}$. For historical reasons, it is quite common to specify the energy cutoff not in atomic units, but in rydberg; 1 Ry = 13.6 eV = 0.5 a.u. or hartree. Therefore, if the energy cutoff is given in Ry, $G_{cut} = 2\sqrt{E_{cut}[\text{Ry}]}$.

The truncation of the basis set leads to an error in the computed physical quantities, but this error can be easily handled by increasing the cutoff. Since this implies an increase in the size of the basis set without modifying the Hamiltonian, then the energy should decrease variationally with increasing E_{cut}. This is different from other types of basis sets where basis functions are not orthogonal (e.g. Gaussians). There, increasing the number of basis functions can lead to overcompleteness of the basis set, and does not necessarily imply a decrease in energy, especially if the basis set has not been carefully optimized or is not well-tempered (see Section 8.4.6).

Ideally, the Fourier series representation ensures that *any* periodic function can be represented using an expansion like that in (8.21). However, the introduction of a finite cutoff in reciprocal space (G_{cut}) puts a restriction to the type of functions that can be represented. In fact, the relation

$$|n_1\mathbf{b}_1 + n_2\mathbf{b}_2 + n_3\mathbf{b}_3| = G_{cut} \tag{8.34}$$

defines the set of the largest integer numbers (n_1, n_2, n_3) required to carry out the PW expansion up to the chosen energy cutoff. Therefore, for any direction in the BZ there is a maximum number of $\mathbf{G}$-vectors given by

$$n_i \geq \frac{G_{cut}}{|\mathbf{b}_i|} = \frac{G_{cut}|\mathbf{a}_i|}{2\pi}, \tag{8.35}$$

where the equality is strictly valid only for orthogonal lattice vectors.

Since large $\mathbf{G}$-vectors are associated with short-range features in real space, cutting off the vectors of modulus larger than G_{cut} implies the loss of information at a short scale in real space. This scale is given by the usual Fourier transform relation

$$\Delta x \, G_{cut} = 2\pi, \tag{8.36}$$

where Δx is the minimum meaningful distance between two points in real space. This is for a one-dimensional case; the three-dimensional case is a straightforward extension. By comparing expressions (8.35) and (8.36), it can be seen that this distance is given by

$$\Delta x = \frac{|\mathbf{a}_i|}{n_i}. \tag{8.37}$$

The meaning of the above relation is that points in real space closer than Δx do not carry additional information. Therefore, this is equivalent to a discrete real-space representation where the wave function is defined on a grid of points separated by the interval Δx. This is precisely the program carried out by the fast Fourier transform technique (FFT) (Press et al., 1992). In the FFT method, the grid of **G**-vectors in reciprocal space has an associated grid of $N_i = 2n_i$ points in real space. The factor 2 arises because the cutoff in reciprocal space is the modulus of the maximum **G**-vector. Therefore, the **G**-vectors run between $-G_{\text{cut}}$ and G_{cut}, and there are $2n_i$ of these. The number of grid points in real space is given by

$$N_i \geq \frac{1}{\pi}\sqrt{2E_{\text{cut}}}|\mathbf{a}_i|, \tag{8.38}$$

or, if E_{cut} is expressed in Rydberg units,

$$N_i \geq \frac{2}{\pi}\sqrt{E_{\text{cut}}[\text{Ry}]}|\mathbf{a}_i|. \tag{8.39}$$

This relation is valid in the three spatial directions. The inequality is to ensure that the number of grid points verifies that the **G**-vectors run *at least* up to G_{cut}.

The number of **G**-vectors in reciprocal space within the sphere of radius G_{cut} depends on the shape of the supercell, but can be approximately calculated by considering that the volume of such a sphere is given by

$$N_G \approx \frac{4\pi}{3}\left(\frac{N_{\text{grid}}}{2}\right)^3 = \frac{4}{3\pi^2}\Omega[\text{bohr}^3]E_{\text{cut}}^{3/2}[\text{Ry}], \tag{8.40}$$

where N_{grid} is an average number of grid points and Ω is the volume of the supercell. Here an important advantage of operating in reciprocal, rather than real, space can be observed; the number of grid points in real space corresponds to the three-dimensional body defined by the lattice vectors, which contains approximately $N_R \approx N_{\text{grid}}^3 \approx 6N_G/\pi \approx 2N_G$ points, i.e. real space requires twice the number of basis functions as reciprocal space.

The key issue when using a PW basis set is *how to choose the cutoff*. The main idea here is that the wave functions respond to the atomic potentials. If the true nuclear potential Z/r is retained, as in all-electron calculations, then the core wave functions exhibit sharp peaks close to the nucleus, and the valence wave functions have $n-1$ nodes, with n the principal quantum number. Both cases require a large number of PWs because the spatial variation is fast. This is the reason why all-electron PW calculations are simply avoided, and the bare nuclear potential is replaced by a smooth pseudopotential. Core states are eliminated, and valence states are pseudized and made nodeless inside the core region. Therefore, being reasonably smooth potentials, there is a spatial scale where the wave functions become smooth enough that decreasing the grid spacing or, equivalently, increasing the energy cutoff, does not introduce any additional relevant information. In that case, it is said that the system has *converged* in PWs. The required cutoff can be estimated directly from the atomic pseudopotentials, but in general it has to be adjusted by trial and error for the particular system under study. The case of the bare Coulomb potential is special, because the PW expansion never really converges when increasing the cutoff. However, physical properties of interest (other than the energy) are converged at a finite cutoff, as in the case of the hydrogen molecule, where 60 Ry is sufficient to reproduce the interatomic distance and vibrational frequency.

The electronic density is constructed by squaring the wave functions in real space. When this operation is written as a convolution in Fourier space, it becomes clear that the density in reciprocal space includes plane waves up to $|\mathbf{G}| = 2G_{\text{cut}}$. Therefore, there is another energy cutoff associated with the density: $E_D = 4E_{\text{cut}}$. This means that the real-space grid for the density is finer than that of the wave functions, although it does not carry more information. It can be made coarser without losing information, but this has to be done carefully.

If the BZ of the supercell is sampled in more than one $\mathbf{k}$-point, then the number of $\mathbf{G}$-vectors $N_G^{\mathbf{k}}$ depends on $\mathbf{k}$ through the relation $E_{\text{cut}} = |\mathbf{k} + \mathbf{G}_{\text{cut}}|^2/2$. Typically, the variation of $N_G^{\mathbf{k}}$ with $\mathbf{k}$ is very small because $|\mathbf{k}| < |\mathbf{G}_{\text{min}}|$. It is basically the effect of de-centering a large sphere in an amount proportional to $1/N$. The number of basis functions changes discontinuously with E_{cut} because the $\mathbf{G}$ vectors are ordered in shells of equal modulus $|\mathbf{G}|$, so that the number of PWs slightly depends on $\mathbf{k}$, especially if the cutoff is not too high.

8.2.3 Advantages and disadvantages of the plane wave expansion

The main advantages of using a truncated PW basis set are the following.

(i) The kinetic term in the one-electron Hamiltonian is diagonal in reciprocal space, while the potential term is local in real space (pseudopotential methods introduce a non-local component of the potential, but this will be treated in detail in Section 9.3). This feature can be exploited to speed up the calculations by transforming the wave functions and the density back and forth from real to reciprocal space and vice versa, and calculating the kinetic and potential contributions in the representation where they are diagonal. The transformation can be done very efficiently by using fast

Fourier transform (FFT) techniques, whose associated computational cost scales like $M \log M$, where M is the number of PWs.
(ii) The calculation of the energy and its derivatives (forces, stress) is analytic and quite simple.
(iii) The PW basis functions do not move together with the atoms, and they represent all regions of space with the same accuracy. There are no additional forces on the nuclei arising from the derivation of the basis functions. This is because the Hellmann–Feynman theorem can be strictly applied only when the basis set is fully converged, or when the basis functions do not depend on the nuclear coordinates. Using atom-centered basis sets, the correction for the finiteness of the basis is very important, and gives rise to the so-called Pulay forces (Pulay, 1969).

On the other hand, the disadvantages are the following.

(i) For systems of low dimensionality, such as molecules, wires, or surfaces, an important computational effort is used to represent the vacuum that fills the supercell, which is not particularly relevant. This is very different from the case of localized basis sets.
(ii) We have already mentioned that charged systems cannot be studied within a standard PW approach, unless some kind of compensating background is added. This is valid not only for solids, but also for molecular systems and charged surfaces and wires. In addition, even in the case of neutral molecules, if they have a dipole moment, a spurious electrostatic interaction between periodic images in adjacent cells appears. This is because PBC are automatically embedded into the PW scheme. Dipole–dipole interactions decay as R^{-3}, which is not particularly slow. Therefore, in order to minimize these effects one has to choose a simulation cell that is sufficiently large, with the undesired consequence of an increase in the number of plane waves. The same happens in the case of non-polar molecules if they have a non-zero quadrupole moment, although in that case the quadrupole–quadrupole interaction decays more rapidly, as R^{-5}.
(iii) Systems where the wave functions vary rapidly close to the nuclei require a very high energy cutoff (many PW components). Localized bases are much better in this because they are normally tuned to reproduce atomic wave functions. This is especially important in hydrogen, first-row elements, and transition metals.

8.3 Other floating basis sets

Some of the inconveniences of the PW basis set can be eased out by the use of a different floating basis set, i.e. not linked to atomic nuclei, but more efficient to represent the features of the electronic orbitals. There have been numerous attempts in this direction, which were more or less successful. However, it was always difficult to beat the simplicity and efficiency of PW calculations.

One interesting step forward proposed by Gygi in the early nineties was to maintain the PW representation but introducing a space-dependent metric tensor.

This allows for an accumulation of real-space grid points in the atomic region, while making the grid coarser in interstitial or vacuum regions (Gygi, 1992). Within this approach, the energy cutoff becomes a local spatial function that has to be kept under control to avoid an excessive curvature in the atomic region. The advantage of this approach is that the extremely efficient FFT machinery remains in place, but the number of basis functions is significantly reduced. This method is reminiscent of re-gridding methods used in continuum mechanics. A few groups have adopted this strategy (Gygi and Galli, 1995; Hamann, 1995).

A method that makes use of the analytic advantages of Gaussian functions while eliminating the features of atom-centered basis functions is that of *floating Gaussians* (Sprik and Klein, 1989). In this approach, a set of Gaussian functions is assigned to a set of grid points that is fixed in space and does not move with the atoms. Problems with completeness, linear dependence, and optimization of basis functions are inherited from the more common atom-centered basis sets (see Section 8.4.5 below). Along the same idea of basis functions attached to fixed grid points is the approach proposed by Gan *et al.* (2001), where the basis functions are now spherical waves that are solutions to Laplace's equation.

Finite-difference schemes have been tried, but they proved not to be competitive with the PW approach, even using sophisticated representations of the Laplacian operator that involve up to six neighboring sites (Chelikowsky *et al.*, 1994; see also Skylaris *et al.*, 2002). Wavelet basis sets that are partially localized in both real and reciprocal space appeared to be efficient, but never became too widespread (Arias, 1999). An interesting alternative based on B-spline basis functions of finite support (Hernandez *et al.*, 1997) proved very efficient for large systems, and scaled very well with increasing number of atoms in the system (linearly with N). This approach is implemented in the code CONQUEST (Bowler and Gillan, 2002); apparently the computational complexity involved by the use of this type of basis function is not irrelevant. One of the most recent proposals is to use Lagrange polynomials as basis functions, which has the advantage of being orthogonal, and optimal for numerical integration (Varga *et al.*, 2004). Lagrange functions also allow one to select an efficient grid mapping according to the peculiarities of the system.

8.4 Atom-centered basis sets

For molecular systems the plane wave methodology is not the most efficient one for the reasons pointed out in Section 8.2. In the quantum chemistry and molecular physics communities the most common approach is to expand the one-electron wave functions, or *molecular orbitals* (MOs), in terms of *atomic orbitals* (AOs). AOs are centered on the atomic nuclei, and can be represented in terms of basis functions with different functional forms, or can be given numerically on a grid. Many times what are called AOs are the basis functions, and not the solutions of the atomic problem.

A suitable basis set should allow for a systematic improvement of its quality when the number of basis functions is incremented. It should also be designed in such a way that only a few basis functions are required to achieve a reasonable accuracy in the electronic distribution. In the case of analytic expressions, the basis functions should be simple enough that the Hamiltonian matrix elements can be easily evaluated, preferably analytically. These requirements lead to truncated basis sets, so that issues of completeness and convergence are bound to be more severe than in PW basis sets, where a single parameter E_{cut} controls the convergence.

The systematic improvement of the quality is better handled if the basis functions are *orthogonal*. An orthogonal basis is simple to construct for a single atomic center, but the efficiency for molecular systems would be quite poor. A natural road to improvement is to consider a *multi-center* expansion, where the basis set is constructed with basis functions centered on every atom in the molecule. This approach introduces information about the characteristics of the electronic distribution, e.g. the fact that core states, or states that are deep in energy, peak close to their corresponding nuclear site. However, in this case orthogonality is usually sacrificed because there is no *a priori* reason why basis functions centered on different atoms should be orthogonal. Just imagine two s-type orbitals sitting on nearest-neighbor sites. This does not exclude the possibility of transforming these basis functions into an orthogonal basis set. A possible orthogonalization procedure suggested by Löwdin consists of transforming the basis functions using the overlap matrix (Löwdin, 1950),

$$\phi'_\alpha(\mathbf{r}) = \sum_\beta S^{-1/2}_{\alpha\beta} \phi_\beta(\mathbf{r}). \tag{8.41}$$

However, any other transformation $\mathbf{X}$ related to $\mathbf{S}$ by a similarity transformation $\mathbf{X}^\dagger \mathbf{S} \mathbf{X} = \mathbf{I}$ would also serve as an orthogonalization scheme. After this transformation, the new, orthogonal basis functions spread over several centers, although they are not necessarily more spread than the original ones. The calculation of the Hamiltonian matrix elements becomes more complicated, but there is no fundamental reason to discard them as a possible basis set. Other more sophisticated orthogonalization schemes where, in addition, the basis functions are required to be localized (*screening transformation*) have been devised in the context of muffin-tin orbitals (Andersen and Jepsen, 1984). While the algebraic screening procedure is more complicated, this approach is very rewarding in terms of efficiency and ease of interpretation.

In contrast to single-center basis sets, non-orthogonal multi-center basis sets require fewer basis functions, but their systematic improvement is compromised. Non-orthogonality can lead to linear dependencies in the basis set, i.e. the same MO can be represented in different ways using different subsets of basis functions, thus leading to numerical instabilities. Therefore, adding more basis functions, if not carefully done, may not lead to a systematic improvement of the quality of the calculation.

A further improvement in the direction of reducing the number of functions in the basis set is to introduce information about the characteristics of the MOs by tailoring the shape of the basis functions appropriately. An important element to take into account when designing a basis set is that the Hamiltonian matrix elements, whether in the Hartree–Fock approximation or in standard DFT approaches, contains two-electron terms. When the one-electron orbitals are expanded in an atom-centered basis set, such integrals involve four basis functions that give rise to a variety of terms, including three- and four-center integrals. The calculation of these terms is the bottleneck of the calculation, as they are computationally the most demanding ones and, in addition, the number of such integrals grows as M^4, with M the size of the basis set. Therefore, the shape of the basis functions should be chosen in such a way that these integrals are easy to calculate.

We shall denote by $\phi_\alpha(\mathbf{r} - \mathbf{R}_I)$, or $\phi_\alpha^I(\mathbf{r})$, a generic basis function centered on atom I. The index α labels the type of basis function, which generally corresponds to a set of indices $\alpha = (n, l, m)$ that indicate angular momentum (l, m) and radial orbital n.

8.4.1 Atomic orbitals

Since the molecular electronic distribution is largely dominated by atomic distributions, a natural choice is to optimize the basis functions to reproduce AOs, and then export them to molecular systems or condensed phases. This scheme is called *linear combination of atomic orbitals* (LCAO). The starting point is to solve the atomic problem, i.e. to find the one-electron eigenstates in the effective Coulomb-like central field. In spherical coordinates the one-electron Schrödinger-like equation reads

$$\left[-\frac{\hbar^2}{2m} \left(\frac{1}{r^2} \frac{\partial}{\partial r} \left(r^2 \frac{\partial}{\partial r} \right) - \frac{\hat{L}^2}{r^2} \right) + v_{\text{eff}}(r) \right] \phi_\alpha(\mathbf{r}) = \varepsilon_\alpha \phi_\alpha(\mathbf{r}), \qquad (8.42)$$

where the $\hat{L}$ is the angular momentum operator, so that

$$\hat{L}^2 = -\frac{1}{\sin\theta} \frac{\partial}{\partial\theta} \left(\sin\theta \frac{\partial}{\partial\theta} \right) - \frac{1}{\sin^2\theta} \frac{\partial^2}{\partial\varphi^2}. \qquad (8.43)$$

Since for a central field $\hat{L}^2$ and $\hat{L}_z$ commute with the Hamiltonian, the AOs factorize into a radial and an angular part (Messiah, 1961)

$$\phi_{nlm}(\mathbf{r}) = \chi_{nl}(r) Y_{lm}(\theta, \varphi), \tag{8.44}$$

where $Y_{lm}(\theta, \varphi)$ are spherical harmonics, i.e. eigenstates of $\hat{L}^2$ and $\hat{L}_z$ with eigenvalues $l(l+1)$ and m, respectively. The mathematical expression of the spherical harmonics can be found in any book on electromagnetism (Jackson, 1975) or quantum mechanics (Messiah, 1961). Spherical harmonics are orthonormal functions in the space of angular variables (θ, ϕ). A useful expression is obtained by absorbing part of the normalization factor into the definition of the functions, $C_{lm}(\theta, \phi) = \sqrt{4\pi/(2l+1)} Y_{lm}(\theta, \phi)$, in what is called the Racah normalization. It is often useful to use *real spherical harmonics*, which are appropriate combinations of Y_{lm} and $Y_{l,-m}$ that convert the complex exponentials into sines and cosines (Finnis, 2003). The number of independent real spherical harmonics for each l is still $2l+1$, but they are not eigenstates of $\hat{L}_z$. The algebra of angular momenta is well known and it is documented in any book on quantum mechanics (Messiah, 1961).

By replacing the above expression for the AOs into the Schrödinger-like equation, we obtain the following differential equation in the radial variable:

$$\frac{d^2 u_{nl}(r)}{dr^2} - \frac{2m}{\hbar^2} \left(v_{\text{eff}}(r) + \frac{l(l+1)}{r^2} - \epsilon_{nl} \right) u_{nl}(r) = 0, \tag{8.45}$$

where the radial part of the AO is given by $\chi_{nl}(r) = u_{nl}(r)/r$. It is useful to indicate here the solutions of Laplace's equation in spherical coordinates, as these play an important role in some schemes for evaluating the electrostatic contribution to the energy and potential. They are also a starting point for the construction of atom-centered basis sets, as they are zero-energy solutions of Schrödinger's equation in the absence of a potential, for angular momentum l. There are two types of solution:

$$R_{lm}(\mathbf{r}) = r^l C_{lm}(\theta, \phi),$$
$$I_{lm}(\mathbf{r}) = r^{-(l+1)} C_{lm}(\theta, \phi), \tag{8.46}$$

where the first solution is regular at the origin and the second is divergent at the origin, but regular at infinity.

8.4.2 Numerical basis sets

The most straightforward basis set is constructed by choosing as basis functions the AOs that are solutions of the atomic problem. This can be done at the desired level of theory, but it is convenient to be consistent with that used for the molecular system or the condensed phase, e.g. Hartree–Fock or DFT-LDA.

Even if, qualitatively, the shape of the AOs is known, it is not trivial to find an analytic functional form that represents them accurately enough. Below we will introduce basis sets made of decaying exponentials and Gaussians. Nevertheless, normally, these basis functions are not accurate enough to represent the AOs using a single basis function, and a set of them is required. Therefore, if the strategy is to reduce as much as possible the number of basis functions, a possible road is to abandon the idea of analyticity, and represent directly the AOs numerically on a grid, as they come out from the atomic calculation. In fact, there are computer codes such as Dmol (Delley, 1990) that use numerical AOs as basis sets.

The main disadvantage of this approach is that the Hamiltonian and overlap matrix elements have to be computed fully numerically. This is more costly from the computational point of view, especially the calculation of three- and four-center Coulomb integrals, but the extra cost can be overcome if the size of the basis set is significantly reduced by carefully tailoring the basis functions. In the case of DFT-based methods, this limitation has been partly relieved by using auxiliary basis sets (Andzelm and Wimmer, 1992; Soler *et al.*, 2002), but it still remains a delicate issue.

The fireball basis set

An attractive modification is to generate the basis functions as solutions of the atomic problem, but with the additional constraint that they become strictly zero beyond some localization radius (Sankey and Niklewski, 1989). This is the *fireball* basis set. The advantage is that overlap, kinetic, and nuclear attraction matrix elements are strictly zero for centers beyond a certain distance. This kind of approach, which was originally envisaged for non-self-consistent calculations, was adopted for the SIESTA self-consistent DFT code (Soler *et al.*, 2002). In addition, since SIESTA is a pseudopotential code, the basis functions are chosen to be the solutions of the pseudo-atomic problem, i.e. where the bare Coulomb potential is replaced by the pseudopotential. These basis functions or *pseudo-atomic orbitals* (PAOs), are expected to provide an optimal representation of the orbitals close to the origin, although their quality decays with the distance from the atomic sites. The interstitial regions, however, are populated with many other basis functions that are centered on other atomic sites, thus adding sufficient variational freedom.

The localization radius R_{loc} can be different for different atomic species and angular momenta. A finite R_{loc} implies a positive shift in the energy of the orbitals (E_{shift}) due to the compression imposed by the modified boundary condition. A good way of generating a well-balanced basis set is to choose the various R_{loc}

by fixing the same E_{shift} for all the atomic species and angular momenta. With this prescription, possible spurious charge transfers due to different shifts in the pseudo-atomic eigenvalues are minimized, which is particularly important for small basis sets. The energy shift assumes the role of a variational parameter; the smaller E_{shift} is, the more the basis functions approach the unconfined atomic orbitals, thus lowering the total energy of the system. The optimal value of E_{shift} depends on the characteristics of the system. For simple and noble metals, where the electrons are quite spread, a large value of E_{shift}, typically of the order of 200 meV, can be used quite safely. Molecular and semiconducting systems in general require a lower shift, which can be as small as 20 meV. Let us recall that the smaller E_{shift} is, the larger is the computational cost.

Recently, a variant of this scheme was introduced, where the strict localization condition is replaced by a soft localization condition, which preserves the strict localization beyond a certain radius, but the orbitals approach this radius smoothly (Junquera et al., 2001). In addition, an automatic procedure for optimizing the localization radii based on a simplex minimization of the total energy has been proposed (Anglada et al., 2002). This procedure is supposed to provide basis functions that can be transferred more safely from one environment to another. For example, the oxygen basis set optimized for an oxide like MgO can then be used for other oxides like perovskites ($BaTiO_3$). Optimization of the basis set can result in significantly smaller localization radii, thus improving the computational performance.

8.4.3 Hydrogenic and Λ-functions

In the pursuit of analytic basis functions, the first step is to use the solutions of Eq. (8.45) when $v_{\text{eff}}(r) = -Z/r$, i.e. hydrogenoid atoms. This leads to radial wave functions composed by a decaying exponential multiplied by a Laguerre polynomial. A basis set of *hydrogenoid functions*, although orthogonal, is not particularly good because it is not complete (it requires unbound continuum states) and the orbitals quickly become too spread. An alternative is to fix the exponent, which leads to the *Laguerre* or Λ-*functions* (Trivedi and Steinborn, 1982):

$$\Lambda_{nl}(r, \zeta, n) = N_{nl}(\zeta) x^l L_{n-l-1}^{2l+2}(x) e^{-x/2}, \tag{8.47}$$

with $x = 2\zeta r$, L an associated Laguerre polynomial, and $N(\zeta)$ a normalization constant. In this case orthogonality is a consequence of using a single exponent for *all* the AOs. An advantage of these basis functions is that there is an expansion theorem for the Λ-functions centered on one site in terms of those centered on a different site in terms of a structure matrix, similar to the case of Hankel and Bessel functions (Lin and Harris, 1993). This simplifies the calculation of Hamiltonian and overlap matrix elements between different sites. Being orthogonal, the on-site overlap matrix elements are strictly zero.

However, as mentioned before, while a particular exponent may represent well a $2s$ or a $2p$ orbital, it cannot simultaneously represent reasonably well a $1s$ orbital that decays much faster and peaks much closer to the nucleus. A possible solution is to use different exponents for different orbitals, at the price of destroying orthogonality between basis functions with different exponent. At this point there is no particular advantage in retaining the complicated form of the Laguerre polynomials.

8.4.4 Slater-type orbitals (STO)

In order to simplify the structure of the radial wave function we notice that, in an orbital corresponding to the principal quantum number n, the leading power of the polynomial that accompanies the exponential is r^{n-1}. Therefore, basis functions that retain this behavior are likely to represent the orbitals quite well in the large r region. Following this reasoning, in 1930 Slater proposed a basis set that today is still alive and very healthy (Slater, 1930a, 1932). The basis functions receive the name of *Slater-type orbitals*, or STO, and are written as

$$\chi_{nl}^{\text{STO}}(r, \zeta, n) = \frac{(2\zeta)^{3/2}}{\sqrt{(2n+1)!}} (2\zeta r)^{n-1} e^{-\zeta r}, \tag{8.48}$$

where the coefficient in front ensures their normalization. For a fixed value of ζ, the STOs generated by systematically increasing the value of n form a complete but non-orthogonal basis set. As in the case of Laguerre functions, fixed-exponent STOs have very poor convergence properties. The problem is that the exponential factor decays always in the same way, and thus STOs cannot efficiently represent orbitals with different principal quantum numbers, e.g. $1s$ and $2s$. Extra flexibility can be added by allowing for variable exponents ζ_{nl}. This permits us to tune the location of the peak and the decay of the basis functions. At this stage, not only is orthogonality lost, but also the completeness of the basis set is compromised. It is easy to construct a sequence of exponents that does not lead to a complete basis set, but it has been proven that completeness is ensured if the sequence has a finite accumulation point. For example, the sequence $\zeta_n = n^{-1}$, which corresponds to hydrogenoid atoms, does not verify the above condition, while the alternative $\zeta_n = 1 + n^{-1}$ does.

The usual approach to the generation of STO basis sets is to choose a sequence of basis functions with variable exponents for each atomic orbital. A single STO may be sufficient for some AOs, but in general it is not. One of the main reasons is that STOs are nodeless functions, so that, except for the lowest-lying states of the various angular momenta (1s, 2p, 3d, etc.), linear combinations of several STOs are required to reproduce the nodal structure. In practice, such linear combinations are also needed to represent the lowest-lying, nodeless AOs accurately. Therefore, an STO basis set consists of a set of exponents for the 1s state, another for the 2s, and so on for all the occupied states. The exponents are optimized to represent accurately the corresponding one-electron atomic wave functions at the required level of theory, whether Hartree–Fock, DFT-LDA, or beyond.

8.4.5 Gaussian-type orbitals (GTO)

Unfortunately, the analytic expressions involved in the computation of three- and four-center Coulomb integrals using Slater-type orbitals are too complicated, although some progress in that direction has been made recently (Watson *et al.*, 2003). Therefore, one has to resort to numerical integration techniques that are computationally more costly than simple analytic expressions. At this stage, one could use the true atomic orbitals given in numerical form directly, and the computational cost would be the same, or even smaller, because fewer basis functions would be required.

An alternative solution was proposed by Boys (1950), who introduced the widely popular Gaussian-type orbitals (GTO) to replace the exponentials. Gaussians have the enormous advantage that all the integrals can be performed analytically, even four-center terms. Analogously to STOs, the exponents of the Gaussian functions are variationally optimized to match the corresponding atomic wave functions. The natural representation for the GTO is in polar coordinates:

$$\phi_{nlm}(\mathbf{r}) = \frac{2(2\alpha_{nl})^{3/4}}{\pi^{1/4}} \sqrt{\frac{2^l}{(2l+1)!!}} \left(\sqrt{2\alpha_{nl}}\,r\right)^l e^{-\alpha_{nl}r^2} Y_{lm}(\theta, \varphi), \qquad (8.49)$$

but a more convenient form for the computation of Hamiltonian matrix elements are the *Cartesian* GTOs, given by

$$\phi_{ijk}(\mathbf{r}) = \chi_i(x)\chi_j(y)\chi_k(z), \qquad (8.50)$$

where

$$\chi_i(x) = \left(\frac{2\alpha}{\pi}\right)^{1/4} \sqrt{\frac{(4\alpha)^i}{(2i-1)!!}} \, x^i e^{-\alpha x^2}, \qquad (8.51)$$

and analogous expressions for $\chi_j(y)$ and $\chi_k(z)$. The advantage of Cartesian GTOs is that molecular integrals factorize into the three Cartesian variables.

Polar (or spherical) Gaussians have a well-defined angular momentum l given by the spherical harmonic function and the factor r^l in (8.49). For a given value of l, the allowed values of m are $m = 0, \pm 1, \ldots, \pm l$. These $2l+1$ functions, e.g. the three p functions or the five d functions, can be constructed with a subset of Cartesian Gaussians. Since the product of the Cartesian coordinates should appear to the same power l of the radial coordinate in the polar Gaussians, then it should be $i + j + k = l$. There is a difference, however, which is that the number of Cartesian Gaussians for a given angular momentum l is $n_l = (l+2)(l+1)/2$, which for $l \geq 2$ exceeds the number of independent polar components. For instance, for $l = 2$, $n_l = 6$ as opposed to the five spherical harmonics $Y_{2\pm2}$, $Y_{2\pm1}$, and Y_{20}. Therefore, for $l \geq 2$, when Cartesian Gaussians are transformed into polar ones, there are some additional combinations of different symmetry. In the case of d orbitals the only such function is $(x^2 + y^2 + z^2)$, which has s character, but for f orbitals there are three additional p-like combinations. These may cause problems of linear dependence, and sometimes are eliminated from the basis set.

An extremely important property of Gaussian functions is that the product of two Gaussians with different centers can be written also as a Gaussian with different center and exponent. This is called the *Gaussian product formula* or *rule*. We first define the unnormalized *primitive* Cartesian Gaussians of exponent α and center $\mathbf{R}_I$ as

$$G_{ijk}(\mathbf{r}, \alpha, \mathbf{R}_I) = (x - x_I)^i (y - y_I)^j (z - z_I)^k e^{-\alpha(\mathbf{r} - \mathbf{R}_I)^2}. \tag{8.52}$$

We next consider the simplest version of the product of two s-type primitive Gaussians ($i = j = k = 0$) of exponents α and β centered at $\mathbf{R}_I$ and $\mathbf{R}_J$, respectively. In that case the product formula says:

$$G_s(\mathbf{r}, \alpha, \mathbf{R}_I) G_s(\mathbf{r}, \beta, \mathbf{R}_J) = e^{-\gamma(\mathbf{R}_I - \mathbf{R}_J)^2} G_s(\mathbf{r}, \kappa, \mathbf{R}_K), \tag{8.53}$$

with

$$\kappa = \alpha + \beta, \qquad \gamma = \frac{\alpha\beta}{\alpha + \beta}, \qquad \mathbf{R}_K = \frac{\alpha \mathbf{R}_I + \beta \mathbf{R}_J}{\alpha + \beta}. \tag{8.54}$$

In the general case of two primitive Cartesian Gaussians of powers (i_1, j_1, k_1) and (i_2, j_2, k_2), respectively, we have

$$G_{i_1 j_1 k_1}(\mathbf{r}, \alpha, \mathbf{R}_I) G_{i_2 j_2 k_2}(\mathbf{r}, \beta, \mathbf{R}_J) = e^{-\gamma(\mathbf{R}_I - \mathbf{R}_J)^2} e^{-\kappa(\mathbf{r} - \mathbf{R}_K)^2} (x - x_I)^{i_1} (x - x_J)^{i_2}$$
$$\times (y - y_I)^{j_1} (y - y_J)^{j_2} (z - z_I)^{k_1} (z - z_J)^{k_2}. \tag{8.55}$$

It is more convenient to transform the Cartesian factors in such a way that the product can be expressed in terms of Cartesian Gaussians centered at $\mathbf{R}_K$. This is achieved by using the binomial expansion

$$(x - x_I)^{i_1} = [(x - x_K) + (x_K - x_I)]^{i_1} = \sum_{i=0}^{i_1} \binom{i_1}{i} (x - x_K)^i (x_K - x_I)^{i_1 - i}, \tag{8.56}$$

so that the product of pairs of Cartesian factors is

$$(x-x_I)^{i_1}(x-x_J)^{i_2} = \sum_{i=0}^{i_1+i_2} (x-x_K)^i C_{ix}^{i_1,i_2}(IJK), \quad (8.57)$$

with

$$C_{ix}^{i_1,i_2}(IJK) = \sum_{l_1=0}^{i_1} \sum_{l_2=i-l_1}^{i_2} \binom{i_1}{l_1}\binom{i_2}{l_2}(x_K-x_I)^{i_1-l_1}(x_K-x_J)^{i_2-l_2}. \quad (8.58)$$

Equivalent expressions apply to the other Cartesian coordinates, y and z.

In summary, the product of any two Cartesian Gaussians can be written as a linear combination of Cartesian Gaussians with a different center:

$$G_{i_1 j_1 k_1}(\mathbf{r}, \alpha, \mathbf{R}_I) G_{i_2 j_2 k_2}(\mathbf{r}, \beta, \mathbf{R}_J) = e^{-\gamma(\mathbf{R}_I-\mathbf{R}_J)^2} \sum_{i=0}^{i_1}\sum_{j=0}^{j_1}\sum_{k=0}^{k_1} C_{ix}^{i_1,i_2}(IJK)$$

$$\times C_{jy}^{j_1,j_2}(IJK) C_{kz}^{k_1,k_2}(IJK)\, G_{ijk}(\mathbf{r}, \kappa, \mathbf{R}_K). \quad (8.59)$$

8.4.6 Size of the basis set

If a single basis function per atomic orbital is used, this is called a *minimal* or *single-ζ* (SZ) basis set. Minimal basis sets can be remarkably accurate for isolated atoms, but they are not flexible enough to describe molecular systems where electrons participate in chemical bonding and become polarized. In that case, additional sets of basis functions of the same type as those already present are required to improve accuracy. If only one additional set per orbital is considered, it becomes a *double-ζ* basis set (DZ). The exponents of such sets are usually variationally optimized, e.g. to minimize the Hartree–Fock energy. Still more flexibility can be obtained by adding more basis functions of the same type, thus leading to *triple-ζ* (TZ), *quadruple-ζ*, and, in general, *n-tuple-ζ* basis sets. Since core states are quite insensitive to the environment, multiple basis functions for core AOs are very rare. In general the *splitting* of the basis set is applied only to the valence orbitals. These are called *split valence basis sets*, and sometimes are denoted by including a prefix V (for valence) in the name, e.g. VDZ.

Yet, adding more functions of the same symmetry is not enough because these cannot accurately represent the deformation of the AOs due to the presence of neighboring atoms. This requires functions of different angular momenta, and perturbation theory indicates that the most important contributions will arise from angular momenta $l \pm 1$. Therefore, to account for *polarization* effects, the solution is to include a set of basis functions of the first angular momentum that is unoccupied in the isolated atom. Sometimes the first two or three unoccupied

8.4 Atom-centered basis sets

angular momenta are added as polarization functions. Hence, when a DZ basis set is supplemented with polarization functions, it becomes a *double-ζ plus polarization* basis, denoted by DZP, or DZVP. In a similar way the TZP basis and, in general, nZP basis sets are constructed. If two sets of polarization functions per angular momentum are included, then the basis set is denoted by nZ2P, e.g. the TZ2P basis set consists of three basis functions for the occupied states, and two basis functions for polarization orbitals. Polarized basis sets are considered good general purpose basis sets for describing chemical bonding in neutral molecules. A cost-effective and reasonably accurate basis set for semi-quantitative calculations is the DZP. The presence of polarization functions in a basis set can also be indicated by an asterisk. The rule of thumb for improving the quality of basis sets says to *first double and then polarize*.

Negatively charged molecules (anions), due to the enhanced Coulomb repulsion between the electrons, usually tend to produce quite spread AOs. To represent these AOs, standard basis functions are not sufficient, and additional basis functions with small exponents have to be included. These are called *diffuse functions*, and are indicated by a plus sign "+". One + if it is a single diffuse function, ++ if it is two, and so on. These functions are also useful to describe non-bonding interactions.

Most basis sets have been developed by optimizing the exponents with respect to Hartree–Fock, numerically exact atomic calculations performed using radial grids. These basis functions, however, may not be good enough when used for MP2 or CI calculations. Dunning has developed basis sets from correlated calculations (Dunning, 1989), which are called *correlation consistent*, and are indicated by a prefix "cc" before the description of the basis set. An example is the cc-pVDZ basis, which corresponds to the DZP basis in Hartree–Fock. If diffuse functions are also included, then the prefix "aug" is added, e.g. aug-cc-pVTZ corresponds to TZP+.

The optimization of exponents becomes problematic when large basis sets are considered, because the energy function is quite flat and presents many local minima. The best solution is to reduce the number of fitting parameters, and this can be done, e.g., by noticing that in variationally optimized basis sets, the ratio between successive exponents is approximately constant. Taking this ratio as effectively constant, i.e. $\zeta_n = \alpha \beta^n$, reduces notably the number of parameters and facilitates the optimization procedure. Basis sets constructed in this way are called *even-tempered*, and the coefficient β depends on the chemical species and on the size of the basis set. A possible variant is to distribute the exponents in such a way as to have the valence region represented with more flexibility than the core region close to the nucleus. These are called *well-tempered* basis sets, and use a similar but modified formula for the exponents (see Helgaker *et al.*, 2000, Chapter 8).

8.4.7 Contracted Gaussians (CGTO)

Gaussian functions do not behave properly at the origin. In fact, they are solutions of the harmonic oscillator, and not of the atomic problem. In particular, they are too smooth at the origin, where they arrive with zero derivative. This is in marked contrast with the behavior of atomic orbitals, which are better represented by exponentials with a *cusp* at the origin (non-zero derivative). However, the simplicity of the algebra of Gaussians makes them the favorite basis set.

The problem is how to recover the cusp condition using GTO. One possible way is to represent the STO as a *fixed* linear combination of GTO (Hehre *et al.*, 1969),

$$\chi_{nl}^{\text{STO}}(r, \zeta) = \sum_{i=1}^{k} a_i G_{nl}(r, \zeta^2 \alpha_i), \tag{8.60}$$

where $G_{nl}(r)$ is the radial part of the polar Gaussians defined in (8.49), and the exponents α_i are obtained by a least squares fitting procedure. Basis functions of this type are called STO-kG, with k the number of GTOs used to represent the STO. A typical example is the STO-3G minimal basis set, where one STO made of three GTOs is assigned to each occupied AO.

Clearly, this approach is limited by the accuracy of the STO. A more flexible road is to construct fixed linear combinations of *primitive* Gaussians (PGTOs) where the coefficients and the exponents are optimized to reproduce the atomic ground state orbitals at the desired level of theory, e.g. Hartree–Fock. These basis functions, written as

$$G^{\text{CGTO}}(r, \{\alpha_i\}) = \sum_{i=1}^{k} a_i G^{\text{PGTO}}(r, \alpha_i), \tag{8.61}$$

are known as *contracted* GTOs, and a_i are the *contraction coefficients*. Since these coefficients are fixed, contracted basis functions are most useful to represent core orbitals. Contracted GTOs are also used to improve the shape of the valence basis functions, but this is not as important as for the core orbitals because the valence MOs usually differ quite significantly from the AOs of the neutral atom. In that case, better than optimizing the CGTO, it is more important to include additional basis functions of the same angular momentum (e.g. double-ζ). The number of contracted functions typically varies between one and ten. A big advantage of CGTOs is that they count as a single basis function, and thus the dimension of the Hamiltonian matrix is enormously reduced with respect to uncontracted GTOs.

The notation used for contraction schemes is as follows: $(9s5p1d)/[4s2p1d]$ indicates that nine s-type, five p-type, and one d-type primitive GTOs are used to construct four s-type, two p-type, and one d-type basis functions. In general, the innermost primitives that represent the core states are contracted, while the more diffuse ones are left uncontracted. In the case above, corresponding to the Dunning–Huzinaga basis set for first-row atoms (Huzinaga, 1965; Dunning, 1989), the first six s-type primitives are contracted to represent a $1s$ core orbital, while the other three are uncontracted. Similarly, out of the five p-type primitives, four are used for a first CGTO and the remaining one is an uncontracted GTO. The above is then a DZP basis set.

More efficient basis sets of the split valence type, where the core shells are represented by a single CGTO and multiple CGTOs are also considered for the valence orbitals, are the most widely used for general purposes. In the standard notation there is a first digit that indicates the number of primitives used for the single core state, and then the contraction scheme for the valence CGTO is specified. The number of valence digits indicates the size of the basis set, whether DZ, TZ, or other. Thus, the 3-21G basis is made of three contracted PGTOs for the $1s$ core state, and then two CGTOs for the $2s$ and $2p$ states, the first one made of two contracted PGTOs, and the second being a single uncontracted primitive (Binkley et al., 1980). A more accurate and widely used DZ basis set is the 6-31G, which supplemented with a single set of $3d$ polarization primitives becomes the famous 6-31G* basis set (Hehre et al., 1972).

The most popular basis sets have been optimized for Hartree–Fock calculations, and it is not necessarily good to use them for DFT-LDA or GGA calculations. In fact, core states are rather different in these two approaches, basically due to the lack of cancelation of the Coulomb self-interaction in the usual DFT-based approaches. Within split valence schemes, the inconsistency of the basis set is worse for the core states that are represented by a single CGTO. For these orbitals there is no flexibility at all, and the results show this unless the basis sets are re-optimized to fit the appropriate orbitals. This has been done for the case of GGA density functional calculations by Godbout et al. (1992).

8.4.8 Basis set superposition error (BSSE)

To calculate the interaction energy between two fragments of a system, e.g. the dissociation energy, the standard approach is to calculate the energy of the whole system and subtract the energies of each one of the fragments. Normally, the energy of the fragments is calculated with the basis functions associated with the atoms *only* in that fragment. However, the energy of the whole system uses the basis functions of *both* fragments. Therefore, there is a systematic error in the interaction energies because the energy of each one of the fragments calculated using the full basis, which is consistent with that of the whole system, is lower than the one calculated with only its own basis set. The result is that interaction energies are overestimated.

A straightforward approximate way of correcting this problem is to calculate the energy of the fragments using the augmented basis, i.e. its own basis functions together with those of the other fragment (ghost orbitals) (Boys and Bernardi, 1970). This is known as *counterpoise correction*. In the limit of a complete basis set the counterpoise correction vanishes, but for truncated basis sets the effect may be quite important, especially for weakly bound systems such as van der Waals complexes and, to a lesser extent, hydrogen-bonded systems.

8.5 Mixed basis sets

An attractive alternative to floating and atom-centered basis sets is to use a combination of both, i.e. a basis set composed of a minimal set of atom-centered basis functions supplemented with plane waves or some other alternative. This is called a *mixed basis set*, and it has the advantages of both worlds. It can be very efficient because most of the features of the MOs are already captured by the small number of atom-centered basis functions, while the remaining details are taken care of by the plane waves. The PW cutoff, however, does not need to be very high because the behavior of the MO close to the atomic nuclei, which is usually the reason why a large number of PWs is needed, is already well covered by the atom-centered basis functions. On the other hand, the completeness issue that is important for atom-centered basis sets is sorted out by the PW expansion, which by construction is complete.

Mixed basis sets also experience the disadvantages of both worlds, in particular the double effort in programming, the inefficient representation of vacuum regions and the artificial imposition of periodic boundary conditions for molecular systems, and also possible problems with linear dependencies of the basis functions if the PW cutoff is increased too much. A mixed basis set of Gaussians and plane waves was implemented long ago by Louie *et al.* (1979), and it was very important due to the computational limitations of the time. More recently, this approach has been revived by Ho *et al.* (1992), and proved to be quite efficient and unbiased.

8.6 Augmented basis sets

As mentioned at the beginning of this chapter, the generalization of atomic structure methods to molecular or condensed phases is feasible. In principle, this problem could be faced by dividing up the system into space-filling non-overlapping elementary volumes, e.g. unit cells, and solving the electronic problem in each volume, imposing the boundary condition that the one-electron wave functions and their derivatives are continuous at the interface between two such volumes.

8.6 Augmented basis sets

The problem is that spheres cannot accomplish the task of filling the whole space without leaving voids; the geometrical figures that fulfill this construction are, instead, rather involved polyhedra (see, e.g., Kittel, 1996). Imposing boundary conditions on a polyhedral unit cell is a complicated operation. The first approach to this problem was proposed by Bloch (1928), and it constitutes the basis for the empirical tight-binding method (see Section 10.1). It basically consists of representing the electronic states as a superposition of atomic wave functions centered on different atomic sites, and treating the interaction between them via perturbation theory (see also Slater, 1934). An improvement over Bloch's approach was the *cellular* method, based on imposing the appropriate boundary conditions for the wave function at the boundaries of the unit cell (Wigner and Seitz, 1933). The wave function is expanded in atomic orbitals, and the expansion coefficients are determined by imposing the boundary conditions at an arbitrary number of boundary points. The practical approach suggested by Wigner and Seitz was to do this at high-symmetry points such as the mid-points of the faces of the cells, where the slope of the wave function has to be zero by symmetry. This is somehow equivalent to approximating the polyhedron with a sphere (the Wigner–Seitz sphere).

A possible strategy to further improve over the cellular method is to partition the system into atomic spheres and interstitial regions (I), as shown in Fig. 8.1. The spherical regions are usually called *muffin-tin* (MT) spheres, and their radius is the muffin-tin radius (R_{MT}). By introducing the interstitial regions explicitly, one can take into account the complicated polyhedral shape of the unit cell, but still maintain the desirable feature of solving the spherically symmetric atomic problem inside MT spheres. The atomic-like solutions have to be matched in

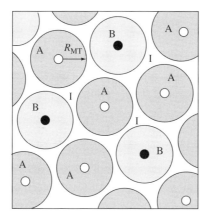

Fig. 8.1 Schematic representation of the partitioning of a system into atomic spheres (shaded areas) of radii R_{MT} and interstitial regions (I). Atoms of two different types, A and B, are shown.

amplitude and derivative to a set of interstitial basis functions at the surface of the spheres. The rationale for such an approach is that inside the MT spheres the potential is spherically symmetric to a very good extent, and then the basis functions in this region can be obtained by integrating the radial Schrödinger equation outwards up to R_{MT}. This strategy is called *augmentation*; the basis functions in the interstitial regions are augmented with atomic-like functions inside the MT spheres.

A general augmented basis set looks like

$$\phi_\alpha(\mathbf{r},\varepsilon) = \begin{cases} \sum_{Ilm} C^I_{\alpha lm}(\varepsilon) \chi^I_l(|\mathbf{r}_I|,\varepsilon) Y_{lm}(\hat{\mathbf{r}}_I), & \text{for } |\mathbf{r}_I| < R_{\mathrm{MT}} \\ \phi_\alpha(\mathbf{r},\varepsilon), & \text{otherwise,} \end{cases} \quad (8.62)$$

where $\mathbf{r}_I = \mathbf{r} - \mathbf{R}_I$, and the functions $\chi^I_l(r,\varepsilon)$ are the regular solutions of the radial Schrödinger equation for atom I at energy ε, i.e. the solutions of

$$\left\{ -\frac{\hbar^2}{2m} \frac{d^2}{dr^2} + \frac{l(l+1)}{2r^2} + v_I(r) \right\} r\chi^I_l(r,\varepsilon) = \varepsilon r \chi^I_l(r,\varepsilon), \quad (8.63)$$

subject to the appropriate boundary conditions on the surface of the spheres. The potential $v_I(r)$ is the sphericalized potential inside the Ith MT sphere, i.e. the Kohn–Sham potential constructed with the electronic density, which is also assumed to be spherically symmetric inside the MT spheres. As in the case of atoms, due to the spherical symmetry this equation can be integrated numerically outwards from the origin and up to R_{MT} for any energy value, thus generating a radial atomic-like wave function for each angular momentum l. The coefficients $C^I_{\alpha lm}(\varepsilon)$ are determined by enforcing the matching conditions on the surface of the spheres and, in general, they depend on energy. Notice that this type of basis function is extended to all space, and depends on the location, type, and radius of *all* the MT spheres. The index α labels the different basis functions.

For a single interstitial basis function this problem can be faced in the same way as the atomic problem, i.e. by searching for the eigenvalues that ensure the continuity and differentiability of the wave functions on the surface of *all* the MT spheres. The lowest eigenvalues will resemble closely the atomic core states, while higher eigenvalues should depart from the atomic valence states in order to describe bonding. While some approaches adopt, indeed, such a minimal basis set, for the sake of generality we shall assume here that the number of basis functions in the interstitial regions is not necessarily one per state. Notice that, since the basis functions are energy-dependent, the secular equation for the target system is not a standard eigenvalue equation, but a set of highly non-linear homogeneous equations in ε, of the form $\mathbf{M}(\varepsilon)\mathbf{c} = 0$, where c_{ij} are the expansion coefficients of the wave functions, $\varphi_i(\mathbf{r}) = \sum_j c_{ij} \phi_j(\mathbf{r})$. The allowed energies are those for which a set of coefficients c_{ij} exist, such that the wave functions are continuous and differentiable everywhere.

Any further advance requires us to specify the basis functions in the interstitial regions, $\bar{\phi}_\alpha(\mathbf{r}, \varepsilon)$. Since in those regions the potential is rather flat, the basis functions can be taken to be solutions of the Schrödinger equation for electrons in a constant potential. There are mainly two strategies in this respect: floating basis functions such as plane waves, and atom-centered basis functions. The first class is unbiased, and can be made increasingly accurate (and computationally heavier) by increasing the number of basis functions. The *augmented plane wave* method (APW), first proposed by Slater (1937), is the paradigm of this class of all-electron methods. The second class is more efficient; it requires only one or a few selected basis functions per angular momentum and per atom, but it is more biased and in general is mathematically more involved than the APW method. The most finished example of a method of this class is the *linear combination of muffin-tin orbitals* (LCMTO) proposed by Andersen (1971).

8.6.1 Augmented plane waves

In the APW method the basis functions are

$$\phi_\mathbf{G}^\mathbf{k}(\mathbf{r}) = \begin{cases} \sum_{llm} A_{lm}^{\mathbf{k},I}(\mathbf{G}, \varepsilon) \chi_l^I(|\mathbf{r}_I|, \varepsilon) Y_{lm}(\hat{\mathbf{r}}_I), & \text{for } |\mathbf{r}_I| < R_{\text{MT}}^I, \\ e^{i(\mathbf{k}+\mathbf{G})\cdot\mathbf{r}}/\sqrt{\Omega}, & \text{otherwise,} \end{cases} \quad (8.64)$$

where $\mathbf{G}$ are the reciprocal lattice vectors of the plane wave expansion, $\mathbf{k}$ are wave vectors within the first Brillouin zone, Ω is the volume of the cell, $Y_{lm}(\hat{\mathbf{r}}_I)$ are spherical harmonic functions referred to the center of the sphere located at $\mathbf{R}_I$, $\chi_l^I(r, \varepsilon)$ are the regular solutions of the radial Schrödinger equation for energy ε, and $A_{lm}^{\mathbf{k},I}(\mathbf{G}, \varepsilon)$ are the expansion coefficients that are determined by the matching and normalization conditions. To impose the matching conditions on the surface of the spheres, the complex exponentials have to be expanded in atom-centered spherical harmonics. For a sphere centered at $\mathbf{R}_I$ the PW at the surface of the sphere is given by

$$e^{i(\mathbf{k}+\mathbf{G})\cdot(\mathbf{r}_s+\mathbf{R}_I)} = e^{i(\mathbf{k}+\mathbf{G})\cdot\mathbf{R}_I} \sum_{l=0}^{\infty} (2l+1) i^l P_l(\cos\theta) j_l(|\mathbf{k}+\mathbf{G}|r_s), \quad (8.65)$$

where the z-axis has been chosen in the direction of vector $\mathbf{k}+\mathbf{G}$ so that the spherical harmonic simplifies to a Legendre polynomial $P_l(\cos\theta)$. The variables r_s and θ are spherical coordinates with origin at the center of the sphere centered on atom I. Therefore, using that $r_s = R_{\text{MT}}$ on the surface of the sphere, the continuity condition leads to

$$A_{lm}^{\mathbf{k},I}(\mathbf{G}, \varepsilon) = (2l+1) i^l e^{i(\mathbf{k}+\mathbf{G})\cdot\mathbf{R}_I} \frac{j_l(|\mathbf{k}+\mathbf{G}|R_{\text{MT}}^I)}{\chi_l^I(R_{\text{MT}}^I, \varepsilon)}. \quad (8.66)$$

In the first APW method (Slater, 1937) this was the only matching condition enforced, and the idea was to achieve differentiability by gradually smoothing the discontinuity in the derivatives of the wave functions of the solid by varying the energy ε, as is usually done in atomic calculations. This approach proved computationally too demanding for those times, thus leading Slater and Saffren to propose an elegant alternative (Slater, 1953; Saffren and Slater, 1953). In this second attempt, continuity was imposed on the logarithmic derivative of the basis functions. This, together with the normalization condition, ensures the continuity of basis functions and derivatives. Incidentally, these same conditions are imposed when constructing pseudopotentials in order to ensure that the scattering phase shifts of the pseudo-cores coincide with those of the full cores (see Chapter 7). Similar approaches, where a few atomic basis functions are carefully chosen for the matching, were proposed later on by Marcus (1967) and Koelling (1970). In particular, Marcus' paper contains a suggestion of the linearization approach later developed by Andersen (1975).

Interestingly, this second method by Slater proved harder than the original 1937 APW method, and it was eventually this latter that developed into practical computational schemes in the group at MIT gathered around Slater himself (Saffren, 1960). A detailed description of the method, which was widely used until the development of linear methods, can be found in the book by Loucks (1967). This method requires an expansion in angular momenta up to some $l_{\max}$ of the order of 10 or larger, and then finding the zeroes of the determinant of the secular equation as a function of ε. A modern version of the APW scheme (Soler and Williams, 1990) is based on directly augmenting the electronic wave functions instead of the individual plane waves. In addition, this method allows for the large l plane wave components to penetrate the spheres, at variance with other methods that make the large l components zero at the surface of the spheres. This allows for a significant reduction of the cutoff in the angular momentum expansion to values of $l_{\max} = 2$ or 3.

8.6.2 Augmented atom-centered basis sets

The use of spherically symmetric basis functions centered on the atomic positions also for the interstitial regions has a number of advantages over floating basis sets. For a start, the matching at the surface of the MT spheres is notably simplified because the two sets of functions have the same angular dependence. Therefore, augmentation can be imposed on every angular momentum individually, by requesting the continuity of the radial logarithmic derivative. Second, this construction leads naturally to localized basis functions that can be identified with particular atoms, while augmented PWs have similar weights in all the atoms in the system (see Williams *et al.*, 1979, Fig. 1). The idea of such a basis set was first suggested by Andersen (1971) (see also Andersen and Kasowski, 1971), and further developed by several authors, largely in Andersen's group, but also in other locations. By now there are several reviews on this class of methods (see e.g. Andersen, 1984; Andersen *et al.*, 1992, 2000).

The development of one particular type of atom-centered basis function, the muffin-tin orbitals (MTO), is closely linked to the KKR method (Korringa, 1947; Kohn and Rostoker, 1954), where the electronic structure problem is formulated in terms of Green's functions within the muffin-tin approximation (MTA). In the MTA, the potential is taken to be spherically symmetric inside a sphere inscribed into the Wigner–Seitz (WS) cell, and outside it is approximated by a constant, with the zero of the potential shifted to make this constant equal to zero. An account of the KKR method is given in Section 9.1, but for our purposes here the important fact is that the KKR wave functions for the solid are expressed as a one-center expansion in partial waves, i.e. an expansion in spherical harmonics around a single point in the cell. Inside the WS sphere the wave function is simply expressed as a linear combination of the solutions of the atomic problem subject to appropriate boundary conditions. Outside the sphere the potential is not spherically symmetric. A partial wave expansion can still be used in that region, although other alternatives such as plane waves have also been proposed (Ham and Segall, 1961). Under these circumstances, the KKR method reduces the electronic structure problem to the determination of the roots of the determinant of a secular equation, where the matrix elements contain two separate contributions: a set of energy-dependent *structure constants*, and the logarithmic derivatives of the radial atomic orbitals. Due to the one-center character of the expansion, the dimension of the KKR matrix is given by the maximum angular momentum included in the partial wave expansion.

Muffin-tin orbitals were introduced by Andersen (1971) with the aim of improving the understanding of the KKR method, and to analyze the radius of convergence of the one-center partial-wave expansion (Andersen *et al.*, 1992). The wave functions of the solid are expressed as a multi-center expansion around each of the atomic spheres, much in the spirit of atom-centered basis sets described in Section 8.4. Outside the spheres the basis orbitals are continued with atom-centered functions that will be described below. A crucial result is that the one-center KKR wave functions can be equivalently written as a multi-center linear combination of MTOs, with the same expansion coefficients. However, MTOs have the advantage that the partial-wave expansion can be cut off at smaller values of angular momenta (Andersen and Kasowski, 1971). In addition, MTOs lead to a formulation in the usual form of a Schrödinger equation, which is more convenient for introducing non-spherical corrections beyond the MTA. Since they were first introduced, methods based on MTOs have departed from the KKR philosophy, following their own route.

The MTO for a single sphere is constructed in the following way: outside the sphere it is the solution of the Helmholtz equation

$$\left[\frac{\hbar^2}{2m}\nabla^2 + \kappa^2\right]\phi_{lm}(\mathbf{r}, \kappa) = 0 \tag{8.67}$$

in spherical coordinates. The Helmholtz equation is just Schrödinger's equation for a constant potential V_{MT}, which is a reasonable approximation for the interstitial regions. The coefficient κ^2 is the kinetic energy of the spherical wave, and is related to the energy by

$$\kappa^2 = \varepsilon - V_{MT}. \tag{8.68}$$

The solutions of this equation are of the form

$$\phi_{lm}(\mathbf{r}, \kappa) = \tilde{\chi}_l(r, \kappa) Y_{lm}(\hat{\mathbf{r}}), \tag{8.69}$$

where $\tilde{\chi}_l(r, \kappa)$ are spherical Bessel $j_l(\kappa r)$ or Neumann $n_l(\kappa r)$ functions. If κ^2 is negative, the latter becomes a Hankel function $h_l(\kappa r)$, which decreases exponentially like $h_l(\kappa r) \sim e^{-\kappa r}$ at long distances. The function $h_l(\kappa r)$ is called the *envelope* function, and it is regular at infinity, while $j_l(\kappa r)$ is regular at the origin. Therefore, the only acceptable solutions outside and inside the spheres are Hankel and Bessel functions, respectively. Notice that the functions $R_l(r)$ and $I_l(r)$ in (8.46), introduced within the context of atomic orbitals are just Bessel and Hankel functions for $\kappa = 0$, corresponding to $V_{MT} = 0$.

As in the case of plane waves, a spherical Hankel function $h_l(\kappa r)$ cannot be matched in amplitude and slope to a regular solution of the radial Schrödinger equation $\chi_l(r, \varepsilon)$ at an arbitrary energy. Only when κ corresponds to an eigenvalue (or a resonance) can this matching be accomplished (Heine, 1967). From the scattering point of view, the solutions of Schrödinger's equation for a single MT potential are the partial waves

$$\phi_{\kappa lm}^{\text{single}}(\mathbf{r}, \varepsilon) = \begin{cases} \chi_l(r, \varepsilon) Y_{lm}(\hat{\mathbf{r}}), & \text{for } r \leq R_{MT}, \\ [h_l(\kappa r) - j_l(\kappa r) \cot \eta_l(\varepsilon)] Y_{lm}(\hat{\mathbf{r}}), & \text{for } r > R_{MT}. \end{cases} \tag{8.70}$$

This expression can be interpreted as an incoming partial wave of angular momentum lm propagating into the sphere, where it is scattered by the potential. This produces an outgoing partial wave which is phase-shifted with respect to the incoming wave by the quantity $\eta_l(\varepsilon)$. The phase shift is determined by the potential inside the sphere, through the continuity of the logarithmic derivative at R_{MT}:

$$\cot \eta_l(\varepsilon) = \frac{h_l(\kappa R_{MT})}{j_l(\kappa R_{MT})}\left(\frac{D_\chi^l(\varepsilon) - D_h^l(\kappa)}{D_\chi^l(\varepsilon) - D_j^l(\kappa)}\right), \tag{8.71}$$

where D^l_χ, D^l_h, and D^l_j are the logarithmic derivatives of χ_l, h_l, and j_l at the muffin-tin radius, e.g.

$$D^l_\chi(\varepsilon) = \left[\frac{\partial \ln |\chi_l(r,\varepsilon)|}{\partial \ln r}\right]_{R_{\mathrm{MT}}} = \left[\frac{r\chi'_l(r,\varepsilon)}{\chi_l(r,\varepsilon)}\right]_{R_{\mathrm{MT}}}. \tag{8.72}$$

These functions are not very convenient as basis functions because the spherical Bessel functions $j_l(\kappa r)$ diverge at infinity when $\kappa^2 \leq 0$. However, the same matching conditions can be realized by adding $j_l(\kappa r) \cot \eta_l(\varepsilon)$ to the partial wave, so that this potentially divergent term is confined within the sphere, where the Bessel function is regular. The muffin-tin orbitals are thus defined as:

$$\phi_{\kappa l m}(\mathbf{r},\varepsilon) = \begin{cases} [\chi_l(r,\varepsilon) + j_l(\kappa r)\cot \eta_l(\varepsilon)] Y_{lm}(\hat{\mathbf{r}}), & \text{for } r \leq R_{\mathrm{MT}}, \\ h_l(\kappa r) Y_{lm}(\hat{\mathbf{r}}), & \text{for } r > R_{\mathrm{MT}}, \end{cases} \tag{8.73}$$

and are regular everywhere in space. Up to this point κ is directly related to the energy by (8.68). However, at this stage one can eliminate this dependence by making κ into an energy-independent parameter that characterizes the behavior of the basis function outside the sphere. The wave function for a solid or a molecule can then be written as a linear combination of MTOs, thus leading to the LCMTO method (Andersen, 1971).

It is interesting to remark here the generality of this idea, which consists of choosing *energy-independent* atom-centered functions as basis functions in the interstitial regions, as in LCAO approaches. In fact, in analogy with the APW method, this method can be viewed as an augmented LCAO approach. The functions in the interstitials can be MTOs, Slater-type orbitals, Gaussian-type orbitals, or any other atom-centered basis function. Which ones are better suited depends on several factors that include, e.g., how easy it is to enforce the matching to the atomic functions on the surface of the spheres, and how computationally costly it is to calculate the Hamiltonian and overlap matrix elements. While STOs and GTOs are certainly quite attractive possibilities, most approaches use MTOs, which are particularly well suited for the augmentation procedure, as will be shown below.

At this stage it is important to observe that the envelope function corresponding to an MTO centered on atom I does not decay fast enough to avoid reaching the spheres corresponding to other atoms J with a significant amplitude. This means that the *tail* of this MTO contributes to the wave function inside the other spheres. Quite conveniently, there is a *one-range* addition theorem for Hankel functions centered on site $\mathbf{R}_I$, which allows us to express them in terms of Bessel functions centered on another site $\mathbf{R}_J$. In one-range addition theorems, the variables $\mathbf{r}$ and $\mathbf{R}$ are completely separated and space does not need to be divided into two regions (two-range addition theorems), such as in the expansion of the Coulomb potential.

To simplify the notation, we shall condense the orbital and magnetic quantum numbers into a single index $L = (l, m)$. Therefore, if $H_L(\mathbf{r}) = h_l(r)Y_L(\hat{\mathbf{r}})$ and $J_L(\mathbf{r}) = j_l(r)Y_L(\hat{\mathbf{r}})$, then the addition theorem for the tails indicates that

$$H_L(\kappa, \mathbf{r} - \mathbf{R}_I) = \sum_{L'} B_{JL', IL}(\kappa) J_{L'}(\kappa, \mathbf{r} - \mathbf{R}_J), \qquad (8.74)$$

where the coefficients

$$B_{JL', IL}(\kappa) = 4\pi \sum_{L''} C_{LL'L''} i^{-l+l'-l''} h_{l''}^*(\kappa|\mathbf{R}_I - \mathbf{R}_J|)Y_{L''}^*(\hat{\mathbf{R}}_{IJ}) \qquad (8.75)$$

are, besides a factor of κ, the same *structure constants* that appear in the KKR method. The quantities

$$C_{LL'L''} = \int Y_L(\hat{\mathbf{r}})Y_{L'}^*(\hat{\mathbf{r}})Y_{L''}(\hat{\mathbf{r}})d^2\hat{\mathbf{r}} \qquad (8.76)$$

are the Gaunt coefficients. For a crystalline system, periodicity can be simply built into the structure constants by means of Bloch sums of the type (8.13), i.e.

$$B_{JL', IL}(\kappa, \mathbf{k}) = \sum_{\mathbf{T}} e^{i\mathbf{k}\cdot\mathbf{T}} B_{JL', (I+T)L}(\kappa), \qquad (8.77)$$

where the sum runs over the lattice translations $\mathbf{T}$.

Therefore, if the wave functions for the target system are written as a linear combination of energy-dependent muffin-tin orbitals:

$$\varphi_n(\mathbf{r}, \varepsilon) = \sum_{I=1}^{P} \sum_{L=0}^{L_{\max}} C_{n\kappa L}^I \phi_{\kappa L}^I(\mathbf{r}, \varepsilon), \qquad (8.78)$$

then in every sphere there is a multitude of tails arising from the MTOs centered on other spheres. According to (8.74), each of these tails contributes to the wave function inside the foreign sphere with Bessel functions of various angular momenta. The information on these contributions is encoded in the matrix of the structure constants $B_{JL', IL}(\kappa)$. Inside the spheres the solutions of Schrödinger's equation should equal the atomic orbitals $\chi_l(r, \varepsilon)$, in accord with the one-center partial wave expansion of KKR. Therefore, for a given sphere the sum of the contributions arising from the tails of all the foreign MTOs should be exactly canceled by the term $j_l(\kappa r)\cot\eta_l^i(\varepsilon_i)$ belonging to its own MTO. This is known as *tail cancelation*, and leads precisely to the KKR equations

$$\det\left|B_{JL', IL}(\kappa) + \cot\eta_l^I(\kappa, \varepsilon)\delta_{IJ}\delta_{LL'}\right| = 0, \qquad (8.79)$$

where $\cot\eta_l^I(\kappa, \varepsilon)$ are the phase shifts at every sphere given by Expression (8.71). Therefore, the solutions of the non-linear KKR secular equation are the energies for which tail cancelation is possible.

It is here where Hankel functions tend to be superior to other choices of envelope functions such as STOs (Davenport, 1984) and GTOs, although the one-range expansion of Λ-functions, i.e. exponentials multiplied by Laguerre polynomials, has also shown good convergence properties (Trivedi and Steinborn, 1982; Lin and Harris, 1993). In general, however, one-range expansions require a large number of terms and two-range expansions are numerically unstable. As a consequence, complete tail cancelation cannot be achieved efficiently, and this leads to systematic errors. Amongst all the possible sensible spherical orbitals outside the spheres, Hankel functions are particularly efficient and numerically stable.

In the LCMTO method κ retains its original meaning in terms of the energy, i.e. $\kappa^2 = \varepsilon - V_{MT}$, but the expression is still valid if it is made into an independent parameter. The advantage of this approximation is that the structure constants become energy-independent, and the calculation is simplified, because the step of re-calculating the structure constants at every iteration, when searching for the roots of the determinant, is eliminated. In the best possible scenario, a single parameter κ may be sufficient to describe the wave functions in the interstitial regions. While this is convenient in terms of computational efficiency, one could also have *double-κ* or *triple-κ* basis sets (Methfessel, 1988), as in multiple-ζ basis sets of atomic orbitals. Typically, it is sufficient to carry the expansion in angular momenta up to $l_{max} = 2$ or $l_{max} = 3$, according to the chemical species present in the system. At a first glance, the MTO basis set may appear a sophisticated version of a linear combination of atomic orbitals. This is actually not the case. The AO basis functions are completely fixed and depend on a single atom. Due to the matching condition at the surface of the spheres, MTOs are flexible basis functions that adapt to changes in the potential of all the spheres.

The LCMTO matrix elements, however, still retain an implicit energy dependence through the phase shifts. This dependence can be eliminated by replacing the energy-dependent solutions of the radial Schrödinger equation inside the spheres with atomic orbitals corresponding to one or more fixed energies. There are different possible ways of doing this. One of these could be to express the atomic wave function as a linear combination of solutions corresponding to the same logarithmic derivative on the surface of the MT spheres, as proposed by Slater (1953) within the APW context. Approaches of this type have been proposed by Koelling (1970) and Williams *et al.* (1979) (ASW method), but a more convenient route proved to be the energy linearization of the atomic wave function.

8.6.3 Linear methods

The advantage of having an energy-independent basis set is that the secular equation becomes a linear eigenvalue problem that can be solved using standard linear algebra procedures. When the basis set is energy-dependent, the solution must be obtained by calculating the determinant of the secular equation as a function of energy, and finding its roots. This operation can be numerically unstable, and also expensive if the secular equation has to be solved for a large number of energies. In addition, the phase shifts develop a singularity whenever a node of the radial function falls on a sphere boundary, as can be seen in Expression (8.66). These two aspects are common to the APW and LCMTO schemes. However, if the basis functions inside the spheres are fixed, then neither of these two problems arises because the energy dependence disappears, and the basis function can be chosen such as to avoid nodes in inconvenient places. The key question is how to choose flexible and efficient radial basis functions.

One possible way is to linearize the solutions of the radial Schrödinger equation in the energy (Andersen, 1975), by expanding $\chi_l(r, \varepsilon)$ in Taylor series around a *reference* energy ε_ν that is representative of the appropriate energy range:

$$\chi_l(r, \varepsilon) = \chi_l(r, \varepsilon_\nu) + (\varepsilon - \varepsilon_\nu)\dot{\chi}_l(r, \varepsilon_\nu) + \mathcal{O}(\varepsilon - \varepsilon_\nu)^2, \qquad (8.80)$$

where $\dot{\chi}_l(r, \varepsilon) = \partial \chi_l(r, \varepsilon)/\partial \varepsilon$ is the energy derivative of the radial atomic orbital, which can be shown to be orthogonal to $\chi_l(r, \varepsilon)$. This suggests the idea of replacing $\chi_l(r, \varepsilon)$ in the basis functions by a linear combination of the two linearly independent functions $\chi_l(r, \varepsilon_\nu)$ and $\dot{\chi}_l(r, \varepsilon_\nu)$, so as to eliminate the energy dependence from the basis. In the linearized APW method (LAPW), the basis functions inside the augmentation spheres become (Andersen, 1975)

$$\phi_{\mathbf{G}}^{\mathbf{k}}(\mathbf{r}) = \sum_{I=1}^{P} \sum_{L=1}^{l_{\max}} \left[A_L^{\mathbf{k},I}(\mathbf{G}) \chi_l^I(|\mathbf{r}_I|, \varepsilon_\nu) + B_L^{\mathbf{k},I}(\mathbf{G}) \dot{\chi}_l^I(|\mathbf{r}_I|, \varepsilon_\nu) \right] Y_L(\hat{\mathbf{r}}_I), \qquad (8.81)$$

where the expansion coefficients are (Koelling and Arbman, 1975):

$$A_L^{\mathbf{k},I}(\mathbf{G}) = \frac{4\pi R_{\mathrm{MT}}^{I\,2}}{\sqrt{\Omega}} i^l Y_L^*(\mathbf{k}+\mathbf{G}) Y_L(\hat{\mathbf{r}}_I) a_l^{\mathbf{k},I}(\mathbf{G}), \qquad (8.82)$$

$$a_l^{\mathbf{k},I}(\mathbf{G}) = j_l'(|\mathbf{k}+\mathbf{G}|R_{\mathrm{MT}})\dot{\chi}_l^I(R_{\mathrm{MT}}, \varepsilon_\nu) - j_l(|\mathbf{k}+\mathbf{G}|R_{\mathrm{MT}})\dot{\chi}_l'^I(R_{\mathrm{MT}}, \varepsilon_\nu),$$

and

$$B_L^{\mathbf{k},I}(\mathbf{G}) = \frac{4\pi R_{\mathrm{MT}}^{I\,2}}{\sqrt{\Omega}} i^l Y_L^*(\mathbf{k}+\mathbf{G}) Y_L(\hat{\mathbf{r}}_I) b_l^{\mathbf{k},I}(\mathbf{G}), \qquad (8.83)$$

$$b_l^{\mathbf{k},I}(\mathbf{G}) = j_l(|\mathbf{k}+\mathbf{G}|R_{\mathrm{MT}})\chi_l'^I(R_{\mathrm{MT}}, \varepsilon_\nu) - j_l'(|\mathbf{k}+\mathbf{G}|R_{\mathrm{MT}})\chi_l^I(R_{\mathrm{MT}}, \varepsilon_\nu),$$

with $j_l'(\kappa r)$ and $\chi_l'^I(r, \varepsilon_\nu)$ the radial derivatives of the spherical Bessel functions and the atomic orbitals, respectively.

Therefore, in addition to eliminating the energy dependence, the linearization procedure also removes the singularities that appear in the APW method whenever a node of the radial atomic wave function falls on the surface of the MT sphere. This is a consequence of asking for the continuity of both basis function and derivative, a feature that is absent in Slater's APW method, where the derivatives are discontinuous during the iterative procedure, and the eigenstates are obtained precisely as those energies for which the discontinuity disappears.

Similarly, in the linearized MTO method (LMTO) the basis functions inside the spheres are (Andersen, 1975)

$$\phi_{\kappa L}(\mathbf{r}) = \sum_{I=1}^{P} \sum_{L'=1}^{L_{max}} \left[\Pi_{L,L'}^{I}(\kappa) \chi_{L'}(\mathbf{r}_I, \varepsilon_\nu) + \Omega_{L,L'}^{I}(\kappa) \dot{\chi}_{L'}(\mathbf{r}_I, \varepsilon_\nu) \right], \qquad (8.84)$$

where the matrices Π and Ω are determined from the tail cancelation condition in all the spheres. They contain the coefficients of the expansion of the MTO in spherical harmonics and, in addition, they enclose information on the potential through the values of χ_l and $\dot{\chi}_l$ (and thus the logarithmic derivative) on the surface of the spheres. Explicit expressions for Π and Ω can be found in Andersen (1975), Andersen *et al.* (2000), or in the book by Skriver (1984). A generalization of this linearization idea is the NMTO method, where the atomic orbitals are expanded to higher orders and at several reference energies (Andersen *et al.*, 2000).

References

Andersen, O. K. (1971). In *Computational Methods in Band Theory*, J. F. J. P. M. Marcus and A. R. Williams, eds. New York, Plenum Press, 178.
  (1975). Linear methods in band theory. *Phys. Rev. B* **12**, 3060–3083.
  (1984). Linear methods in band theory. In *The Electronic Structure of Complex Systems*, P. Phariseau and W. M. Temmerman, eds. New York, Plenum Press, 11–66.
Andersen, O. K. and Jepsen, O. (1984). Explicit, first-principles tight-binding theory. *Phys. Rev. Lett.* **53**, 2571–1574.
Andersen, O. K. and Kasowski, R. V. (1971). Electronic states as linear combinations of Muffin Tin orbitals. *Phys. Rev. B* **4**, 1064–1069.
Andersen, O. K., Postnikov, A. V., and Savrasov, S. Y. (1992). The muffin-tin-orbital point of view. In *Applications of Multiple Scattering Theory to Materials Science*, W. H. Butler, P. H. Dederichs, A. Gonis, and R. L. Weaver, eds. MRS Symposium Proceedings, vol. 253. Pittsburgh, Materials Research Society, 37–70.
Andersen, O. K., Saha-Dasgupta, T., Tank, R. W., Arcangeli, C., Jepsen, O., and Krier, G. (2000). Developing the MTO formalism. In *Electronic Structure and the Physical Properties of Solids: the Uses of the LMTO Method*, H. Dreyssé, ed. Lecture Notes in Physics. Berlin, Springer Verlag.
Andzelm, J. and Wimmer, E. (1992). Density functional Gaussian-type-orbital approach to molecular geometries, vibrations, and reaction energies. *J. Chem. Phys.* **96**, 1280–1303.
Anglada, E., Soler, J. M., Junquera, J., and Artacho, E. (2002). Systematic generation of finite-range atomic basis sets for linear-scaling calculations. *Phys. Rev. B* **66**, 205101.

Arias, T. A. (1999). Multiresolution analysis of electronic structure: semicardinal and wavelet bases. *Rev. Mod. Phys.* **71**, 267–311.

Binkley, J. S., Pople, J. A., and Hehre, W. J. (1980). Self-consistent molecular orbital methods. XXI. Small split-valence basis sets for first-row elements. *J. Am. Chem. Soc.* **102**, 939–947.

Bloch, F. (1928). Über die Quantenmechanik der Elektronen in Kristallgittern. *Z. Phys.* **52**, 555–560.

Bowler, D. and Gillan, M. (2002). Recent progress in linear scaling *ab initio* electronic structure techniques. *J. Phys. Condens. Matter* **14**, 2781–2798.

Boys, S. F. (1950). Electron wave functions I. A general method for calculation for the stationary states of any molecular system. *Proc. Roy. Soc. London* **200**, 542–554.

Boys, S. F. and Bernardi, F. (1970). The calculations of small molecular interaction by the difference of separate total energies. Some procedures with reduced error. *Mol. Phys.* **19**, 553–566.

Chelikowsky, J. R., Troullier, N., and Saad, Y. (1994). Finite-difference pseudopotential method: electronic structure calculations without a basis. *Phys. Rev. Lett.* **72**, 1240–1243.

Davenport, J. W. (1984). Linear augmented-Slater-type-orbital method for electronic-structure calculations. *Phys. Rev. B* **29**, 2896–2904.

Delley, B. (1990). An all-electron numerical method for solving the local density functional for polyatomic molecules. *J. Chem. Phys.* **92**, 508–517.

Dunning, T. H. (1989). Gaussian basis sets for use in correlated molecular calculations. I. The atoms boron through neon and hydrogen. *J. Chem. Phys.* **90**, 1007–1023.

Finnis, M. W. (2003). *Interatomic forces in condensed matter*. Oxford Series on Materials Modelling, Oxford, Oxford University Press.

Gan, C. K., Haynes, P. D., and Payne, M. C. (2001). First-principles density-functional calculations using localized spherical-wave basis sets. *Phys. Rev. B* **63**, 205109.

Godbout, N., Salahub, D. R., Andzelm, J., and Wimmer, E. (1992). Optimization of Gaussian-type basis sets for local spin density functional calculations. Part I. Boron through Neon, optimization technique and validation. *Can. J. Chem.* **70**, 560–571.

Gygi, F. (1992). Adaptive riemannian metric for plane-wave electronic-structure calculations. *Europhys. Lett.* **19**, 617–620.

Gygi, F. and Galli, G. (1995). Real-space adaptive-coordinate electronic-structure calculations. *Phys. Rev. B* **52**, R2229–R2232.

Hall, G. G. (1951). The molecular orbital theory of chemical valency. VIII - A method of calculating ionization potentials. *Proc. Roy. Soc. London* **205**, 541–552.

Ham, F. S. and Segall, B. (1961). Energy bands in periodic lattices – Green's function method. *Phys. Rev.* **124**, 1786–1796.

Hamann, D. R. (1995). Application of adaptive curvilinear coordinates to the electronic structure of solids. *Phys. Rev. B* **51**, 7337–7340.

Hehre, W. J., Stewart, R. F., and Pople, J. A. (1969). Self-consistent molecular-orbital methods. I. Use of Gaussian expansions of Slater-type atomic orbitals. *J. Chem. Phys.* **51**, 2657–2664.

Hehre, W. J., Ditchfield, R., and Pople, J. A. (1972). Self-consistent molecular orbital methods. XII. Further extensions of Gaussian-type basis sets for use in molecular orbital studies of organic molecules. *J. Chem. Phys.* **56**, 2257–2261.

Heine, V. (1967). *s-d* interaction in transition metals. *Phys. Rev.* **153**, 673–682.

Helgaker, T., Jørgensen, P., and Olsen, J. (2000). *Molecular Electronic-Structure Theory*. Chichester, Wiley.

Hernandez, E., Gillan, M. J., and Goringe, C. M. (1997). Basis functions for linear-scaling first-principles calculations. *Phys. Rev. B* **55**, 13485–13493.

Ho, K. M., Elsässer, C., Chan, C. T., and Fähnle, M. (1992). First-principles pseudopotential calculations for hydrogen in 4d transition metals. I. Mixed-basis method for total energies and forces. *J. Phys. Condens. Matter* **4**, 5189–5206.

Huzinaga, S. (1965). Gaussian-type functions for polyatomic systems. I. *J. Chem. Phys.* **42**, 1293–1302.

Jackson, J. D. (1975). *Classical Electrodynamics*, 2nd edn. New York, Wiley.

Junquera, J., Paz, O., Sánchez-Portal, D., and Artacho, E. (2001). Numerical atomic orbitals for linear-scaling calculations. *Phys. Rev. B* **64**, 235111.

Kittel, C. (1996). *Introduction to Solid State Physics*, 7th edn. New York, Wiley.

Koelling, D. D. (1970). Alternative augmented-plane-wave technique: theory and application to copper. *Phys. Rev. B* **2**, 290–298.

Koelling, D. D. and Arbman, G. O. (1975). Use of energy derivative of the radial solution in an augmented plane wave method: application to Cu. *J. Phys. F: Metal Phys.* **5**, 2041–2054.

Kohn, W. and Rostoker, N. (1954). Solution of the Schrödinger equation in periodic lattices with an application to metallic lithium. *Phys. Rev.* **94**, 1111–1120.

Korringa, J. (1947). On the calculation of the energy of a Bloch wave in a metal. *Physica* **13**, 392–400.

Lin, Z. and Harris, J. (1993). A localized-basis scheme for molecular dynamics. *J. Phys. Condens. Matter* **5**, 1055–1080.

Loucks, T. L. (1967). *The Augmented Plane Wave Method*. New York, Benjamin.

Louie, S. G., Ho, K. M., and Cohen, M. L. (1979). Self-consistent mixed-basis approach to the electronic structure of solids. *Phys. Rev. B* **19**, 1774–1782.

Löwdin, P. O. (1950). On the non-orthogonality problem connected with the use of atomic wave functions in the theory of molecules and crystals. *J. Chem. Phys.* **18**, 365–375.

Marcus, P. M. (1967). *Int. J. Quantum Chem.* **1**, 567.

Messiah, A. (1961). *Quantum Mechanics*. Amsterdam, North Holland.

Methfessel, M. (1988). Elastic constants and phonon frequencies of Si calculated by a fast full-potential linear-muffin-tin-orbital method. *Phys. Rev. B* **38**, 1537–1540.

Pople, J. A. and Nesbet, R. K. (1954). Self-consistent orbitals for radicals. *J. Chem. Phys.* **22**, 571–572.

Press, W. H., Teukolsky, S. A., Vetterling, W. T., and Flannery, B. P. (1992). *Numerical Recipes. The Art of Scientific Computing*. Cambridge, Cambridge University Press.

Pulay, P. (1969). Ab initio calculation of force constants and equilibrium geometries in polyatomic molecules. I. Theory. *Mol. Phys.* **17**, 197–204.

Roothaan, C. C. J. (1951). New developments in molecular orbital theory. *Rev. Mod. Phys.* **23**, 69–89.

Saffren, M. M. (1960). *Bull. Am. Phys. Soc.* **5**, 298.

Saffren, M. M. and Slater, J. C. (1953). An augmented plane-wave method for the periodic potential problem. II. *Phys. Rev.* **92**, 1126–1128.

Sankey, O. F. and Niklewski, D. J. (1989). *Ab initio* multicenter tight-binding model for molecular dynamics simulations and other applications in covalent systems. *Phys. Rev. B* **40**, 3979–3995.

Skriver, H. (1984). *The LMTO Method*. New York, Springer.

Skylaris, C.-K., Diéguez, O., Haynes, P. D., and Payne, M. C. (2002). Comparison of variational real-space representations of the kinetic energy operator. *Phys. Rev. B* **66**, 073103.

Slater, J. C. (1930a). Atomic shielding constants. *Phys. Rev.* **36**, 57–64.
  (1930b). Note on Hartree's method. *Phys. Rev.* **35**, 210–211.
  (1932). Analytic atomic wave functions. *Phys. Rev.* **42**, 33–43.

(1934). The electronic structure of metals. *Rev. Mod. Phys.* **6**, 209–281.
(1937). Wave functions in a periodic potential. *Phys. Rev.* **51**, 846–851.
(1953). An augmented plane wave method for the periodic potential problem. *Phys. Rev.* **92**, 603–608.
Soler, J. M. and Williams, A. R. (1990). Augmented-plane-wave forces. *Phys. Rev. B* **42**, 9728–9731.
Soler, J. M., Artacho, E., Gale, J. D., García, A. I., Junquera, J., Ordejón, P., and Sánchez-Portal, D. (2002). The SIESTA method for *ab initio* order-N materials simulation. *J. Phys. Condens. Matter* **14**, 2745–2780. www.siesta.org.
Sprik, M. and Klein, M. L. (1989). Adiabatic dynamics of the solvated electron in liquid ammonia. *J. Chem. Phys.* **91**, 5665–5671.
Trivedi, H. P. and Steinborn, E. O. (1982). Numerical properties of a new translation formula for exponential-type functions and its application to one-electron multicenter integrals. *Phys. Rev. A* **25**, 113–127.
Varga, K., Zhang, Z., and Pantelides, S. T. (2004). Lagrange functions: a family of powerful basis sets for real-space order-N electronic structure calculations. *Phys. Rev. Lett.* **93**, 176403.
Watson, M. A., Handy, N. C., and Cohen, A. J. (2003). Density functional calculations, using Slater basis sets, with exact exchange. *J. Chem. Phys.* **119**, 6475–6481.
Wigner, E. and Seitz, F. (1933). On the constitution of metallic sodium. *Phys. Rev.* **43**, 804–810.
Williams, A. R., Kübler, J., and Gelatt Jr., C. D. (1979). Cohesive properties of metallic compounds: augmented-spherical-wave calculations. *Phys. Rev. B* **19**, 6094–6118.

9

Electronic structure methods

Once the level of theory (DFT-LDA, Hartree–Fock, or other) has been chosen, the differences between electronic structure methods are essentially due to the choice of basis set. Pseudopotentials may or may not be part of the package. The main difference is the replacement of the bare Coulomb potential of the nucleus by a softer potential and some technical issues regarding the angular dependence of the pseudopotential, but the abundance of electronic structure methods in the market is mostly due to the quest for the ultimate basis set.

The central and computationally most intensive aspect of an electronic structure calculation is the self-consistent solution of the one-electron eigenvalue equation. This involves the calculation of the Kohn–Sham or the Fock matrix elements and the corresponding energy. In addition, geometry optimization and molecular dynamics simulations require the calculation of forces on the nuclear degrees of freedom. In solid-state applications, the optimization of lattice parameters and constant-pressure molecular dynamics simulations require also the calculation of the stress tensor.

In this chapter we shall describe how the Hamiltonian and the total energy are calculated in practice, in the most widely used methods. We start in Section 9.1 by introducing the KKR method as an approach derived from multiple scattering theory, where basis sets expansions are bypassed by using Green's function techniques. Section 9.2 is devoted to describing the most relevant aspects of all-electron schemes based on augmentation spheres. These are considered amongst the most accurate approaches because they do not approximate the behavior of core electrons through pseudopotentials, and the basis sets are flexible and adjust themselves according to the eigenvalue energies. These methods have invariably been applied within the realm of DFT. Not as accurate, but in general more economical and still very reliable and robust, is the pseudopotential plane wave (PPW) method, which has been used for a large proportion of the DFT electronic structure calculations – and most first-principles molecular dynamics simulations – for condensed phases to date. Therefore, in Section 9.3 we shall focus our attention on the computation of Hamiltonian, energy, forces, and stress within this scheme. Next, we address the main issues that arise when using generic atom-centered basis sets (or *local orbitals*), and finally, in Section 9.5, we describe the use of the Gaussian basis set that for many decades has been the way of choice for molecular calculations, especially at the Hartree–Fock and post-Hartree–Fock levels, but also more recently within DFT-GGA. This latter is usually an all-electron scheme, but can be, and indeed has been, used in conjunction with pseudopotentials or effective core potentials, especially for heavy atoms with a large number of chemically inert core electrons.

9.1 Multiple scattering methods: the KKR approach

One of the difficult issues in electronic structure calculations is to take into account the polyhedral shape of the unit cell. This is a difficult boundary value problem. A way to avoid it is by using Green's function techniques, as in the KKR method, where the differential equation is transformed into an integral equation (Korringa, 1947; Kohn and Rostoker, 1954). The Green's function for the lattice is the solution to the following differential equation:

$$\left[\frac{\hbar^2}{2m}\nabla^2 + E\right] G(\mathbf{r}, \mathbf{r}', E) = \delta(\mathbf{r} - \mathbf{r}'), \tag{9.1}$$

with the boundary conditions

$$G(\mathbf{r}_s, \mathbf{r}', \mathbf{k}, E) = e^{i\mathbf{k}\cdot\mathbf{T}} G(\mathbf{r}, \mathbf{r}', \mathbf{k}, E),$$

$$\frac{\partial G(\mathbf{r}_s, \mathbf{r}', \mathbf{k}, E)}{\partial n} = -e^{i\mathbf{k}\cdot\mathbf{T}}\frac{\partial G(\mathbf{r}, \mathbf{r}', \mathbf{k}, E)}{\partial n}, \tag{9.2}$$

where $\mathbf{r}$ and $\mathbf{r}_s$ are conjugate points on the boundary of the polyhedron. Under these conditions, the Green's function for a certain energy E and $\mathbf{k}$-point in the Brillouin zone can be written as

$$G(\mathbf{r}, \mathbf{r}', \mathbf{k}, E) = -\frac{1}{4\pi}\sum_{\mathbf{T}}\frac{e^{i\kappa|\mathbf{r}-\mathbf{r}'-\mathbf{T}|}}{|\mathbf{r}-\mathbf{r}'-\mathbf{T}|} e^{i\mathbf{k}\cdot\mathbf{T}}, \tag{9.3}$$

where the $\mathbf{T}$ vectors run over the Bravais lattice of the solid, $\kappa = \sqrt{E}$ for $E > 0$, and $\kappa = i\sqrt{-E}$ for $E < 0$. This expression can also be written as a plane wave expansion, which automatically satisfies the periodicity conditions of the lattice.

In order to avoid the use of boundary conditions for the wave function, the KKR method replaces the differential version of the Schrödinger equation with the equivalent integral equation over the volume of the atomic polyhedron:

$$\varphi(\mathbf{r}, E) = \int_\Omega G(\mathbf{r}, \mathbf{r}', E) v(\mathbf{r}') \varphi(\mathbf{r}', E)\, d\mathbf{r}', \tag{9.4}$$

where the wave vector index **k** is understood, and has been dropped for the sake of clarity. The above equation can also be obtained by applying the variational principle to the quantity

$$\Lambda = \int_\Omega \varphi^*(\mathbf{r}, E) v(\mathbf{r}) \left[\varphi(\mathbf{r}, E) - \int_\Omega G(\mathbf{r}, \mathbf{r}', E) v(\mathbf{r}') \varphi(\mathbf{r}', E) \, d\mathbf{r}' \right] d\mathbf{r}' \quad (9.5)$$

with respect to the wave function $\varphi^*(\mathbf{r}, E)$. The variational principle $\delta\Lambda = 0$ leads to a secular equation in the energy that results in the desired energy dispersion relations, or bands, $E(\mathbf{k})$. This program can be carried out, but it requires the calculation of six-dimensional integrals with the additional caveats that the Green's function is singular for $\mathbf{r} = \mathbf{r}'$, and that the region of integration is the complicated atomic polyhedron.

At this point, the original KKR method introduces the approximation of considering the potential as spherically symmetric inside atom-centered spheres inscribed into the atomic polyhedron, and constant in the interstitial region. Furthermore, this constant is made zero by choosing the energy reference appropriately. This is the muffin-tin approximation (MTA) mentioned before, which is common with Slater's 1937 APW method. Since the potential is zero in the interstitials, the integrals are confined to the atomic spheres. Inside the spheres, the solutions are expanded in one-center partial waves:

$$\varphi(\mathbf{r}, E) = \sum_{l=0}^{l_{max}} \sum_{m=-l}^{l} C_{lm} \bar{\chi}_l(r, E) Y_{lm}(\hat{\mathbf{r}}), \quad (9.6)$$

where the functions $\bar{\chi}_l(r, E)$ are regular solutions of the radial Schrödinger equation for energy E in the presence of the spherical MT potential. If there is more than one atom in the unit cell, then there is a set of coefficients C_{lm}^l for each atom. Using the MTA and Green's theorem, the volume integrals in (9.5) can be transformed into surface integrals on the surface of the spheres. The integrand contains the trial wave functions, the Green's functions, and their derivatives, all evaluated at the muffin-tin radius.

By expanding the Green's function in spherical harmonics around the two centers $\mathbf{r}$ and $\mathbf{r}'$, and using the condensed notation $L = (l, m)$, we obtain:

$$G(\mathbf{r}, \mathbf{r}', \mathbf{k}, E) = \sum_L \sum_{L'} Y_L(\hat{\mathbf{r}}) Y_{L'}^*(\hat{\mathbf{r}}')$$
$$\times \left[B_{L,L'}(\mathbf{k}, E) j_l(\kappa r) j_{l'}(\kappa r') + \kappa \delta_{L,L'} j_l(\kappa r) n_l(\kappa r') \right], \quad (9.7)$$

where $B_{L,L'}(\mathbf{k}, E)$ are quantities that depend only on the type of Bravais lattice, the wave vector, and the energy. They are completely independent of the chemical species, and contain all the information on the space group of the crystal (Ham and Segall, 1961). The functions $j_l(x)$ and $n_l(x)$ are the spherical Bessel and Neumann functions. For negative energy, $n_l(x)$ is replaced by the spherical Hankel function $h_l(x)$. The structure constants (B) are calculated by considering all the periodic replicas of the unit cell, normally using the Ewald summation technique (see Section 9.3.1). Now, if the wave function $\varphi(\mathbf{r}, E)$ and the Green's function are replaced into the expression obtained from (9.5) in terms of surface integrals, then the matrix elements of Λ in the basis of spherical harmonics are

$$\Lambda_{L,L'} = \left[(B_{L,L'} j'_{l'} + \kappa \delta_{L,L'} n'_{l'}) - (B_{L,L'} j_{l'} + \kappa \delta_{L,L'} n_{l'}) D_\chi^{l'} \right]$$
$$\times \left(D_\chi^l j_l - j'_l \right), \tag{9.8}$$

where D_χ^l are the logarithmic derivatives of the radial atomic functions $\bar{\chi}_l(r)$, given by (8.72), and j'_l and n'_l are the derivatives of $j_l(x)$ and $n_l(x)$. All the Bessel and Neumann functions are evaluated at $x = \kappa R_{\mathrm{MT}}$.

Minimization of Λ, when the wave function is expanded in spherical harmonics as in (9.6), leads to the set of linear equations $\sum_{L'} \Lambda_{L,L'} C_{L'} = 0$, which has a non-trivial solution only when the determinant is zero:

$$\det \left| B_{L,L'}(\mathbf{k}, E) + \kappa(E) \delta_{LL'} \frac{n'_l - n_l D_\chi^l(E)}{j'_l - j_l D_\chi^l(E)} \right| = 0. \tag{9.9}$$

This is the KKR secular equation, which contains structure-dependent quantities that are independent of the chemical species, and the logarithmic derivatives that carry information on the potential. Notice that the same structure constants appear in augmentation methods based on atom-centered basis functions for the interstitials, such as the LCMTO and the ASW methods. Improvements of the KKR methods beyond the MTA have been proposed by, amongst others, Williams (1970). For a review of Green's function methods, see Gonis (1992).

9.2 All-electron methods based on augmentation spheres

Augmented basis sets that divide the space into muffin-tin (MT) spheres and interstitial regions have been described in Section 8.6. A basis set is selected for the interstitial region, and the basis functions are augmented with solutions of the spherically symmetric atomic problem within the MT spheres. There are two main approaches of this kind: either the wave functions in the interstitials are expanded in a floating basis set like plane waves, or they are expanded in atom-centered basis functions like muffin-tin orbitals.

Once this choice has been made, the next step is to decide on how to treat the potential when calculating the Hamiltonian matrix elements in the augmented basis. Matrix elements contain a contribution from the spheres, which is calculated using the atomic wave functions, and another from the interstitials. Different technical issues arise according to the choice of interstitial basis functions but, nevertheless, there are some common features.

First, we notice that for isolated atoms the potential is spherically symmetric. In a molecule or a condensed phase this potential is modified by the presence of neighboring atoms; the molecular (or crystal) field breaks the spherical symmetry. However, due to the strong attraction of the nucleus, within the spheres the non-spherical contribution to the potential constitutes a small fraction of the full potential. Therefore, it makes sense to think of the potential inside the spheres as the sum of a major spherical, or muffin-tin, contribution, and a small non-spherical correction v_{ns}. In the interstitials there is only a non-muffin-tin contribution v_{i}. Therefore, the full potential can be written as

$$v(\mathbf{r}) = \sum_{I=1}^{P} v_I(|\mathbf{r}-\mathbf{R}_I|)\theta(|\mathbf{r}-\mathbf{R}_I|) + v_{\mathrm{nmt}}(\mathbf{r}), \qquad (9.10)$$

where $\theta(x)$ is the step function that assumes the value one inside the sphere and zero outside. The last term, $v_{\mathrm{nmt}} = v_{\mathrm{i}} + v_{\mathrm{ns}}$ is the sum of the interstitial potential and the non-spherical contribution from the spheres. The same can be said about the electronic density:

$$\rho(\mathbf{r}) = \sum_{I=1}^{P} \rho_I(|\mathbf{r}-\mathbf{R}_I|)\theta(|\mathbf{r}-\mathbf{R}_I|) + \rho_{\mathrm{nmt}}(\mathbf{r}). \qquad (9.11)$$

This separation into spherically-symmetric and non-muffin-tin contributions is very useful, because the matrix elements involving the spherical components of the potential can be reduced to simple one-dimensional integrals. In augmented PW methods the calculation of the interstitial potential and density is quite straightforward, but for atom-centered interstitial functions it becomes more delicate.

The potential in the interstitials, which correspond to bonding regions, tends to be rather flat (see Figure 3 in Jepsen *et al.*, 1982). Therefore, it makes sense, as first suggested by Slater (1937), to approximate the non-muffin-tin potential by a constant potential. Moreover, the potential reference can be shifted so that this constant is zero, i.e. $v_{\mathrm{nmt}}^{\mathrm{MTA}}(\mathbf{r}) = 0$. This is called the *muffin-tin approximation* (MTA), and in this case the only contribution of the potential to the Hamiltonian matrix elements arises from the integrals within the spheres.

A variant of the MTA is the *atomic sphere approximation* (ASA), first introduced by Andersen (1973). In the ASA the spheres are *inflated* by choosing the MT radius equal to the Wigner–Seitz radius, so that they become space-filling and slightly overlapping (around 10%), thus eliminating the interstitial regions. Therefore, the matrix elements of the potential contain a main contribution from the spheres, and a small correction from the overlapping regions that, in principle, should be discounted. In the ASA this correction due to the overlap is neglected.

It has been shown that, without correcting for overlapping or interstitial regions, the ASA is more accurate an approximation than the MTA (Andersen, 1973).

In more recent times these approximations have been replaced by computationally more intensive methods that take into account the *full potential*. The MTA and the ASA, however, have made an extremely important contribution towards the understanding of the band structures of materials, and remain a cornerstone of augmentation methods. The ASA, in particular, has led to the concept of *canonical bands*, i.e. electronic bands that depend only on the crystal structure, given by the eigenvalues of the diagonal blocks of the structure matrix $B^{\mathbf{k}}_{lm,l'm'}\delta_{l,l'}$, where the l, l' coupling is neglected. The information on the specific system is introduced by means of a unitary transformation of the Hamiltonian that depends on the potential parameters (Andersen, 1984), and leads to a system-dependent deformation of the bands.

9.2.1 Augmented plane waves: APW, LAPW, and PAW

In the original APW method the basis functions are plane waves matched to solutions of the atomic problem inside the MT spheres, as indicated in (8.64). An alternative is to expand the wave functions of the target system in plane waves, and to augment these directly instead of each individual PW component (Soler and Williams, 1990). The first approach requires a large number of angular momenta, of the order of $l_{\max} \approx 10$ or more, to achieve convergence (Loucks, 1965). A detailed description can be found in Loucks (1967). In the Soler–Williams approach only the projections of the wave functions onto the atom-centered spherical harmonics up to some $l_{\max}$ are matched, while the remaining contributions, i.e. the projections of angular momentum $l > l_{\max}$, are allowed to penetrate the spheres without forcing any matching condition. This approach can be argued to be better than just ignoring large l components above $l_{\max}$, as this latter implies that the PW amplitude should be set to zero on the surface of the spheres, while this amplitude is usually largest there (Soler and Williams, 1989). This is probably the reason why fewer angular momenta, of the order of $l_{\max} = 2$ or 3 depending on the chemical species present in the system, are required in this approach.

The original APW method is based on the MTA, i.e., it neglects any non-spherical contributions to the potential. The wave functions of the target system are written as linear combinations of the augmented PWs basis functions, and the energies are obtained by solving a generalized eigenvalue equation with a non-unity overlap matrix, due to the non-orthogonality of the basis functions for different values of **G**. Since the PWs are orthonormal when integrating over the entire cell, it is more convenient to calculate the matrix elements by first assuming that the PW part extends to all space, and then discount the contribution from the spheres together with the augmentation part. Using this strategy, the overlap matrix is written:

$$\langle \phi_\mathbf{G}^\mathbf{k} | \phi_{\mathbf{G}'}^\mathbf{k} \rangle = \delta_{\mathbf{G},\mathbf{G}'} - \frac{4\pi}{\Omega} \sum_{I=1}^{P} R_{\mathrm{MT}}^{I\,3} e^{i(\mathbf{G}-\mathbf{G}')\cdot\mathbf{R}_I} \frac{j_1(|\mathbf{G}-\mathbf{G}'|R_{\mathrm{MT}}^I)}{|\mathbf{G}-\mathbf{G}'|R_{\mathrm{MT}}^I}, \qquad (9.12)$$

where the first term is the usual orthogonality relation for PW, and the second term is the subtraction of the integrals inside the spheres, which were transformed into surface integrals with the aid of Green's theorem. The Hamiltonian matrix elements are

$$\langle \phi_\mathbf{G}^\mathbf{k} | \hat{\mathcal{H}} | \phi_{\mathbf{G}'}^\mathbf{k} \rangle = \frac{\hbar^2 |\mathbf{k}_\mathbf{G}|^2}{2m} \delta_{\mathbf{G},\mathbf{G}'} \qquad (9.13)$$

$$- \frac{4\pi}{\Omega} \frac{\hbar^2}{2m} \mathbf{k}_\mathbf{G} \cdot \mathbf{k}'_\mathbf{G} \sum_{I=1}^{P} R_{\mathrm{MT}}^{I\,3} e^{i(\mathbf{G}-\mathbf{G}')\cdot\mathbf{R}_I} \frac{j_1(|\mathbf{G}-\mathbf{G}'|R_{\mathrm{MT}}^I)}{|\mathbf{G}-\mathbf{G}'|R_{\mathrm{MT}}^I}$$

$$+ \frac{4\pi}{\Omega} \sum_{I=1}^{P} R_{\mathrm{MT}}^{I\,2} e^{i(\mathbf{G}-\mathbf{G}')\cdot\mathbf{R}_I} \sum_{l=0}^{\infty} (2l+1) P_l(\cos\theta_{\mathbf{G},\mathbf{G}'}) h_{\mathbf{G},\mathbf{G}'}^{l,I,\mathbf{k}}$$

with

$$h_{\mathbf{G},\mathbf{G}'}^{l,I,\mathbf{k}} = j_l(k_\mathbf{G} R_{\mathrm{MT}}^I) j_l(k_{\mathbf{G}'} R_{\mathrm{MT}}^I) \left[\frac{\chi_l'(R_{\mathrm{MT}}^I, E)}{\chi_l(R_{\mathrm{MT}}^I, E)} - \frac{j_l'(k_{\mathbf{G}'} R_{\mathrm{MT}}^I)}{j_l(k_{\mathbf{G}'} R_{\mathrm{MT}}^I)} \right]. \qquad (9.14)$$

Here, $\mathbf{k}_\mathbf{G} = \mathbf{k} + \mathbf{G}$, $j_l(x)$ are spherical Bessel functions, $\chi_l(r, E)$ are the solutions of the radial Schrödinger equation inside the spheres, and j_l' and χ_l' are the derivatives with respect to the radial coordinate. The ratio $r\chi_l'(r, E)/\chi_l(r, E)$, i.e. the logarithmic derivative, contains the information on the potential inside the spheres. The second term in that expression, $j_l'(kr)/j_l(kr)$, appears due to the discontinuity of the kinetic energy across the surface of the spheres, because in the APW method the derivatives of the basis functions are discontinuous at the MT surface, except when E is an eigenvalue (Loucks, 1967).

In the LAPW method these logarithmic derivatives are linearized with respect to energy (Andersen, 1975), thus resulting in the modified basis functions given by (8.81), (8.82), and (8.83). The overlap and Hamiltonian matrix elements within the MTA are only slightly modified with respect to the original APW method (Koelling and Arbman, 1975). The LAPW overlap matrix elements are

$$\langle \phi_\mathbf{G}^\mathbf{k} | \phi_{\mathbf{G}'}^\mathbf{k} \rangle_{\mathrm{LAPW}} = \langle \phi_\mathbf{G}^\mathbf{k} | \phi_{\mathbf{G}'}^\mathbf{k} \rangle_{\mathrm{APW}}$$

$$+ \frac{4\pi}{\Omega} \sum_{I=1}^{P} R_{\mathrm{MT}}^{I\,4} e^{i(\mathbf{G}-\mathbf{G}')\cdot\mathbf{R}_I} \sum_{l=0}^{\infty} (2l+1) P_l(\cos\theta_{\mathbf{G},\mathbf{G}'}) s_{\mathbf{G},\mathbf{G}'}^{l,I,\mathbf{k}} \qquad (9.15)$$

with

$$s_{\mathbf{G},\mathbf{G}'}^{l,I,\mathbf{k}} = a_l^{\mathbf{k},I}(\mathbf{G}) a_l^{\mathbf{k},I}(\mathbf{G}') + b_l^{\mathbf{k},I}(\mathbf{G}) b_l^{\mathbf{k},I}(\mathbf{G}') N_l^I. \qquad (9.16)$$

Here, $N_l^I = \int_0^{R_{MT}^I} r^2 |\dot{\chi}_l^I(r)| \, dr$ is an appropriate normalization constant, and $a_l^{k,I}(G)$ and $b_l^{k,I}(G')$ are given in (8.82) and (8.83), respectively.

The first two terms in (9.13) derive from the plane wave part of the basis functions, and are thus unchanged in the LAPW approach. To take into account the linearization of the atomic basis functions, the only requirement is to modify the quantity $h_{G,G'}^{l,I,k}$ in (9.14) to

$$h_{G,G'}^{l,I,k} = E_l^I s_{G,G'}^{l,I,k} + a_l^{k,I}(G) b_l^{k,I}(G'), \tag{9.17}$$

where E_l^I is the linearization energy for angular momentum l and atom I. The linearization energy can, in principle, be allowed to depend on the atomic species, and even on the specific atom. In general, however, it is chosen to be atom-independent, and located approximately in the center of the energy band of predominantly l character, as indicated by the empirical Wigner–Seitz rule (Andersen, 1973; Blaha et al., 1990). This rule establishes that the two edges of the band can be determined by requesting either the atomic wave function $\chi_l(r, E_1)$ or its radial derivative $\chi'_l(r, E_2)$ to be zero at the MT radius.

Going beyond the MTA requires us to choose a representation for the non-muffin-tin potential and density. It is convenient to write v_{nmt} as the sum of two contributions: one in the interstitial regions (v_i), which is expanded in plane waves, and another, non-spherical contribution inside the spheres (v_{ns}), which is expanded in spherical harmonics or, for periodic cell calculations, in *lattice harmonics* (Bloch sums of spherical harmonics):

$$v(\mathbf{r}) = \begin{cases} \sum_{lm} v_{lm}(r) Y_{lm}(\theta, \phi) & \text{inside the spheres } (v_{ns}), \\ \sum_G \tilde{v}(G) e^{i G \cdot r} & \text{in the interstitials } (v_i). \end{cases} \tag{9.18}$$

The s component, $v_{00}(r) Y_{00}(\theta, \phi)$, is just the muffin-tin potential. An analogous expression applies to the electronic density. Schemes that avoid any kind of shape approximation for the potential and density are called *full-potential*, in this case full-potential LAPW, or F-LAPW.

The calculation of the matrix elements of the potential in the interstitial region, common to the APW and LAPW methods, proceeds in the same way as for the overlap matrix elements, by integrating over the entire cell and then discounting the contributions from the spheres. Therefore,

$$\langle \phi_G^k | \hat{v}_i | \phi_{G'}^k \rangle = \tilde{v}(G - G') - \frac{4\pi}{\Omega} \sum_{I=1}^P R_{MT}^{I\,3} e^{i(G-G') \cdot R_I} \tilde{v}^I(G - G'), \tag{9.19}$$

with

$$\tilde{v}^I(G - G') = \sum_{G''} e^{i G'' \cdot R_I} \tilde{v}(G'') \left(\frac{\cos(x'') \sin(x)/x - \cos(x) \sin(x'')/x''}{x^2 - x''^2} \right), \tag{9.20}$$

where $x = |\mathbf{G} - \mathbf{G}'|R_{\mathrm{MT}}^I$ and $x'' = G''R_{\mathrm{MT}}^I$. Since the non-spherical contribution within the spheres is much smaller than the interstitial contribution, the approach of retaining the latter while ignoring the former has been frequently used. This intermediate approach is called *warped muffin-tin* (Koelling *et al.*, 1970).

In the LAPW method, the matrix elements of the non-spherical contribution within the spheres are easier to calculate than in the APW method. They assume the following expression (Koelling and Arbman, 1975):

$$\langle \phi_{\mathbf{G}}^{\mathbf{k}} | \hat{v}_{\mathrm{ns}} | \phi_{\mathbf{G}'}^{\mathbf{k}} \rangle = \frac{16\pi^2}{\Omega} \sum_{I=1}^{P} R_{\mathrm{MT}}^{I\,4} e^{i(\mathbf{G}-\mathbf{G}')\cdot \mathbf{R}_I} v_{\mathrm{ns}}^{I,\mathbf{k}}(\mathbf{G},\mathbf{G}') \qquad (9.21)$$

with

$$v_{\mathrm{ns}}^{I,\mathbf{k}}(\mathbf{G},\mathbf{G}') = \sum_{L,L',L''} C_{L,L',L''} \int_0^{R_{\mathrm{MT}}^I} r^2 \left[a_{l'}^{\mathbf{k},I}(\mathbf{G})\chi_{l'}(r) + b_{l'}^{\mathbf{k},I}(\mathbf{G})\dot{\chi}_{l'}(r) \right]$$

$$\times v_{l'm'}(r) \left[a_{l''}^{\mathbf{k},I}(\mathbf{G})\chi_{l''}(r) + b_{l''}^{\mathbf{k},I}(\mathbf{G})\dot{\chi}_{l''}(r) \right] dr, \qquad (9.22)$$

where the notation $L = (l,m)$ has been used, and $C_{L,L',L''}$ are the Gaunt coefficients defined in (8.76).

In the context of density-functional calculations it is also necessary to solve Poisson's equation in order to obtain the Hartree potential, as part of the Kohn–Sham potential. The option of Fourier transforming the whole charge density, as is done in the plane wave method (see Section 9.3) is out of the question here because the number of PWs required to represent the density inside the spheres would be prohibitive. An alternative method to obtain the Hartree potential in the interstitial region was developed by Wimmer *et al.* (1981). The charge density inside the spheres is replaced with a smoother pseudo-charge density that has the same multi-polar expansion, thus reducing significantly the required number of PWs. Next, the potential inside the spheres is obtained by solving Poisson's equation for the original charge density, with the potential on the surface of the spheres as boundary condition, e.g. using Green's function techniques. The boundary potential is obtained by expanding the PW components of the potential in the interstitials into lattice harmonics. The exchange-correlation potential is also represented in PW and lattice harmonics, with the expansion coefficients determined by least-squares fits.

Recently, an approach that lies half-way between the LAPW and the pseudopotential PW methods, and is reminiscent of the orthogonalized plane wave (OPW) method (Herring, 1940), has been introduced by Blöchl (1994) under the name of *projected augmented waves* (PAW). This approach retains the all-electron character, but it uses a decomposition of the all-electron wave function in terms of a *smooth* pseudo-wave function, and a rapidly varying contribution localized within the core region (the muffin-tin spheres, in the augmentation language). The true and pseudo-wave functions are related by a linear transformation:

$$|\varphi_{AE}^n\rangle = |\varphi_{PS}^n\rangle + \sum_{I=1}^{P}\sum_{l,m}\sum_{i}(|\Phi_{AE}^{iIlm}\rangle - |\Phi_{PS}^{iIlm}\rangle)\langle \tilde{p}_i^{Ilm}|\varphi_{PS}^n\rangle, \qquad (9.23)$$

where $\Phi_{AE}^{iIlm}(\mathbf{r})$ are all-electron partial waves centered on atom I obtained for a reference atom, and $\Phi_{PS}^{iIlm}(\mathbf{r})$ are pseudo-atomic partial waves that coincide with the all-electron one outside a cutoff radius and match continuously inside. The projector functions verify the relation $\langle \tilde{p}_i^{Ilm}|\Phi_{PS}^{jIlm}\rangle = \delta_{ij}$. The sums run over all the atomic sites I, angular momenta (l, m), and projector functions i. For a single projector this is basically the OPW method. Using the above expression for the wave functions, the orthogonality relation of projectors and pseudo-atomic partial waves splits the electronic density naturally into three disjoint contributions: a soft pseudo-charge density arising from $\varphi_{PS}^n(\mathbf{r})$, and two localized charge densities involving $\Phi_{AE}^{iIlm}(\mathbf{r})$ and $\Phi_{PS}^{iIlm}(\mathbf{r})$, respectively. Similar partitions can be obtained for the potential and the energy (Blöchl, 1994; Kresse and Joubert, 1999). All the expressions involving the pseudized quantities are evaluated on a Cartesian grid using plane waves, while the expressions involving localized quantities are evaluated using radial grids. No cross terms appear that require both grids. Furthermore, although it is not strictly necessary, the PAW method freezes the core orbitals to those of a reference configuration and works only with the valence wave functions, exactly as pseudopotential methods. Therefore, all the pseudopotential machinery is available for the PAW method, which just has to be supplemented with contributions from spherical regions. Typically, two reference energies are chosen, and the projectors determined according to the procedure described in Section 7.3. The pseudo-atomic wave functions can be constructed in the same way as is done in the context of pseudopotential generation (see Section 7.2).

The expansion of the wave functions in PWs makes the APW, LAPW, and PAW methods very flexible and unbiased. Forces on the nuclear degrees of freedom can be calculated, and thus geometry optimization and molecular dynamics simulations are possible (Soler and Williams, 1990; Yu *et al.*, 1991). A comparison of LAPW, PAW, and pseudopotential methods can be found in Holtzwarth *et al.* (1997). The PAW method seems to be one of the most powerful approaches at present, combining the efficiency of the pseudopotential PW method with the accuracy provided by augmentation. Interestingly, it has been shown that the PAW and the ultrasoft pseudopotential method of Vanderbilt are closely related (Kresse and Joubert, 1999).

9.2.2 Muffin-tin orbitals: LCMTO and LMTO

In atom-centered augmented methods the wave functions in the interstitial region are expanded in exponentially decaying spherical Hankel functions centered on the atomic positions. These are matched to atomic wave functions at the surface of muffin-tin spheres. In the *linear combination of muffin-tin orbitals* method (LCMTO), the atomic wave functions are re-calculated at every self-consistent iterative step according to the value of the energy, until they satisfy a non-linear secular equation (Andersen and Wooley, 1973). It has been shown that the LCMTO method is equivalent to the KKR Green's function method (Korringa, 1947; Kohn and Rostoker, 1954). The linearization of the matching condition eliminates the energy dependence of the basis, thus leading to a linear method similar to LAPW, which receives the name of *linear muffin-tin orbitals* (LMTO). The linearized version of the MTOs was originally developed by Andersen (1975), and is quite a popular all-electron method because it is very fast. The fastness stems from the fact that the basis functions can be finely tuned so that a small number of them is enough to have an accurate description of the system. The price for this is a more complicated mathematical form. In this section we shall not give the explicit mathematical formulation of the LCMTO and LMTO methods. For this, the reader is referred to the many reviews (see, e.g., Andersen *et al.*, 2000, and references therein) and books (Skriver, 1984). Instead, we discuss some general aspects of these methods.

The LCMTO and LMTO methods have been extensively used within the muffin-tin (MTA) and atomic sphere (ASA) approximations, where the non-spherical contributions to the potential are neglected, thus leading to a secular equation of the form (8.79). The value of κ, i.e. the decay of the spherical Hankel function outside the spheres, is a small number that represents the kinetic energy of the envelope function in the flat interstitial region. At this point, the mathematical framework is notably simplified by choosing $\kappa = 0$ (Andersen, 1975). In that case the envelope function becomes an algebraically long-range decaying function given by the irregular solution of the Laplace equation $I_{lm}(\mathbf{r}) = r^{-(l+1)} C_{lm}(\theta, \phi)$. The LMTO-ASA scheme is very fast, as it requires only nine orbitals per atom for *spd*-systems, and the matrix elements are all calculated in terms of radial integrals in the inflated spheres. The LMTO-ASA can be corrected for the neglected overlapping region. A simple way is to take the potential as constant in that region (combined-correction), as proposed by Andersen (1975). For the energetics and band structures of close-packed crystalline materials, the LMTO-ASA plus combined correction method is very accurate indeed. The error in the kinetic energy has been estimated to be of the order of 10^{-3} hartree/atom (Andersen *et al.*, 1992).

MTO methods within the MTA and ASA approximations have difficulties in describing open (loosely packed) structures, e.g. the diamond structure. The interstitial region is poorly described by the Hankel functions, unless empty spheres, i.e. MT spheres with zero charge, are added in the voids to fill the empty space. This works well for symmetric structures where the voids also occupy high-symmetry positions. Nevertheless, the introduction of empty spheres makes the comparison of different structures at the energetic level more difficult, and accurate forces on the nuclear degrees of freedom cannot be computed, thus closing the road to geometry optimization and molecular dynamics. In order to address these points, the above approximations have to be abandoned and the potential has to be treated more accurately. Since the late eighties, several full-potential LMTO (FP-LMTO) approaches have been proposed (Methfessel *et al.*, 1989; Savrasov and Savrasov, 1992), which allow for an accurate computation of energies and forces, independently of the packing characteristics of the system. These schemes allow for several κ panels. Typically, between one and three are used.

The bottleneck of FP-LMTO schemes is the calculation of the matrix elements of the potential in the interstitial regions. There are basically two ways to proceed at this stage: either the potential in the interstitial is expanded in some suitable basis set of auxiliary atom-centered functions, or its Fourier representation in plane waves (or in a grid) is given. This latter approach is quite convenient, as it allows us to use the FFT machinery for solving Poisson's equation, it is simpler to compute the exchange-correlation potential, and it avoids the step of fitting the density and potential to a combination of auxiliary basis functions. Additional advantages regarding the coarseness of the FFT grid and in the size of the basis set are gained by a projection procedure in the intra-sphere region similar to the PAW method (Methfessel *et al.*, 2000). Other alternatives consist of expressing the products of envelope functions in an auxiliary basis, or performing one-range expansions of the Hankel functions into another site so as to transform two- and three-center integrals into one-center ones.

9.3 The pseudopotential plane wave method (PPW)

The Kohn–Sham equations in a PW basis set are given by Eq. (8.31), where the Kohn–Sham potential replaces the external potential. This introduces the self-consistency element because $v_{\text{KS}}[\rho]$ depends on the solutions of the equation through the electronic density,

$$\sum_{\mathbf{G}'} \left[\frac{\hbar^2}{2m} |\mathbf{k}+\mathbf{G}|^2 \delta_{\mathbf{G},\mathbf{G}'} + \tilde{v}_{\text{KS}}[\rho](\mathbf{G}-\mathbf{G}') \right] C_{j\mathbf{k}}(\mathbf{G}') = \epsilon_{j\mathbf{k}} C_{j\mathbf{k}}(\mathbf{G}). \qquad (9.24)$$

An important issue in PW calculations is that the size of the basis set can be enormous. As an illustration, for a cubic supercell of 10 bohr side and a cutoff of 40 Ry, according to Expression (8.40) the number of plane waves is about 34 000 per **k**-point. The actual number of independent PW components is smaller than that. If the **k**-point in the Brillouin zone is a high-symmetry point, e.g. the Γ-point (**k** = **0**), the wave functions in real space $\varphi_\mathbf{k}(\mathbf{r})$ can be taken as real so that $C_{j\mathbf{k}}(-\mathbf{G}) = C_{j\mathbf{k}}^*(\mathbf{G})$. Therefore, only half of the PW wave components are really necessary in those cases. For perfect crystalline structures the number of independent PW components can be reduced further by using the symmetry group of the crystal. Nevertheless, it remains a considerable number.

The immediate consequence of having to deal with a large number of basis functions is that the dimension of the matrix to diagonalize, and consequently the number of matrix elements to calculate, is very large. Therefore, PW methods that explicitly calculate and diagonalize the full Hamiltonian matrix tend to be computationally too costly. In fact, most modern PW codes are based on the iterative diagonalization of the Hamiltonian matrix. In iterative methods, matrix elements are never calculated explicitly. The only quantity that has to be computed is the action of the Hamiltonian on a trial wave function, i.e. $\hat{\mathcal{H}}\varphi^{(\mathbf{k})}(\mathbf{r})$, which can be viewed as a *generalized force* on the electronic orbitals. A description of various iterative diagonalization methods is given in Chapter 11.

In a PW basis set, advantage can be taken from the fact that *the kinetic term of the Hamiltonian is local in reciprocal space*, while *the potential term is local in real space*, although this last assertion is not strictly true when PW basis sets are used in conjunction with non-local pseudopotentials. In a DFT calculation, the Kohn–Sham potential is written

$$\hat{v}_{KS}[\rho] = \left\{v_{PS}^{loc}(\mathbf{r}, \mathbf{R}) + v_H[\rho(\mathbf{r})] + \mu_{XC}[\rho(\mathbf{r})]\right\}\hat{I} + \sum_{l=0}^{l_{max}} \Delta v_{PS}^l(r)\hat{P}_l. \quad (9.25)$$

The first term in the RHS represents an effective local potential that includes the local pseudopotential contribution and the Hartree and exchange-correlation terms arising from the electron–electron interaction. This part of $v_{KS}(\mathbf{r})$ is local in real space. The second term is the non-local pseudopotential contribution, which is local neither in real nor in reciprocal space.

The advantage of a substantial part of the potential being local in real space is that the action of the Hamiltonian onto the wave function can be simply calculated by multiplying the potential with the wave function in real space, and then Fourier transforming the product to reciprocal space. Once there, the kinetic contribution given by (8.29), which is local in reciprocal space, is added. The non-local pseudopotential contribution is most often computed in reciprocal space, although real-space implementations have also been proposed (King-Smith *et al.*, 1991). One of the key technical issues that makes PW calculations feasible is the possibility of using the fast Fourier transform method (FFT) to transform back and forth between real and reciprocal space. If the Fourier transform of the potential, the wave functions, or the density had to be calculated for each **G** component by integrating the real-space quantity on a grid, then the computational cost would scale like $\mathcal{O}(M^2)$, with M the number of PWs. The enormous advantage of the FFT technique is that the Fourier transform is calculated by a decimation method that reduces the cost of this part of the calculation to operations that scale like $\mathcal{O}(M \log M)$, with a smaller pre-factor (Press et al., 1992). In what follows we describe how the different energy terms and Hamiltonian matrix elements are calculated in PPW electronic structure codes.

9.3.1 Energy functional

In PW calculations under periodic boundary conditions, the actual system is infinitely periodically repeated. Of course, the energy of the infinite system is infinite but, since all the replicas of the simulation cell are identical, then the energy of the system *per cell* is a well-defined quantity that can be calculated through Kohn–Sham's expression:

$$E_{\text{KS}}[\rho] = \frac{1}{N_{\text{cell}}} \left(T_e[\rho] + E_{\text{H}}[\rho] + E_{\text{PS}}^{\text{loc}} + E_{ii} + E_{\text{XC}}[\rho] + E_{\text{PS}}^{nl} \right), \qquad (9.26)$$

where the factor N_{cell} in the denominator has been introduced to normalize the energy to a single simulation cell. Here, E_{ii} and $E_{\text{PS}}^{\text{loc}} + E_{\text{PS}}^{nl}$ are the ion–ion and electron–ion interaction energies, respectively.

Local electrostatic energy

The three local electrostatic contributions, E_{H}, $E_{\text{PS}}^{\text{loc}}$, and E_{ii}, diverge when taken individually. The reason for this, which represents a problem from the computational point of view, is that each individual term corresponds to the electrostatic interaction of a non-neutral charge distribution with an infinite set of periodic images of another non-neutral charge distribution.

This can be clearly seen by considering a single point charge q in a periodically repeated one-dimensional box of side a. The electrostatic energy of the infinite system is given by

$$E_{\text{es}}(q, q) = \frac{1}{2} \sum_{i=1, i \neq j}^{N_{\text{cell}}} \sum_{j=1}^{N_{\text{cell}}} \frac{q^2}{|x_i - x_j|} = N_{\text{cell}} \left(\frac{q^2}{2a} \right) \sum_{j=1}^{N_{\text{cell}}} \frac{1}{j}, \qquad (9.27)$$

where the last term involves the sum of the geometric series, which grows as N_{cell}. Therefore, the electrostatic energy scales as N_{cell}^2 so that even when it is normalized to a single cell, it still diverges as N_{cell}. The same happens if we consider a charge of opposite sign, and if we calculate the interaction between the two opposite charges of equal magnitude. The former leads to an electrostatic energy $E_{es}(-q,-q) = E_{es}(q,q)$, but for the latter the sign of the energy is negative and the double counting factor 1/2 is not present. Therefore this electrostatic contribution will have the form

$$E_{es}(q,-q) = -\sum_{i=1}^{N_{cell}}\sum_{j=1}^{N_{cell}} \frac{q^2}{|x_i-x_j|} = -N_{cell}\left(\frac{q^2}{a}\right)\sum_{j=1}^{N_{cell}} \frac{1}{j+\Delta}, \quad (9.28)$$

with $\Delta = d/a$ the distance between the oppositely charged particles in the cell, in units of the lattice constant a.

When the three terms are taken together the electrostatic energy of the system becomes

$$E_{es} = E_{es}(q,q) + E_{es}(-q,-q) + E_{es}(q,-q)$$

$$= N_{cell}\left(\frac{q^2}{a}\right)\sum_{j=1}^{N_{cell}}\left(\frac{1}{j} - \frac{1}{j+\Delta}\right)$$

$$\approx N_{cell}\left(\frac{q^2\Delta}{a}\right)\sum_{j=1}^{N_{cell}} \frac{1}{j^2}, \quad (9.29)$$

where the last expression has been obtained by expanding $1/(j+\Delta)$ in Taylor series and retaining the leading term in $1/j$. The overall sum is now convergent for $N_{cell} \to \infty$. The key issue to ensure the regularity of the energy per cell is that the charge distribution in the simulation cell must be neutral. The above reasoning can be generalized straightforwardly to continuous charge distributions, and combinations of continuous distributions and point-like charges, as in the present case.

Coming back to our problem, the contribution of the three local electrostatic terms to the Kohn–Sham energy is

$$E_{es}[\rho] = E_H[\rho] + E_{PS}^{loc}[\rho] + E_{ii} = \frac{1}{2}\int\int \frac{\rho(\mathbf{r})\rho(\mathbf{r}')}{|\mathbf{r}-\mathbf{r}'|}\,d\mathbf{r}\,d\mathbf{r}'$$

$$+ \sum_{s=1}^{N_s}\sum_{I=1}^{P_s}\int \rho(\mathbf{r})v_{PS}^{loc,s}(|\mathbf{r}-\mathbf{R}_I|)\,d\mathbf{r}$$

$$+ \frac{1}{2}\sum_{I=1}^{P}\sum_{J=1,J\neq I}^{P} \frac{Z_I Z_J}{|\mathbf{R}_I-\mathbf{R}_J|}, \quad (9.30)$$

where $\rho(\mathbf{r})$ is the electronic charge distribution, s in the second term labels the different atomic species, with N_s the number of atoms of species s, and $v_{PS}^{loc,s}(\mathbf{r})$ is the local component of the pseudopotential corresponding to that species, whose long-distance behavior is purely coulombic. Z_I is the valence charge of the ionic core I, and $\mathbf{R}_I$ its location. The electronic charge density is taken positive by convention. Therefore, the charge associated with the ionic cores (or nuclei) is negative. This strange convention is due purely to computational reasons, because it is more convenient to work with arrays of positive numbers for the electronic density.

According to the preceding digression, all three terms diverge for an infinite system. However, the total charge distribution of valence electrons plus ionic cores is neutral, and then the electrostatic energy of a single cell should be finite. In order to sort out this problem, we consider a neutralizing, continuous auxiliary charge distribution $\rho_i(\mathbf{r})$ associated with the nuclear subsystem (and thus negative), and add and subtract the electrostatic self-interaction energy of this distribution,

$$\tilde{E}_{ii} = \frac{1}{2} \int \int \frac{\rho_i(\mathbf{r})\rho_i(\mathbf{r}')}{|\mathbf{r}-\mathbf{r}'|} d\mathbf{r} \, d\mathbf{r}'. \quad (9.31)$$

After some elementary algebra the electrostatic energy can be re-written as

$$E_{\text{es}}[\rho] = \frac{1}{2} \int \int \frac{\rho_T(\mathbf{r})\rho_T(\mathbf{r}')}{|\mathbf{r}-\mathbf{r}'|} d\mathbf{r} \, d\mathbf{r}'$$

$$+ \int \rho(\mathbf{r}) \left(\sum_{s=1}^{N_s} \sum_{I=1}^{P_s} v_{\text{PS}}^{\text{loc},s}(|\mathbf{r}-\mathbf{R}_I|) - \int \frac{\rho_i(\mathbf{r}')}{|\mathbf{r}-\mathbf{r}'|} d\mathbf{r}' \right) d\mathbf{r}$$

$$+ \frac{1}{2} \left(\sum_{I=1}^{P} \sum_{J=1, J \neq I}^{P} \frac{Z_I Z_J}{|\mathbf{R}_I - \mathbf{R}_J|} - \int \int \frac{\rho_i(\mathbf{r})\rho_i(\mathbf{r}')}{|\mathbf{r}-\mathbf{r}'|} d\mathbf{r} \, d\mathbf{r}' \right), \quad (9.32)$$

where $\rho_T(\mathbf{r}) = \rho(\mathbf{r}) + \rho_i(\mathbf{r})$ is a neutral charge distribution. With this rearrangement, all the three terms give an energy contribution that is finite in the simulation cell.

At this point it is important to establish a convention for the *practical* definition of Fourier transforms. The inverse FT (reciprocal to real space) is defined as

$$f(\mathbf{r}) = FT^{-1}[\tilde{f}(\mathbf{G})] = \sum_{\mathbf{G}} \tilde{F}(\mathbf{G}) e^{i\mathbf{G}\cdot\mathbf{r}}, \quad (9.33)$$

while the direct FT (real to reciprocal space) is

$$\tilde{f}(\mathbf{G}) = FT[f(\mathbf{r})] = \frac{1}{\Omega} \int_{\Omega} f(\mathbf{r}) e^{-i\mathbf{G}\cdot\mathbf{r}} d\mathbf{r} \approx \frac{1}{N_R} \sum_{n=1}^{N_R} f(\mathbf{r}_n) e^{-i\mathbf{G}\cdot\mathbf{r}_n}, \quad (9.34)$$

where N_R is the number of points in the real-space grid, and the last sum runs over the grid points.

The real-space calculation of the Coulomb double integrals is an extremely expensive operation. However, a second glance reveals that these are simple convolution products. Therefore, using the fact that the Fourier transform of a convolution is the product of the Fourier transforms,

$$\int \frac{\rho_T(\mathbf{r}')}{|\mathbf{r}-\mathbf{r}'|} \, d\mathbf{r}' = FT^{-1}\left[\tilde{\rho}_T(\mathbf{G})\frac{4\pi}{G^2}\right] = \sum_{\mathbf{G}} \tilde{\rho}_T(\mathbf{G})\frac{4\pi}{G^2}e^{i\mathbf{G}\cdot\mathbf{r}}, \quad (9.35)$$

where $4\pi/G^2$ is the Fourier transform of the Coulomb potential. When this expression is multiplied by the charge density and integrated, the result is

$$\frac{1}{2}\int\int \frac{\rho_T(\mathbf{r})\rho_T(\mathbf{r}')}{|\mathbf{r}-\mathbf{r}'|} \, d\mathbf{r} \, d\mathbf{r}' = N_{\text{cell}}\frac{\Omega}{2}\sum_{\mathbf{G}\neq 0}\frac{4\pi}{G^2}\tilde{\rho}_T(\mathbf{G})\tilde{\rho}_T(-\mathbf{G}). \quad (9.36)$$

With this transformation the double integral in real space is evaluated as a single sum over PW components, while the Fourier transform of the density can be efficiently done via FFTs. The divergent $\mathbf{G} = \mathbf{0}$ term in the sum is avoided because the $\mathbf{G} = \mathbf{0}$ component of the charge density, given by

$$\tilde{\rho}_T(\mathbf{G}=\mathbf{0}) = \frac{1}{\Omega}\int_\Omega \rho_T(\mathbf{r}) \, d\mathbf{r} = \frac{Q_T}{\Omega}, \quad (9.37)$$

vanishes when the distribution $\rho_T(\mathbf{r})$ is neutral (Q_T is the integrated charge).

The second term ($\tilde{E}_{\text{PS}}^{\text{loc}}$ for short) can also be calculated more efficiently by Fourier transforming the density and the local pseudopotential. Using the convolution theorem again we have

$$\int \rho(\mathbf{r})v_{\text{PS}}^{\text{loc},s}(|\mathbf{r}-\mathbf{R}_I|) \, d\mathbf{r} = \Omega\sum_{\mathbf{G}} e^{i\mathbf{G}\cdot\mathbf{R}_I}\tilde{v}_{\text{PS}}^{\text{loc},s}(G)\tilde{\rho}(-\mathbf{G}). \quad (9.38)$$

We now define the atomic *structure factor* of species s as

$$S_s(\mathbf{G}) = \sum_{I=1}^{P_s} e^{-i\mathbf{G}\cdot\mathbf{R}_I^s}, \quad (9.39)$$

where $\mathbf{R}_I^s$ are the positions of the atoms belonging to that species. All the information on the nuclear positions is contained in $S_s(\mathbf{G})$. This atomic structure factor is not to be confused with what is usually known as structure factor in statistical mechanics, i.e. the quantity measured in scattering experiments that is obtained by Fourier transforming the pair correlation function. The latter is a statistical average, while the present one corresponds to a single, frozen configuration.

Combining the two expressions above we obtain for $\tilde{E}_{\text{PS}}^{\text{loc}}$:

$$\tilde{E}_{\text{PS}}^{\text{loc}}[\rho] = N_{\text{cell}}\Omega\sum_{\mathbf{G}}\left[\sum_{s=1}^{N_s} S_s(\mathbf{G})\tilde{v}_{\text{PS}}^{\text{loc},s}(G) - \frac{4\pi}{G^2}\tilde{\rho}_i(\mathbf{G})\right]\tilde{\rho}(-\mathbf{G}), \quad (9.40)$$

where the sum in the first term on the RHS runs over all the different atomic species in the system. The distinction between atomic species is required because usually the local pseudopotential contribution depends on the species, although this is not strictly necessary.

Here the $\mathbf{G}=0$ term requires careful consideration. As before, it does not diverge, but it assumes a finite value instead of zero. The $\mathbf{G}=0$ component of the structure factor is $S_s(0) = P_s$, the number of atoms belonging to species s. Since the auxiliary charge distribution $\rho_i(\mathbf{r})$ has been tailored to neutralize the electronic charge, the $\mathbf{G}=0$ component of its Fourier transform is $\tilde{\rho}_i(0) = \Omega^{-1} \int \rho_i(\mathbf{r})\,d\mathbf{r} = -Q/\Omega$, where $Q = Ne$ is the total electronic charge (a positive number) in the simulation cell.

This, however, is not sufficient. In fact, since $\tilde{\rho}_i(G)$ is multiplied by $4\pi/G^2$ one should keep terms up to the order G^2, and only then let $G \to 0$. By doing that, and taking into account that the auxiliary charge density cannot be asymmetric, we have $\tilde{\rho}_i(G) = \tilde{\rho}_i(0) + \tilde{\rho}_i''(0)G^2/2$. Therefore, apart from the multiplicative factor N_{cell}, the $\mathbf{G}=0$ term in (9.40) can be written as

$$\lim_{G \to 0} \left[\Omega \sum_{s=1}^{N_s} P_s \tilde{v}_{\text{PS}}^{\text{loc},s}(G) + \frac{4\pi Q}{G^2} \right] \tilde{\rho}(G) - 2\pi\Omega\tilde{\rho}_i''(0)\tilde{\rho}(0). \qquad (9.41)$$

The Fourier transform of the local part of the pseudopotential can be written as

$$\tilde{v}_{\text{PS}}^{\text{loc},s}(G) = \frac{1}{\Omega} \int_{\Omega} v_{\text{PS}}^{\text{loc},s}(r) e^{-i\mathbf{G}\cdot\mathbf{r}}\,d\mathbf{r}$$

$$= \frac{1}{\Omega} \int_{\Omega_c} \left(v_{\text{PS}}^{\text{loc},s}(r) + \frac{Z_s}{r} \right) e^{-i\mathbf{G}\cdot\mathbf{r}}\,d\mathbf{r} + \tilde{v}_c(G), \qquad (9.42)$$

where Ω_c is the volume of the pseudized region, Z_s is the valence charge of species s (a positive value), and we have used the fact that the pseudopotential is identical to $-Z_s/r$ beyond the cutoff radius, so that its Fourier transform is given by $\Omega\tilde{v}_c(G) = -4\pi Z_s/G^2$. For $\mathbf{G} = 0$ the integral in the first term is a finite quantity that we denote by $\Delta\tilde{v}_{\text{PS}}^{\text{loc},s}$, while the second term is simply the divergent $\tilde{v}_c(0)$. Therefore, the $\mathbf{G}=0$ term in (9.40) becomes

$$\sum_{s=1}^{N_s} P_s \Delta\tilde{v}_{\text{PS}}^{\text{loc},s} \tilde{\rho}(0) + \lim_{G \to 0} \left[-\sum_{s=1}^{N_s} P_s Z_s + Q \right] \frac{4\pi}{G^2} \tilde{\rho}(0) - 2\pi\Omega\tilde{\rho}_i''(0)\tilde{\rho}(0), \qquad (9.43)$$

where $Q_s = P_s Z_s$ is the total charge associated with species s. It is clear that the second term vanishes because $\sum_s Q_s = Q$, thus compensating the unpleasant divergence of the Coulomb potential.

We now replace $\tilde{\rho}(0) = Q/\Omega$ to obtain the final expression for the second term in (9.32):

$$\tilde{E}_{\text{PS}}^{\text{loc}}[\rho]/N_{\text{cell}} = \Omega \sum_{\mathbf{G} \neq \mathbf{0}} \left[\sum_{s=1}^{N_s} S_s(\mathbf{G}) \tilde{v}_{\text{PS}}^{\text{loc},s}(G) - \frac{4\pi}{G^2} \tilde{\rho}_i(\mathbf{G}) \right] \tilde{\rho}(-\mathbf{G})$$

$$+ \frac{Q}{\Omega} \sum_{s=1}^{N_s} P_s \Delta \tilde{v}_{\text{PS}}^{\text{loc},s} - 2\pi Q \tilde{\rho}_i''(0). \tag{9.44}$$

Ewald sums

The third term in (9.32) involves the computation of the Coulomb energy of a periodic collection of point-like particles. As discussed before, this type of summation diverges, but the self-interaction of the auxiliary charge distribution $\rho_i(\mathbf{r})$, i.e. the term that is subtracted from the interaction between ionic cores, cancels the divergence. The fact remains that the real-space numerical evaluation of this sum is extremely difficult because it is only conditionally convergent, i.e. it depends on how the terms in the summation are grouped. Coulomb interactions are also long-ranged in reciprocal space, so that Fourier transformation does not solve the problem. The solution was provided in 1917 by Ewald (1921), who proposed a powerful technique that consists of dividing the summation into a real- and a reciprocal-space part, both of which are rapidly convergent. Ewald's method starts from the mathematical identity (Payne et al., 1992):

$$\sum_{L=-\infty}^{\infty} \frac{1}{|\mathbf{R}_1 + L - \mathbf{R}_2|} = \frac{2}{\sqrt{\pi}} \sum_{L=-\infty}^{\infty} \int_{\eta}^{\infty} e^{-|\mathbf{R}_1 + L - \mathbf{R}_2|^2 \sigma^2} d\sigma$$

$$+ \frac{2\pi}{\Omega} \sum_{\mathbf{G}} \int_0^{\eta} e^{-G^2/4\sigma^2} e^{i\mathbf{G} \cdot (\mathbf{R}_1 - \mathbf{R}_2)} \frac{d\sigma}{\sigma^3}, \tag{9.45}$$

which, after some algebra, carries to the following expression for the energy per cell, known as *Ewald sum* (Ihm et al., 1979):

$$E_{ii}/N_{\text{cell}} = \frac{1}{2} \sum_{I=1}^{P} \sum_{J \neq I}^{P} Z_I Z_J \sum_{L=-\infty}^{\infty} \frac{1}{|\mathbf{R}_I + L - \mathbf{R}_J|}$$

$$= \frac{1}{2} \sum_{I=1}^{P} \sum_{J \neq I}^{P} Z_I Z_J \left[\sum_{L=-\infty}^{\infty} \frac{\text{erfc}(|\mathbf{R}_I + L - \mathbf{R}_J| \eta)}{|\mathbf{R}_I + L - \mathbf{R}_J|} \right]$$

$$+ \frac{1}{2\Omega} \sum_{\mathbf{G}} \frac{4\pi}{G^2} e^{-G^2/4\eta^2} \left| \sum_s Z_s S_s(\mathbf{G}) \right|^2 - \frac{\eta}{\sqrt{\pi}} \sum_{I=1}^{P} Z_I^2, \tag{9.46}$$

where erfc(x) is the complementary error function, which decays almost like a Gaussian, and $1/\eta$ is a cutoff distance, which is appropriately chosen to optimize the convergence properties of real- and reciprocal-space sums (Allen and Tildesley, 1987).

This implies that the lattice sums in the first term can be cut off after a certain number $L_{\max}$ of periodically repeated boxes, while the reciprocal-space sum in the second term can also be terminated at some cutoff G_{cut}. In practice, G_{cut} has been already determined by the PW expansion of the electronic wave functions, and the value of η has to be chosen such that the reciprocal-space sum converges within G_{cut}. Typical values of $L_{\max}$ are 0 or 1, except for very small cells and low cutoffs. If η is chosen in such a way that $L_{\max} = 0$, then only the interactions with the closest images of the other atoms are considered. This is known as *minimum image convention*.

When $I = J$, the $L = 0$ term in the first summation should be absent because it would imply that the ionic core I interacts with itself. This is not so if $L \neq 0$, because ionic cores do interact with their images in the periodically repeated boxes. Because of the partition in the Ewald construction, part of the self-interaction has been displaced to the reciprocal-space sum. This contribution can be readily obtained by letting $\mathbf{R}_1 = \mathbf{R}_2$ in (9.45), and then calculating the reciprocal-space sum as an integral. This results in the negative of the last term in (9.46), which is subtracted to compensate for the self-interaction in the previous term, which is not removed.

As it is, this expression diverges because it corresponds to a non-neutral charge distribution. However, this is cured by the self-interaction energy of the auxiliary charge distribution $\rho_i(\mathbf{r})$, which is formally identical to Expression (9.36), so that:

$$\frac{1}{2} \int \int \frac{\rho_i(\mathbf{r})\rho_i(\mathbf{r}')}{|\mathbf{r}-\mathbf{r}'|} \, d\mathbf{r} \, d\mathbf{r}' = N_{\mathrm{cell}} \frac{\Omega}{2} \sum_{\mathbf{G}} \frac{4\pi}{G^2} \tilde{\rho}_i(\mathbf{G}) \, \tilde{\rho}_i(-\mathbf{G}). \tag{9.47}$$

Therefore, when the two terms are taken together as in (9.32), we obtain

$$\tilde{E}_{ii}/N_{\mathrm{cell}} = \frac{1}{2} \sum_{I=1}^{P} \sum_{J \neq I}^{P} Z_I Z_J \left[\sum_{L=-L_{\max}}^{L_{\max}} \frac{\mathrm{erfc}\,(|\mathbf{R}_I + l - \mathbf{R}_J|\eta)}{|\mathbf{R}_I + l - \mathbf{R}_J|} \right]$$

$$+ \frac{1}{2\Omega} \sum_{\mathbf{G} \neq 0} \frac{4\pi}{G^2} \left[e^{-G^2/4\eta^2} \left| \sum_{s} Z_s S_s(\mathbf{G}) \right|^2 - \Omega^2 |\tilde{\rho}_i(\mathbf{G})|^2 \right]$$

$$- \frac{\eta}{\sqrt{\pi}} \sum_{I=1}^{P} Z_I^2 - \frac{\pi Q^2}{2\Omega \eta^2} - 2\pi \Omega \tilde{\rho}_i''(0)\tilde{\rho}_i(0), \tag{9.48}$$

9.3 The pseudopotential plane wave method (PPW)

where the $\mathbf{G} = \mathbf{0}$ term in the reciprocal-space sum has been omitted. In fact, the expansion of the Gaussian around $G = 0$ gives rise to a divergent term $4\pi Q^2/G^2$ plus a regular term $-\pi Q^2/\eta^2$, where $Q = \sum_s Z_s S_s(0)$. The divergent term is exactly canceled by a similar divergence in the second term, $-4\pi\Omega^2 \tilde{\rho}_i^2(0)/G^2$, because $\tilde{\rho}_i(0) = -Q/\Omega$ by construction. This second term also has a regular contribution. The two regular contributions become the last two terms in (9.48).

This suggests that a convenient choice for the auxiliary charge distribution $\rho_i(\mathbf{r})$ is the following:

$$\rho_i(\mathbf{r}) = -\frac{\eta^3}{\pi^{3/2}} \sum_{I=1}^{P} Z_I e^{-2\eta^2 |\mathbf{r}-\mathbf{R}_I|^2}, \tag{9.49}$$

$$\tilde{\rho}_i(\mathbf{G}) = -\frac{1}{\Omega} e^{-G^2/8\eta^2} \left(\sum_{s=1}^{N_s} Z_s S_s(\mathbf{G}) \right),$$

because in this way the two contributions in the second term of (9.48) are identical, and thus cancel out. Moreover, with this choice $\tilde{\rho}_i''(0) = Q/4\eta^2 \Omega$, so that also the last two regular terms in (9.48) cancel, and the only surviving contribution is the real-space sum. This distribution, which is often called the *pseudo-ionic charge density*, can also be viewed as a *smearing* of the point-like ionic charge into a distribution with the shape of an error function. The purpose of this construction is strictly computational convenience, in order to ensure the convergence of the lattice sums. The smearing is then corrected back.

In conclusion, if we now define a modified pseudopotential that includes the contribution of the pseudo-ionic charge density as

$$\tilde{u}_{PS}^{loc,s}(G) = \tilde{v}_{PS}^{loc,s}(G) + \frac{4\pi}{G^2} \frac{Z_s}{\Omega} e^{-G^2/8\eta^2}, \tag{9.50}$$

then the local electrostatic energy per cell is written in the following way:

$$E_{es}[\rho]/N_{cell} = \frac{\Omega}{2} \sum_{\mathbf{G} \neq 0} \frac{4\pi}{G^2} \tilde{\rho}_T(\mathbf{G}) \tilde{\rho}_T(-\mathbf{G})$$

$$+ \Omega \sum_{\mathbf{G} \neq 0} \sum_{s=1}^{N_s} S_s(\mathbf{G}) \tilde{u}_{PS}^{loc,s}(G) \tilde{\rho}(-\mathbf{G})$$

$$+ \frac{1}{2} \sum_{I=1}^{P} \sum_{J \neq I}^{P} Z_I Z_J \left[\sum_{L=-L_{max}}^{L_{max}} \frac{\text{erfc}(|\mathbf{R}_I + \mathbf{L} - \mathbf{R}_J| \eta)}{|\mathbf{R}_I + \mathbf{L} - \mathbf{R}_J|} \right]$$

$$- \frac{\eta}{\sqrt{\pi}} \sum_{I=1}^{P} Z_I^2 - \frac{Q}{\Omega} \left(\sum_{s=1}^{N_s} P_s \Delta \tilde{v}_{PS}^{loc,s} + \frac{\pi Q}{2\eta^2} \right), \tag{9.51}$$

where the last term arises from the regular (non-Coulomb) contribution of the modified local pseudopotential term, $\tilde{E}_{\text{PS}}^{\text{loc}}$, at $\mathbf{G}=\mathbf{0}$. It can also be written as $-Q\left(\sum_{s=1}^{N_s} P_s \Delta\tilde{u}_{\text{PS}}^{\text{loc},s}\right)/\Omega$.

Other energy terms

Apart from the electrostatic energy, the other energy terms in the total energy per cell are the following.

(i) The *exchange-correlation* energy $E_{\text{XC}}[\rho] = \int \rho(\mathbf{r})\epsilon_{\text{XC}}[\rho]\,d\mathbf{r}$, which is calculated by integrating the XC energy density numerically in the real-space grid,

$$\tilde{E}_{\text{XC}}(\mathbf{G}) = \frac{\Omega}{N_R}\sum_{n=1}^{N_R} \rho(\mathbf{r}_n)\,\epsilon_{\text{XC}}[\rho(\mathbf{r}_n)]. \tag{9.52}$$

(ii) The *kinetic energy*, whose expression in real space is:

$$T_e[\rho(\mathbf{r})] = -\frac{\hbar^2}{2m}\sum_{\mathbf{k}\in BZ}\omega_{\mathbf{k}}\sum_{i=1}^{N_{\mathbf{k}}} f_i^{(\mathbf{k})}\int \varphi_i^{(\mathbf{k})*}(\mathbf{r})\nabla^2\varphi_i^{(\mathbf{k})}(\mathbf{r})\,d\mathbf{r}, \tag{9.53}$$

where $f_i^{\mathbf{k}}$ is the occupation number of state i at wave vector $\mathbf{k}$, and $\omega_{\mathbf{k}}$ is the weight of point $\mathbf{k}$ in the BZ averages. This term is computed more efficiently in reciprocal space where the kinetic operator is diagonal:

$$T_e[\rho] = \frac{\hbar^2}{2m}\sum_{\mathbf{k}\in BZ}\omega_{\mathbf{k}}\sum_{i=1}^{N_{\mathbf{k}}} f_i^{(\mathbf{k})}\sum_{\mathbf{G}}|\mathbf{k}+\mathbf{G}|^2\,|C_{i\mathbf{k}}(\mathbf{G})|^2. \tag{9.54}$$

(iii) The *non-local pseudopotential* energy, whose expression will be given in Section 9.3.2.

The electronic density, which is also needed in the calculation, is more easily calculated in real space:

$$\rho(\mathbf{r}) = \sum_{\mathbf{k}\in BZ}\omega_{\mathbf{k}}\sum_{i=1}^{N_{\mathbf{k}}} f_i^{(\mathbf{k})}\left|\varphi_i^{(\mathbf{k})}(\mathbf{r})\right|^2. \tag{9.55}$$

9.3.2 The Kohn–Sham potential

The Kohn–Sham potential can be computed as the functional derivative of the Kohn–Sham energy functional with respect to the density. It comprises the local electrostatic and the exchange-correlation contributions that are local in real space, and the non-local pseudopotential contribution. The local electrostatic part is more easily calculated in reciprocal space and then transformed to real space via FFT, while the exchange-correlation term is calculated directly by evaluating the XC functional on the real-space grid.

Local electrostatic contribution

The local electrostatic part of the Kohn–Sham potential is given by the functional derivative

$$v_{\text{KS}}^{\text{es}}(\mathbf{r}) = \delta E_{\text{es}}[\rho]/\delta\rho(\mathbf{r}), \quad (9.56)$$

By Fourier transforming this identity, we obtain the corresponding expression in reciprocal space:

$$\tilde{v}_{\text{KS}}^{\text{es}}(\mathbf{G}) = \frac{1}{\Omega}\left(\frac{\delta E_{\text{es}}[\rho]}{\delta\rho(-\mathbf{G})}\right) = \frac{4\pi}{G^2}\tilde{\rho}_T(\mathbf{G}) + \sum_{s=1}^{N_s} S_s(\mathbf{G})\tilde{u}_{\text{PS}}^{\text{loc},s}(G), \quad (9.57)$$

because only the first two terms in (9.51) provide a non-zero contribution. If the Hamiltonian matrix element between two plane waves $\tilde{v}_{\text{KS}}^{\text{es}}(\mathbf{G},\mathbf{G}')$ is required, then it is sufficient to replace $\mathbf{G}$ by $\mathbf{G}-\mathbf{G}'$ in the above expression. Since $\rho_T = \rho + \rho_i$, the first term contains a contribution $4\pi\rho_i(\mathbf{G})/G^2$ that exactly cancels an equal contribution in $\tilde{u}_{\text{PS}}^{\text{loc},s}(G)$. The present form, however, is more convenient to show that the term $\tilde{v}_{\text{KS}}^{\text{es}}(0)$ vanishes, thus avoiding the $G=0$ divergence (see below).

The Fourier transform of the local part of the modified pseudopotential, $\tilde{u}_{\text{PS}}^{\text{loc},s}(G)$, depends only on the modulus of $\mathbf{G}$, and is calculated either analytically or numerically by Fourier transforming $u_{\text{PS}}^{\text{loc},s}(r)$ in a one-dimensional grid. Due to the $1/r$ behavior at long distances, it is convenient to use a *logarithmic*, rather than uniform, radial grid to compute $\tilde{u}_{\text{PS}}^{\text{loc},s}(G)$. Logarithmic grids, where the sequence of grid points is defined by $x_{i+1} = Cx_i$, with C a constant, are finer close to the origin where the variation of the potential and wave function is more pronounced, while at long distances they become increasingly coarser.

Potential reference (zero)

In a periodic system it is not possible to set the zero of the potential at infinity, as is done for atomic or molecular systems, because the charge density extends to the whole space. Therefore, the reference has to be chosen in a different way. The $\mathbf{G}=0$ component of the local electrostatic potential can be obtained from Expression (9.57) by taking the limit $G \to 0$:

$$\tilde{v}_{\text{KS}}^{\text{es}}(0) = \lim_{G \to 0}\left\{\frac{4\pi Q_T}{G^2} + \sum_{s=1}^{N_s} P_s\left(\tilde{v}_{\text{PS}}^{\text{loc},s}(G) + \frac{4\pi}{G^2}\frac{Z_s}{\Omega}\right)\right\}. \quad (9.58)$$

In the expression above, Q_T is the total charge in the supercell. It has already been established that charge neutrality is required in order to avoid the $\mathbf{G}=0$ divergence arising from the long-range behavior of the electrostatic potential under PBC. Therefore, $Q_T = 0$ and the first term vanishes. Also, the second term vanishes because for $G \to 0$ it is $\tilde{v}_{\text{PS}}^{\text{loc},s}(G) \to -4\pi Z_s/G^2\Omega$. The zero of

240   *Electronic structure methods*

the electrostatic potential is then fixed by the condition that its average in the simulation cell is zero:

$$\tilde{v}_{KS}^{es}(\mathbf{G}=\mathbf{0}) = \int v_{KS}^{es}(\mathbf{r})\,d\mathbf{r} = 0. \qquad (9.59)$$

Plane waves automatically imply PBC. Hence, it is not possible to study charged systems within a PW approach straightforwardly. For charged systems the customary approach (although not fully justified) is to artificially neutralize the charge in the supercell by adding a uniformly distributed charge of opposite sign. In practice, this simply means that we ignore the $\mathbf{G}=0$ term in the potential and add a background energy to the total energy.

Exchange-correlation

The exchange-correlation potential is local in real space. Therefore, it is calculated by evaluating the appropriate expression $\mu_{XC}[\rho(\mathbf{r})]$, where $\rho(\mathbf{r})$ is the electronic density available at the real-space grid points. In the case of gradient-corrected functionals this approach is still valid, but in PWs the gradient of the density is most conveniently calculated in reciprocal space and then transformed to real space via inverse FFT:

$$\nabla \rho(\mathbf{r}) = \sum_{\mathbf{G}} \rho(\mathbf{G}) \nabla(e^{i\mathbf{G}\cdot\mathbf{r}}) = \sum_{\mathbf{G}} i\mathbf{G}\rho(\mathbf{G}) e^{i\mathbf{G}\cdot\mathbf{r}}, \qquad (9.60)$$

and similarly for the Laplacian, if required. The exchange-correlation Hamiltonian matrix elements in the PW basis set $\mu_{XC}(\mathbf{G},\mathbf{G}')$ are obtained by Fourier transforming the real-space potential via FFT.

Non-local pseudopotential

The only contribution that is somewhat more complicated to compute is the one corresponding to the non-local components of the pseudopotential. For a particular l-component, the matrix elements for an atom of species s located at the origin are:

$$\begin{aligned}
\Delta \tilde{V}_{\mathbf{k}+\mathbf{G},\mathbf{k}+\mathbf{G}'}^{l,s} &= \langle \mathbf{k}+\mathbf{G}|\Delta \tilde{V}_{PS}^{l,s} \hat{P}_l|\mathbf{k}+\mathbf{G}'\rangle \\
&= \sum_{m=-l}^{l} \langle \mathbf{k}+\mathbf{G}|Y_{lm}\rangle \Delta v_{PS}^{l,s}(r) \langle Y_{lm}|\mathbf{k}+\mathbf{G}'\rangle \\
&= 4\pi(2l+1) P_l(\cos\theta_{\mathbf{k}+\mathbf{G},\mathbf{k}+\mathbf{G}'}) \\
&\quad \times \int r^2 j_l(|\mathbf{k}+\mathbf{G}|r) \Delta v_{PS}^{l,s}(r) j_l(|\mathbf{k}+\mathbf{G}'|r)\,dr, \qquad (9.61)
\end{aligned}$$

where $P_l(\cos\theta)$ are Legendre polynomials and $j_l(x)$ are spherical Bessel functions. Notice that, unlike the case of local contributions to the potential, non-local pseudopotential matrix elements depend on the **k**-point in the BZ.

The $M(M+1)/2$ integrals in the last line can be calculated by quadratures. However, it is much more efficient to use the alternative method proposed by Kleinman and Bylander (1982), as discussed in Section 7.3.1. The matrix elements of the separable KB projection operator for the target system are

$$\Delta \tilde{V}^{l,s}_{\mathbf{k+G},\mathbf{k+G'}}(KB) = \sum_{m=-l}^{l}\sum_{I=1}^{P} \alpha_{lm}^s e^{-i\mathbf{G}\cdot\mathbf{R}_I} f_{lm}^{s*}(\mathbf{k+G}) f_{lm}^s(\mathbf{k+G'}) e^{i\mathbf{G'}\cdot\mathbf{R}_I} \quad (9.62)$$

with

$$f_{lm}^s(\mathbf{k+G}) = \langle \Phi_{\mathrm{PS}}^{lm,s} | \Delta\hat{V}_{\mathrm{PS}}^{l,s} | \mathbf{k+G}\rangle$$

$$= \int r^2 \Phi_{\mathrm{PS}}^{lm,s}(r) \Delta v_{\mathrm{PS}}^{l,s}(r) j_l(|\mathbf{k+G}|r)\,dr \quad (9.63)$$

and

$$\alpha_{lm}^s = \langle \Phi_{\mathrm{PS}}^{lm,s} | \Delta\hat{V}_{\mathrm{PS}}^{l,s} | \Phi_{\mathrm{PS}}^{lm,s}\rangle^{-1} = \left(\int r^2 |\Phi_{\mathrm{PS}}^{lm,s}(r)|^2 \Delta v_{\mathrm{PS}}^{l,s}(r)dr\right)^{-1}. \quad (9.64)$$

The advantage of this formulation is obvious: the calculation of the matrix elements is reduced to the evaluation of only the M integrals $f_{lm}^s(\mathbf{k+G})$, instead of the former $M(M+1)/2$.

The non-local contribution of species s to the energy is then given by the following expression:

$$E_{\mathrm{PS}}^{\mathrm{nl}} = \sum_{\mathbf{k}\in BZ} \omega_{\mathbf{k}} \sum_{l=0}^{l_{\max}} \sum_{m=-l}^{l} \sum_{I=1}^{P} \sum_{j=1}^{N_{\mathbf{k}}} f_j^{(\mathbf{k})} \alpha_{lm}^s \left|F_{I,j}^{lm,s}\right|^2, \quad (9.65)$$

with

$$F_{I,j}^{lm,s} = \sum_{\mathbf{G}} e^{i\mathbf{G}\cdot\mathbf{R}_I} f_{lm}^s(\mathbf{k+G}) C_{j\mathbf{k}}(\mathbf{G}). \quad (9.66)$$

The non-local pseudopotential contribution to the Hamiltonian and the energy can also be computed in a *real-space* implementation. In fact, since the pseudopotential components $\Delta\hat{V}_{\mathrm{PS}}^{l,s}$ identically vanish outside the cutoff radius, the cost of computing these terms in real space should scale more favorably than the reciprocal-space implementation. The rationale is that the sums over the M **G**-vectors in the basis set are replaced by integrals confined to the core region that has a finite extent independently of the system size, while M grows linearly with the number of atoms. For increasing system size the reciprocal-space implementation becomes one of the bottlenecks of the PPW calculation, scaling as $\mathcal{O}(N^3)$. The real-space implementation, scaling as $\mathcal{O}(N^2)$, is potentially more efficient.

The road to the real-space implementation, however, presents some technical difficulties that have been addressed by King-Smith, Payne, and Lin. The basic problem is that the wave function in real space is known only at the FFT grid points, so that real-space integrals can only be approximated by sums over these points (King-Smith *et al.*, 1991). In the reciprocal-space implementation these become sums over PW components, which correspond to a proper real-space integration where the regions between grid points are Fourier interpolated. The only approximation is set by the PW cutoff G_{cut}. Therefore, the real-space approach of summing over FFT grid points cannot represent the PW sums accurately enough. The solution proposed by King-Smith *et al.* consists of variationally modifying the Kleinman–Bylander projector operators $|\zeta^l\rangle$ in such a way as to minimize the difference between the real-space sums on the FFT grid and the corresponding reciprocal-space sums computed with the original projectors. Of course, this procedure is not perfect, but the errors can be made quite small, of the order of a few μeV in the energies and a few tens of μeV/Å in the forces. This procedure is implemented in the CASTEP code (Segall *et al.*, 2002).

For the matrix elements of Vanderbilt's ultrasoft pseudopotentials the reader is referred to the original work (Laasonen *et al.*, 1993).

9.3.3 Generalized forces on the electronic orbitals

It has been mentioned that most PPW codes do not construct the Hamiltonian matrix explicitly because they are based on iterative diagonalization techniques. These require only the computation of the action of the Hamiltonian operator on a trial wave function. This product can be viewed as a *generalized force* on the electronic orbitals, which is obtained from the Kohn–Sham energy (9.26) by functional derivation with respect to the Kohn–Sham orbitals:

$$\frac{\delta E_{\text{KS}}[\rho]}{\delta \varphi_i^{(\mathbf{k})*}(\mathbf{r})} = \omega_{\mathbf{k}} f_i^{(\mathbf{k})} \hat{\mathcal{H}}_{\text{KS}} \varphi_i^{(\mathbf{k})}(\mathbf{r}). \tag{9.67}$$

The generalized forces in reciprocal space are obtained by Fourier transforming the above expression:

$$\frac{\delta E_{\text{KS}}[\rho]}{\delta C_{i\mathbf{k}}^*(-\mathbf{G})} = \omega_{\mathbf{k}} f_i^{(\mathbf{k})} \hat{\mathcal{H}}_{\text{KS}} C_{i\mathbf{k}}(\mathbf{G}). \tag{9.68}$$

In what follows, the factors $\omega_{\mathbf{k}}$ and $f_i^{(\mathbf{k})}$ are omitted for the sake of clarity.

9.3 The pseudopotential plane wave method (PPW)

The kinetic contribution is most efficiently calculated in reciprocal space:

$$\hat{T}_e[\rho] C_{i\mathbf{k}}(\mathbf{G}) = \frac{\hbar^2}{2m} |\mathbf{k}+\mathbf{G}|^2 C_{i\mathbf{k}}(\mathbf{G}). \tag{9.69}$$

The electrostatic potential is local (multiplicative) in real space. Therefore, the most efficient way of computing its contribution to the generalized force is by multiplying potential and wave function in real space, and then back Fourier transforming it to reciprocal space.

$$\hat{V}_{\text{KS}}^{\text{es}} C_{i\mathbf{k}}(\mathbf{G}) = \frac{1}{\Omega} \int_\Omega \tilde{v}_{\text{KS}}^{\text{es}}(\mathbf{r}) \varphi_i^{(\mathbf{k})}(\mathbf{r}) e^{-i\mathbf{G}\cdot\mathbf{r}} \, d\mathbf{r}, \tag{9.70}$$

where $v_{\text{KS}}^{\text{es}}(\mathbf{r})$ is the inverse Fourier transform of $\tilde{v}_{\text{KS}}^{\text{es}}(\mathbf{G})$ in Eq. (9.57).

The exchange-correlation contribution is treated in the same way as the electrostatic potential, with the difference that $\mu_{\text{XC}}[\rho(\mathbf{r})]$ is calculated directly in real space, and then back Fourier transformed to reciprocal space:

$$\hat{\mu}_{\text{XC}} C_{i\mathbf{k}}(\mathbf{G}) = \frac{1}{\Omega} \int_\Omega \mu_{\text{XC}}[\rho(\mathbf{r})] \varphi_i^{(\mathbf{k})}(\mathbf{r}) e^{-i\mathbf{G}\cdot\mathbf{r}} \, d\mathbf{r}. \tag{9.71}$$

The non-local pseudopotential contribution is calculated in reciprocal space:

$$\Delta \hat{V}_{\text{PS}}^{\text{nl}} C_{i\mathbf{k}}(\mathbf{G}) = \sum_{s=1}^{N_s} \sum_{l=0}^{l_{\max}} \sum_{\mathbf{G}'} \Delta \tilde{V}_{\mathbf{k}+\mathbf{G},\mathbf{k}+\mathbf{G}'}^{l,s} C_{j\mathbf{k}}(\mathbf{G}') \tag{9.72}$$

$$= \sum_{l=0}^{l_{\max}} \sum_{m=-l}^{l} \sum_{I=1}^{P} \alpha_{lm}^s e^{-i\mathbf{G}\cdot\mathbf{R}_I} f_{lm}^{s*}(\mathbf{k}+\mathbf{G}) F_{I,i}^{lm,s}(\mathbf{k}+\mathbf{G}).$$

Figure 9.1 shows the flow diagram of a typical PPW calculation, as it is realized in the CPMD code (Parrinello *et al.*, 2004). This code uses an iterative diagonalization scheme, so that only the action of the Hamiltonian on the wave functions is required. The operations performed in real space are indicated on the left, while those carried out in reciprocal space are on the right. Going from one representation to another is achieved via FFT.

9.3.4 Forces on the nuclear coordinates

Forces on the nuclear coordinates are required in order to perform geometry optimizations and MD simulations. Within the PPW approach the forces are very simple to calculate and computationally inexpensive. The fact that plane waves do not depend on the nuclear coordinates implies that the so-called Pulay forces (Pulay, 1969), arising from the derivatives of the basis functions with respect to the nuclear coordinates, vanish identically.

In DFT the only terms that include an explicit dependence on the nuclear coordinates are the pseudopotential (local and non-local parts), and the ion–ion

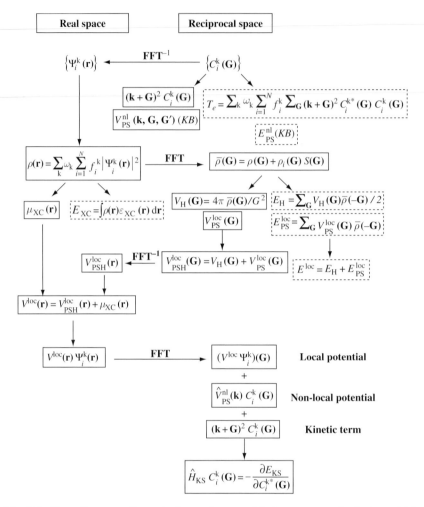

Fig. 9.1 Flow diagram of a typical pseudopotential plane wave calculation within an iterative diagonalization scheme, such as that the one embodied in the CPMD code. Boxes surrounded by dashed lines are contributions to the energy, while those enclosed by solid lines are the different terms of the potential. The transformation between real and reciprocal space is done via fast Fourier transforms.

interaction. All the other terms depend only on the electronic density. The forces on the nuclei of atomic species s, given by $\mathbf{F}_I^s = -\partial E_{KS}/\partial \mathbf{R}_I^s$, have the following expression:

$$\mathbf{F}_I^s = \int \rho(\mathbf{r})(v_{PS}^{loc,s})'(|\mathbf{r} - \mathbf{R}_I|)\frac{\mathbf{r} - \mathbf{R}_I}{|\mathbf{r} - \mathbf{R}_I|} d\mathbf{r}$$

$$+ \frac{Z_I}{2}\sum_{J \neq I} Z_J \frac{\mathbf{R}_I - \mathbf{R}_J}{|\mathbf{R}_I - \mathbf{R}_J|^3} + \mathbf{F}_I^{nl,s}, \quad (9.73)$$

where $(v_{PS}^{loc,s})'$ is the derivative of the local part of the pseudopotential and $\mathbf{F}_I^{nl,s}$ is the non-local contribution to the force.

In the PPW methodology, since we are describing infinite, periodically replicated systems, the forces between the nuclear degrees of freedom have to be computed via the Ewald summation technique. This modifies the form of the interaction in real space, and introduces an extra term in reciprocal space. The force on the nucleus I belonging to species s becomes

$$\mathbf{F}_I^s = \frac{Z_I}{2} \sum_{J \neq I}^{P} Z_J \sum_{L=-L_{max}}^{L_{max}} (\mathbf{R}_I + l - \mathbf{R}_J) \quad (9.74)$$

$$\times \left[\frac{\mathrm{erfc}(|\mathbf{R}_I + l - \mathbf{R}_J|\eta)}{|\mathbf{R}_I + l - \mathbf{R}_J|^3} + \frac{\eta e^{-\eta^2 |\mathbf{R}_I + l - \mathbf{R}_J|^2}}{|\mathbf{R}_I + l - \mathbf{R}_J|} \right]$$

$$+ \Omega \sum_{\mathbf{G} \neq 0} i\mathbf{G} \, e^{-i\mathbf{G} \cdot \mathbf{R}_I} \tilde{u}_{PS}^{loc,s}(\mathbf{G}) \tilde{\rho}(-\mathbf{G})$$

$$+ 2\mathcal{R}e \left\{ \sum_{\mathbf{k} \in BZ} \omega_{\mathbf{k}} \sum_{l=0}^{l_{max}} \sum_{m=-l}^{l} \sum_{i=1}^{N_{\mathbf{k}}} f_i^{(\mathbf{k})} \alpha_{lm}^s F_{I,i}^{lm,s*} D_{I,i}^{lm,s} \right\},$$

where $\mathcal{R}e$ indicates the real part, and

$$D_{I,i}^{lm,s} = \sum_{\mathbf{G}} i\mathbf{G} e^{i\mathbf{G} \cdot \mathbf{R}_I} f_{lm}^s(\mathbf{k} + \mathbf{G}) C_{i\mathbf{k}}(\mathbf{G}). \quad (9.75)$$

The first and second terms arise from the real-space and reciprocal-space parts of the Ewald sum, respectively, and the third term is the non-local pseudopotential contribution. Naturally, the forces on the nuclear degrees of freedom are real quantities. In the above expression this may not appear so obvious, but it can be easily understood by realizing that the $\mathbf{G}$ and $-\mathbf{G}$ contributions to the sum in the third term of (9.74) are complex conjugates, so that the sum of the two is real.

9.3.5 Stress tensor

The expression for the stress tensor in PPW calculations was derived by Nielsen and Martin in 1985. The derivation is based on the application of the variational principle together with a scaling of the wave function (Nielsen and Martin, 1983). An infinitesimal homogeneous deformation $r_i^\alpha \to r_i^\alpha + \sum_\beta \epsilon_{\alpha\beta} r_i^\beta$, with ϵ a symmetric strain tensor, is applied to all the coordinates, nuclear and electronic, so that the ground state wave function is modified to:

$$\Phi_\epsilon(\mathbf{r}) = [\det(\mathbf{1}+\epsilon)]^{-1/2} \Phi[(\mathbf{1}+\epsilon)^{-1}\mathbf{r}], \quad (9.76)$$

where the pre-factor ensures the normalization of the scaled wave function Φ_ϵ. Similarly to the Hellmann–Feynman theorem, the variational principle now requires that the first order variation of the expectation value of the Hamiltonian with respect to the strain tensor vanishes. As a result of this minimization, external and internal stresses are in equilibrium, and this allows us to obtain an expression for the internal stress tensor:

$$T_{\alpha\beta} = -\sum_i \left\langle \Phi \left| \frac{\hat{p}_{i\alpha}\hat{p}_{i\beta}}{m_i} - r_{i\beta}\nabla_{i\alpha}V \right| \Phi \right\rangle, \quad (9.77)$$

which is the quantum form of the virial theorem.

Nielsen and Martin (1985a) extended this construction to density functional theory, where the role of the many-body electronic wave function is played by the electronic density. Here, also, the electronic density experiences the coordinate rescaling. Therefore, the kinetic, Hartree, and exchange-correlation terms of the energy functional contribute to the stress tensor. The expression of the stress within DFT is:

$$T_{\alpha\beta} = -\frac{\hbar^2}{m}\sum_{i=1}^{N}\int \varphi_i^*(\mathbf{r})\nabla_\alpha\nabla_\beta\varphi_i(\mathbf{r})\,d\mathbf{r}$$

$$-\frac{1}{2}\int\int \rho(\mathbf{r})\rho(\mathbf{r}')\frac{(\mathbf{r}-\mathbf{r}')_\alpha(\mathbf{r}-\mathbf{r}')_\beta}{|\mathbf{r}-\mathbf{r}'|^3}\,d\mathbf{r}\,d\mathbf{r}'$$

$$-\frac{1}{2}\sum_{s=1}^{N_s}\sum_{I=1}^{P_s}\int \rho(\mathbf{r})(v_{\text{PS}}^{\text{loc},s})'(|\mathbf{r}-\mathbf{R}_I|)\frac{(\mathbf{r}-\mathbf{R}_I)_\alpha(\mathbf{r}-\mathbf{R}_I)_\beta}{|\mathbf{r}-\mathbf{R}_I|}\,d\mathbf{r}$$

$$-\frac{1}{2}\sum_{I=1}^{P}\sum_{J\neq I}^{P}Z_I Z_J \frac{(\mathbf{R}_I-\mathbf{R}_J)_\alpha(\mathbf{R}_I-\mathbf{R}_J)_\beta}{|\mathbf{R}_I-\mathbf{R}_J|^3}$$

$$+\delta_{\alpha\beta}\int \rho(\mathbf{r})\{\epsilon_{\text{XC}}[\rho(\mathbf{r})]-\mu_{\text{XC}}[\rho(\mathbf{r})]\}\,d\mathbf{r} + T_{\alpha\beta}^{\text{nl}}, \quad (9.78)$$

where the sum in the first term runs over all the occupied Kohn–Sham orbitals and $T_{\alpha\beta}^{\text{nl}}$ is the non-local pseudopotential contribution to the stress.

In a plane wave approach, the real-space scaling procedure $\mathbf{r} \to (1+\epsilon)\mathbf{r}$ can be interpreted as a scaling of the wave vectors $\mathbf{G}$ associated with the plane wave components. To first order in the strain, the transformation reads $\mathbf{G} \to (1-\epsilon)\mathbf{G}$. From the expression of the total energy in a PPW approach, the following expression is obtained for the average stress tensor per unit volume (Nielsen and Martin, 1985b): $\sigma_{\alpha\beta} = \Omega^{-1} T_{\alpha\beta} = \Omega^{-1} \partial E_{\text{KS}}/\partial \epsilon_{\alpha\beta}$:

$$\sigma_{\alpha\beta} = \frac{\hbar^2}{m} \sum_{\mathbf{k} \in BZ} \omega_{\mathbf{k}} \sum_{i=1}^{N_{\mathbf{k}}} f_i^{(\mathbf{k})} \sum_{\mathbf{G}} (\mathbf{k}+\mathbf{G})_\alpha (\mathbf{k}+\mathbf{G})_\beta |C_{i\mathbf{k}}(\mathbf{G})|^2$$

$$+ \frac{1}{2} \sum_{\mathbf{G} \neq 0} \left(\frac{2G_\alpha G_\beta}{G^2} - \delta_{\alpha\beta} \right) \frac{4\pi}{G^2} \tilde{\rho}_T(\mathbf{G}) \tilde{\rho}_T(-\mathbf{G})$$

$$- \sum_{\mathbf{G} \neq 0} \sum_{s=1}^{N_s} S_s(\mathbf{G}) \left[2G_\alpha G_\beta \frac{\partial \tilde{u}_{\text{PS}}^{\text{loc},s}(G)}{\partial(G^2)} + \tilde{u}_{\text{PS}}^{\text{loc},s}(G) \delta_{\alpha\beta} \right] \rho(-\mathbf{G})$$

$$+ \frac{\eta}{2} \sum_{I=1}^{P} \sum_{J \neq I}^{P} \sum_{L=-L_{\text{max}}}^{L_{\text{max}}} Z_I Z_J \, \text{erfc}'(|\mathbf{R}_I + \mathbf{L} - \mathbf{R}_J| \eta)$$

$$\times \frac{(\mathbf{R}_I + \mathbf{L} - \mathbf{R}_J)_\alpha (\mathbf{R}_I + \mathbf{L} - \mathbf{R}_J)_\beta}{|\mathbf{R}_I + \mathbf{L} - \mathbf{R}_J|^3}$$

$$+ \delta_{\alpha\beta} \frac{Q}{\Omega} \left(\sum_{s=1}^{N_s} P_s \Delta \tilde{v}_{\text{PS}}^{\text{loc},s} + \frac{\pi Q^2}{2\Omega \eta^2} \right)$$

$$+ \delta_{\alpha\beta} \int \rho(\mathbf{r}) \{\epsilon_{\text{XC}}[\rho(\mathbf{r})] - \mu_{\text{XC}}[\rho(\mathbf{r})]\} \, d\mathbf{r}$$

$$+ \sum_{\mathbf{k} \in BZ} \omega_{\mathbf{k}} \sum_{l=0}^{l_{\text{max}}} \sum_{m=-l}^{l} \sum_{I=1}^{P} \sum_{j=1}^{N_{\mathbf{k}}} f_j^{(\mathbf{k})} \alpha_{lm}^s F_{I,j}^{lm,s*} \frac{\partial F_{I,j}^{lm,s}}{\partial \epsilon_{\alpha\beta}}, \qquad (9.79)$$

with $\text{erfc}'(x) = d\,\text{erfc}(x)/dx - \text{erfc}(x)/x$, and $\partial F_{I,j}^{lm,s}/\partial \epsilon_{\alpha\beta}$ is calculated from Expression (9.65). The first term is the kinetic contribution, the next three terms arise from the local electrostatic contribution, the fifth term is due to the non-Coulomb part of the modified local pseudopotential at $G = 0$, and the last two are the exchange-correlation and non-local pseudopotential contributions.

9.3.6 Response functions

Higher-order derivatives of the energy allow for the calculation of response functions such as phonon modes, Born effective charges, piezoelectric coefficients, dielectric constants, and infrared and Raman activities. These quantities can also be used to study the thermodynamics of the system within the quasi-harmonic approximation, which allows for the computation of quantities such as the thermal expansion coefficient and the vibrational free energy. All these derive from response functions calculated for virtually negligible displacements from equilibrium. Finite displacements can be dealt with via molecular dynamics or Monte

Carlo simulations (see Chapter 12). Also finite external electric fields can be treated, but this issue is much more delicate, and will not be addressed any further here.

There are mainly two ways of computing phonon modes: by effectively displacing the particles according to the desired pattern in a supercell of the required size (*frozen phonon*), and by considering only displacements of the atoms in the unit cell, but taking into account the appropriate vector in the phonon Brillouin zone (*linear response*). This latter approach makes use of the powerful density functional perturbation theory (DFPT), and is much more efficient than frozen phonons. Another advantage is that it also allows us to consider virtual infinitesimal electric fields (effective charges and dielectric constants). Electron–phonon couplings, required to calculate superconducting parameters, are also accessible. An excellent review of DFPT and its uses has been recently published by Baroni *et al.* (2001).

9.4 Atom-centered basis sets

In an atom-centered basis set, whether atomic orbitals, STO, GTO, or other, the one-electron wave functions are written:

$$\varphi_i(\mathbf{r}) = \sum_{\mu=1}^{M} \sum_{I=1}^{P} c_{i\mu}^{I} \phi_\mu(\mathbf{r} - \mathbf{R}_I), \tag{9.80}$$

where M is the number of basis functions per atom and P is the number of atoms in the system. The basis functions $\phi_\mu(\mathbf{r} - \mathbf{R}_I)$ and the corresponding expansion coefficients $c_{i\mu}^{I}$ now carry two indices: μ for the type of orbital and I for the center. Here we have expanded the notation in (8.1) from a single index to make this double identity more explicit. Now the generalized eigenvalue equation (8.2) becomes

$$\sum_{\beta=1}^{M} \sum_{J=1}^{P} \left(\mathcal{H}_{\alpha\beta}^{IJ} - \varepsilon_j S_{\alpha\beta}^{IJ} \right) c_{j\beta}^{J} = 0, \tag{9.81}$$

where, at variance with orthogonal basis sets like PWs, atom-centered basis sets require the calculation of overlap matrix elements.

In all the matrix elements the basis functions can be centered on different atomic sites, thus giving rise to a variety of integrals, ranging from one to four centers. Three types of integral have to be calculated: overlaps, one-electron integrals (kinetic and nuclear potential), and two-electron integrals (electron–electron interaction). In this section we will drop the wave vector index **k** for the sake of clarity in the notation, but it can be understood as included in the basis function index. The framework presented below is completely general for one-electron schemes, and applies to both Kohn–Sham and Hartree–Fock equations.

9.4.1 Overlap integrals

These have the form:

$$S_{\mu\nu}^{IJ} = \langle \phi_\mu^I | \phi_\nu^J \rangle = \int \phi_\mu^*(\mathbf{r} - \mathbf{R}_I)\phi_\nu(\mathbf{r} - \mathbf{R}_J)\,d\mathbf{r}. \tag{9.82}$$

When both orbitals are centered on the same site, we have $S_{\mu\nu}^{II} = \langle \phi_\mu^I | \phi_\nu^I \rangle$. It may seem at first glance that these terms are one if $\mu = \nu$ and zero otherwise. While the first part is true, the second is not necessarily so if the basis functions have been optimized according to the schemes presented in Section 8.4. Overlaps of basis functions corresponding to different angular momenta are zero but, in general, those for the same angular momentum are not. Imagine an uncontracted Gaussian double-ζ split valence basis set, where two basis functions are used to represent the s valence state. If the exponents of these two functions are α and β, then, apart from normalization factors,

$$S_{\mu\nu}^{II} \propto \int e^{-\alpha(\mathbf{r}-\mathbf{R}_I)^2} e^{-\beta(\mathbf{r}-\mathbf{R}_I)^2}\,d\mathbf{r} = \left(\frac{\pi}{\alpha+\beta}\right)^{3/2} \neq 0. \tag{9.83}$$

9.4.2 Density matrix, Hamiltonian, and total energy

The electronic density can be expressed as:

$$\rho(\mathbf{r}) = \sum_{\mu,\nu=1}^{M} \sum_{I,J=1}^{P} \rho_{\nu\mu}^{JI} \phi_\mu^*(\mathbf{r}-\mathbf{R}_I)\phi_\nu(\mathbf{r}-\mathbf{R}_J), \tag{9.84}$$

where

$$\rho_{\nu\mu}^{JI} = \rho_{\mu\nu}^{IJ*} = \sum_{i=1}^{N} c_{i\nu}^J c_{i\mu}^{I*}. \tag{9.85}$$

$\rho_{\nu\mu}^{JI}$ is usually called the *density matrix*, and it actually corresponds to the basis set representation of the one-body density operator $\rho(\mathbf{r},\mathbf{r}')$, whose trace is the electronic density. The elements of the density matrix between two different centers are called *bond orders*, and give an indication of the strength of the chemical bond between atoms I and J due to the orbitals μ and ν. Occupation numbers can be included in the definition of the density matrix, if required. Interestingly, the operation of reconstructing the real-space representation of the density from the elements of the density matrix can be computationally quite intensive.

The density matrix enters the calculation of the total energy directly as follows. Let us consider, for example, Expression (3.32) for the Hartree–Fock energy

$$E_{\text{HF}} = \sum_{i=1}^{N} E_{ii} + \frac{1}{2}\sum_{i,j=1}^{N}\left(J_{ij} - K_{ij}\right) + V_{nn}. \tag{9.86}$$

The first term is the sum of the one-electron terms of the Hamiltonian:

$$\sum_{i=1}^{N} E_{ii} = \sum_{i=1}^{N} \int \varphi_i^*(\mathbf{r}) \hat{h}_1 \varphi_i(\mathbf{r}) \, d\mathbf{r} = \sum_{i=1}^{N} \sum_{\mu,\nu=1}^{M} \sum_{I,J=1}^{P} c_{i\mu}^{I*} c_{i\nu}^{J} \langle \phi_\mu^I | \hat{h}_1 | \phi_\nu^J \rangle$$

$$= \sum_{\nu=1}^{M} \sum_{J=1}^{P} \left(\sum_{\mu=1}^{M} \sum_{I=1}^{P} \rho_{\nu\mu}^{JI} h_{1\mu\nu}^{IJ} \right) = \sum_{\nu=1}^{M} \sum_{J=1}^{P} \left(\hat{\rho} \hat{h}_1 \right)_{\nu\nu}^{JJ}$$

$$= \text{Tr}(\hat{\rho} \hat{h}_1), \tag{9.87}$$

where the symbol Tr indicates the trace of a matrix, in this case the product of the density and the one-electron Hamiltonian matrices. The one-electron matrix elements are

$$h_{1\mu\nu}^{IJ} = \langle \phi_\mu^I | \left[-\frac{\hbar^2}{2m} \nabla^2 + \sum_{K=1}^{P} u_K(\mathbf{r} - \mathbf{R}_K) \right] | \phi_\nu^J \rangle, \tag{9.88}$$

where u_K is the interaction between the electron and the nucleus at $\mathbf{R}_K$.

The two-electron part of the energy can be calculated in a similar way, but first it is convenient to define the following shorthand notation for the matrix elements between atom-centered orbitals:

$$\langle \mu\gamma | \nu\delta \rangle = \langle \phi_\mu^I \phi_\gamma^K | \hat{v}_2 | \phi_\nu^J \phi_\delta^L \rangle$$

$$= \int \int \frac{\phi_\mu^*(\mathbf{r} - \mathbf{R}_I) \phi_\gamma^*(\mathbf{r} - \mathbf{R}_K) \phi_\nu(\mathbf{r}' - \mathbf{R}_J) \phi_\delta(\mathbf{r}' - \mathbf{R}_L)}{|\mathbf{r} - \mathbf{r}'|} \, d\mathbf{r} \, d\mathbf{r}'. \tag{9.89}$$

With this notation, apart from the factor 1/2, the second term in (9.86) becomes

$$\sum_{i,j=1}^{N} (J_{ij} - K_{ij}) = \sum_{i,j=1}^{N} \sum_{\mu\gamma\nu\delta=1}^{M} \sum_{IKJL=1}^{P} c_{i\mu}^{I*} c_{j\gamma}^{K*} c_{i\nu}^{J} c_{j\delta}^{L} [\langle \mu\gamma | \nu\delta \rangle - \langle \mu\gamma | \delta\nu \rangle]$$

$$= \sum_{\mu\gamma\nu\delta=1}^{M} \sum_{IKJL=1}^{P} \rho_{\nu\mu}^{JI} \rho_{\delta\gamma}^{LK} [\langle \mu\gamma | \nu\delta \rangle - \langle \mu\gamma | \delta\nu \rangle]$$

$$= \sum_{\mu\gamma\nu\delta=1}^{M} \sum_{IKJL=1}^{P} \left(\rho_{\nu\mu}^{JI} \rho_{\delta\gamma}^{LK} - \rho_{\delta\mu}^{LI} \rho_{\nu\gamma}^{JK} \right) \langle \mu\gamma | \nu\delta \rangle. \tag{9.90}$$

Combining the expressions for the one- and two-electron contributions, the Hartree–Fock energy is written in the following way:

$$E_{\text{HF}} = \frac{1}{2} \sum_{\mu,\nu=1}^{M} \sum_{I,J=1}^{P} \rho_{\nu\mu}^{JI} \left(h_{1\mu\nu}^{IJ} + \tilde{F}_{\mu\nu}^{IJ} \right) + V_{nn} = \frac{1}{2} \text{Tr}\left[\hat{\rho} \left(\hat{h}_1 + \hat{\mathcal{F}} \right) \right] + V_{nn}, \quad (9.91)$$

where $\hat{\mathcal{F}}$ is the Fock operator and the factor 1/2 is required to avoid double-counting of the electron–electron interaction. The Fock matrix elements are:

$$F_{\mu\nu}^{IJ} = h_{1\mu\nu}^{IJ} + \sum_{\gamma\delta=1}^{M} \sum_{KL=1}^{P} \rho_{\delta\gamma}^{LK} \left[\langle \mu\gamma|\nu\delta \rangle - \langle \mu\gamma|\delta\nu \rangle \right], \quad (9.92)$$

and the Hartree–Fock equations become the Roothaan–Hall equations:

$$\sum_{\nu=1}^{M} \sum_{J=1}^{P} \left(F_{\mu\nu}^{IJ} - \epsilon_i S_{\mu\nu}^{IJ} \right) c_{i\nu}^{J} = 0. \quad (9.93)$$

The Kohn–Sham case is completely analogous. It only requires the appropriate modification of the expression for the exchange energy. The Kohn–Sham energy is given by

$$E_{\text{KS}} = \frac{1}{2} \sum_{\mu,\nu=1}^{M} \sum_{I,J=1}^{P} \rho_{\nu\mu}^{JI} \left(h_{1\mu\nu}^{IJ} + \mu_{\text{XC}\mu\nu}^{IJ} + H_{\text{KS}\mu\nu}^{IJ} \right) + V_{nn}$$

$$= \frac{1}{2} \text{Tr}\left[\hat{\rho} \left(\hat{h}_1 + \hat{\mu}_{\text{XC}} + \hat{H}_{\text{KS}} \right) \right] + V_{nn}, \quad (9.94)$$

with $\hat{H}_{\text{KS}}$ the Kohn–Sham Hamiltonian with matrix elements

$$H_{\text{KS}\mu\nu}^{IJ} = h_{1\mu\nu}^{IJ} + \mu_{\text{XC}\mu\nu}^{IJ} + \sum_{\gamma\delta=1}^{M} \sum_{KL=1}^{P} \rho_{\delta\gamma}^{LK} \langle \mu\gamma|\nu\delta \rangle. \quad (9.95)$$

Again the factor 1/2 is to compensate for double-counting in the energy expression.

The above expressions can be readily generalized to the case of open-shell systems in the unrestricted Hartree–Fock or Kohn–Sham schemes, where two sets of molecular orbitals, one for each spin projection, are optimized independently. In that case we define the *spin density matrices*

$$\rho_{\mu\nu}^{IJ}(\alpha) = \sum_{i=1}^{N} c_{i\mu}^{I}(\alpha) c_{i\nu}^{J*}(\alpha), \quad (9.96)$$

with $\alpha = \{\uparrow, \downarrow\}$ the two possible spin projections. The full density matrix is written as $\rho_{\mu\nu}^{IJ} = \rho_{\mu\nu}^{IJ}(\uparrow) + \rho_{\mu\nu}^{IJ}(\downarrow)$. With these definitions the Roothaan–Hall equations split into two sets of coupled eigenvalue equations (Pople and Beveridge, 1970):

$$\sum_{\nu=1}^{M} \sum_{J=1}^{P} \left[F_{\mu\nu}^{IJ}(\alpha) - \epsilon_i(\alpha) S_{\mu\nu}^{IJ} \right] c_{i\nu}^{I}(\alpha) = 0, \quad (9.97)$$

where

$$F_{\mu\nu}^{IJ}(\alpha) = h_{1\mu\nu}^{IJ} + \sum_{\gamma\delta=1}^{M} \sum_{KL=1}^{P} \left[\rho_{\delta\gamma}^{LK} \langle\mu\gamma|\nu\delta\rangle - \rho_{\delta\gamma}^{LK}(\alpha)\langle\mu\gamma|\delta\nu\rangle \right], \quad (9.98)$$

and the energy is given by

$$E_{\text{HF}}^{\text{U}} = \sum_{\mu,\nu=1}^{M} \sum_{I,J=1}^{P} \rho_{\nu\mu}^{JI} h_{1\mu\nu}^{IJ}$$

$$+ \frac{1}{2} \sum_{\mu\gamma\nu\delta=1}^{M} \sum_{IKJL=1}^{P} \left(\rho_{\nu\mu}^{JI} \rho_{\delta\gamma}^{LK} - \frac{1}{2} \sum_{\alpha=\uparrow,\downarrow} \rho_{\delta\mu}^{LI}(\alpha) \rho_{\nu\gamma}^{JK}(\alpha) \right) \langle\mu\gamma|\nu\delta\rangle. \quad (9.99)$$

The above expressions can be easily adapted to the case of open-shell systems in the unrestricted Kohn–Sham scheme.

9.4.3 One-electron integrals

The matrix elements of the Fock operator or the Kohn–Sham Hamiltonian are of two types: the kinetic term and the nuclear attraction involve a single electronic state and are called *one-electron integrals*, while the direct (Hartree) and exchange (Fock) terms involve two electronic states interacting via Coulomb forces and are called *two-electron integrals*.

The Hamiltonian matrix elements for a one-electron operator $\hat{\mathcal{A}}$ in an atom-centered basis set are written:

$$\mathcal{A}_{\mu\nu}^{IJ} = \int \phi_\mu^*(\mathbf{r}-\mathbf{R}_I) \mathcal{A}(\mathbf{r},\nabla_\mathbf{r}) \phi_\nu(\mathbf{r}-\mathbf{R}_J)\, d\mathbf{r}. \quad (9.100)$$

In particular, the kinetic contribution to the Hamiltonian matrix is:

$$T_{\mu\nu}^{IJ} = -\frac{\hbar^2}{2m} \int \phi_\mu^*(\mathbf{r}-\mathbf{R}_I) \nabla^2 \phi_\nu(\mathbf{r}-\mathbf{R}_J)\, d\mathbf{r}$$

$$= \frac{\hbar^2}{2m} \int \nabla \phi_\mu^*(\mathbf{r}-\mathbf{R}_I) \cdot \nabla \phi_\nu(\mathbf{r}-\mathbf{R}_J)\, d\mathbf{r}, \quad (9.101)$$

where the second equality was obtained by integrating by parts and using the fact that the basis functions and their first derivatives vanish at infinity. This second form is symmetric in the indices μ and ν, and thus esthetically more satisfactory. For $I = J$ these are one-center integrals that are often called *on-site* terms (Finnis, 2003), while for $I \neq J$ they are two-center integrals.

The contribution arising from the nuclear attraction is written:

$$U_{\mu\nu}^{IJ} = \sum_{K=1}^{P} \int \phi_\mu^*(\mathbf{r}-\mathbf{R}_I) u_K(\mathbf{r}-\mathbf{R}_K) \phi_\nu(\mathbf{r}-\mathbf{R}_J)\, d\mathbf{r}, \quad (9.102)$$

where in general $u_K(\mathbf{r} - \mathbf{R}_K) = 1/|\mathbf{r} - \mathbf{R}_K|$ is the bare Coulomb potential, but it can also be a pseudopotential if the core electrons have been excluded. This is the reason for retaining the index K in u_K, because the pseudopotential would depend on the atomic species. Here we have three different types of integrals depending on whether $I = J = K$ (one center), $I = J \neq K$, $I \neq J = K$, or $I = K \neq J$ (two-center), or $I \neq J \neq K$ (three-center).

9.4.4 Two-electron integrals

The matrix elements of the two-electron Coulomb interaction involve the integrals $\langle \mu\nu|\gamma\delta \rangle$ defined by (9.89). These range from one up to four centers. At first glance, the distinction between Coulomb and exchange integrals disappears for atom-centered basis sets. The only difference is that the matrix elements are contracted with the density matrix in different ways. There is, however, a difference, and that is that the direct Coulomb term can be expressed as a contraction between the density matrix and the four-index tensor $\langle \mu\nu|\gamma\delta \rangle$, while the exchange term cannot. In other words, the Hartree energy can be written as $E_H = (1/2)\text{Tr}(\hat{\rho}\,\hat{v}_2\hat{\rho})$, and the Hartree potential is represented by a partial trace.

Three- and four-center integrals are, in general, difficult to calculate and computationally very costly, unless the basis set used allows for their analytic computation, as is the case of GTOs. However, the limiting factor is really the number of such integrals that have to be calculated, which scales like M^4. In a plane wave basis set the exchange terms are prohibitively expensive, thus almost precluding the possibility of using a PW basis set for Hartree–Fock calculations, except for very small systems (a few atoms).

9.4.5 The Hartree term

In usual DFT methods, where the exchange interaction is approximated by a local or semi-local expression, there is no need to calculate four-center integrals of the Fock type. Since the remaining two-electron integrals that enter the Hartree potential and energy are of the direct Coulomb type, there are alternative roads to simplify their calculation by using the fact that these depend only on the density and not explicitly on the orbitals.

Auxiliary basis sets

One possibility is to expand the density itself in an auxiliary basis set. Here we are back to the problem of choosing a basis set, but now the quantity to expand has some features that are different from those of the electronic orbitals. In fact, the density is a single nodeless function, smoother than the orbitals, and without orthogonalization requirements.

One possible choice is to represent the density in another atom-centered basis set, e.g. Gaussians, as in the code DGAUSS (Godbout *et al.*, 1992; Andzelm and Wimmer, 1992),

$$\rho(\mathbf{r}) = \sum_{\alpha=1}^{M_{\text{aux}}} \sum_{I=1}^{P} d_\alpha^I \varphi_\alpha(\mathbf{r} - \mathbf{R}_I), \qquad (9.103)$$

where the coefficients d_α^I are determined by a least squares fitting procedure (see also Estrín *et al.*, 1993). Using this technique, the Hartree energy can be calculated in terms of one- and two-center integrals in the following way:

$$E_{\text{H}} = \frac{1}{2} \sum_{\alpha,\beta=1}^{M_{\text{aux}}} \sum_{I,J=1}^{P} d_\alpha^I d_\beta^J \int \int \frac{\varphi_\alpha(\mathbf{r}-\mathbf{R}_I)\varphi_\beta(\mathbf{r}'-\mathbf{R}_J)}{|\mathbf{r}-\mathbf{r}'|} \, d\mathbf{r}\, d\mathbf{r}'. \qquad (9.104)$$

Another alternative recently proposed is to represent the density on a real-space uniform grid by means of Expression (9.84), and then calculate the Hartree energy and matrix elements by fast Fourier transform techniques. Obviously, this automatically imposes periodic boundary conditions to the system, which makes this method useful for condensed phases, but introduces a complication for molecules. The codes SIESTA (Soler *et al.*, 2002) and QUICKSTEP (VandeVondele *et al.*, 2005) adopt this type of approach.

Atom-centered expansion of the effective potential

A different strategy is to express the effective (Kohn–Sham) potential defined either in (8.7) or in (8.10) as a sum of atom-centered terms (Finnis, 2003):

$$v_{\text{eff}}(\mathbf{r}) = \sum_{K=1}^{P} \bar{V}^{(K)}(\mathbf{r} - \mathbf{R}_K). \qquad (9.105)$$

The advantage of this expression is that four-center integrals are eliminated, but this approach entails a rather uncontrolled approximation that, depending on the case, may or may not be important. The matrix elements now assume the same form as the nuclear attraction term (9.102):

$$\bar{V}_{\mu\nu}^{IJ} = \bar{V}_{\mu\nu}^{(I)IJ} + \bar{V}_{\mu\nu}^{(J)IJ} + \sum_{K \neq I,J} \bar{V}_{\mu\nu}^{(K)IJ}, \qquad (9.106)$$

where it can be seen that the last term still includes expensive three-center integrals. These are usually neglected in tight-binding or semiempirical schemes (see Sections 10.1 and 10.2), where the two approximations just mentioned are expected to be compensated by a suitable choice of the basis set and by the parameterization of one- and two-center integrals.

9.4.6 Scaling and multi-polar expansion for the Coulomb terms

For atom-centered basis sets all the integrals decay quite fast with the distance between centers, e.g. exponentially for STOs and GTOs. This means that the overlap integrals are negligibly small beyond a certain cutoff distance, which depends on the accuracy required of the calculation. Therefore, the number of significant overlap integrals from a particular center is a fixed number that depends on how quick the basis functions decay, and the number of overlap integrals to be calculated is proportional to the number of atoms N. The same reasoning is valid for the kinetic integrals.

The argument for the Coulomb interaction is somewhat different because integrals of the type (9.102) or (9.89) never really vanish, no matter how far the centers of the basis functions are located; they decay following a power-law behavior. Nevertheless, what does happen is that, at long distance, the matrix elements approach the expression of the classical electrostatic interaction between two charge distributions. The cutoff distance depends on how quick the basis functions decay. An estimate can be obtained from the Schwarz inequality (Helgaker *et al.*, 2000). The number of integrals whose centers are closer than the cutoff distance still scales with N, but the remaining integrals are non-zero and have to be calculated.

An efficient way to evaluate the classical Coulomb interaction between two general, non-overlapping charge distributions is by means of a multi-pole or partial wave expansion. We begin by expanding the Coulomb potential:

$$\frac{1}{|\mathbf{r} - \mathbf{r}'|} = \sum_{l=0}^{\infty} \frac{r_<^l}{r_>^{l+1}} P_l(\cos\theta), \qquad (9.107)$$

where $r_<$ and $r_>$ are referred to some arbitrary origin $\mathbf{r}_0$. The quantities $r_<$ and $r_>$ are the smallest and largest distances among $|\mathbf{r} - \mathbf{r}_0|$ and $|\mathbf{r}' - \mathbf{r}_0|$, and θ is the angle between the two vectors $(\mathbf{r} - \mathbf{r}_0)$ and $(\mathbf{r}' - \mathbf{r}_0)$. $P_l(\cos\theta)$ are Legendre polynomials (see e.g. Jackson, 1975).

Following Finnis (2003), the electrostatic potential due to a charge distribution enclosed within a sphere of volume Ω_s and centered at point $\mathbf{R}$, at a point $\mathbf{r}$ exterior to the sphere, is given by

$$v_H(\mathbf{r}) = \int_{\Omega_s} \frac{\rho(\mathbf{r}')}{|\mathbf{r} - \mathbf{r}'|} \, d\mathbf{r} = \sum_{l,m} Q_{lm}(\mathbf{R}) I_{lm}(\mathbf{r} - \mathbf{R}), \qquad (9.108)$$

where $I_{lm}(\mathbf{r} - \mathbf{R})$ are the irregular solutions of the Laplace equation given by Expression (8.46), which diverge at $\mathbf{r} = \mathbf{R}$ but not around $\mathbf{r}$ outside the sphere. The quantities $Q_{lm}(\mathbf{R})$ are defined as

$$Q_{lm}(\mathbf{R}) = \int_{\Omega_s} R_{lm}(\mathbf{r}' - \mathbf{R}) \rho(\mathbf{r}') \, d\mathbf{r}', \qquad (9.109)$$

where $R_{lm}(\mathbf{r})$ are the regular solutions of the Laplace equation. These are precisely the *multi-pole moments* of the charge distribution calculated with center in $\mathbf{R}$. The functions $I_{lm}(\mathbf{r} - \mathbf{R})$ can be expanded as

$$I_{lm}(\mathbf{r} - \mathbf{R}) = \sum_{l',m'} B_{lm}^{l'm'}(\mathbf{R}) R_{l'm'}(\mathbf{r}) \qquad (9.110)$$

in terms of the regular solutions of the Laplace equation centered at point $\mathbf{r}$, and the coefficients $B_{lm}^{l'm'}(\mathbf{R})$ are the same structure constants that appear in augmentation methods based on atom-centered basis functions, and are given by (8.75) with $\kappa = 0$.

Replacing (9.110) into (9.108) gives the following compact expression for the electrostatic potential in terms of the multi-poles of the charge distribution:

$$v_{\mathrm{H}}(\mathbf{r}) = \sum_{l,m} \sum_{l',m'} Q_{lm}(\mathbf{R}) B_{lm}^{l'm'}(\mathbf{R}) R_{l'm'}(\mathbf{r}). \qquad (9.111)$$

By integrating the electrostatic potential over a second charge distribution centered at $\mathbf{r}$ we finally obtain the expression for the electrostatic energy. If the charge distributions are centered at $\mathbf{R}_I$ and $\mathbf{R}_J$, respectively, the energy assumes the following bilinear form in terms of the multi-pole moments:

$$E_{\mathrm{H}}^{IJ} = \sum_{l,m} \sum_{l',m'} Q_{lm}(\mathbf{R}_I) B_{lm}^{l'm'}(\mathbf{R}_I - \mathbf{R}_J) Q_{l'm'}(\mathbf{R}_J). \qquad (9.112)$$

The accuracy of the multi-pole expansion is controlled by the number of angular momenta included in the sum.

The multi-pole method applies straightforwardly to the calculation of nuclear attraction integrals of the form (9.102). Two-electron Coulomb integrals involve four basis functions, but the four centers become effectively two when defining the overlap distributions:

$$O_{\mu\nu}^{IJ}(\mathbf{r}) = \phi_{\mu}^{*}(\mathbf{r} - \mathbf{R}_I) \phi_{\nu}(\mathbf{r} - \mathbf{R}_J). \qquad (9.113)$$

These distributions generally behave as basis functions centered somewhere between the two centers $\mathbf{R}_I$ and $\mathbf{R}_J$, although, depending on the particular basis set, their specific functional form may not be obvious at all. It is particularly simple in the case of Gaussian basis functions due to the product formula (see Section 8.4.5). In any case, the classical contribution to the electron–electron interaction can be written as a multi-pole expansion, where the multi-pole moments in Eq. (9.112) are calculated using the overlap distributions centered at $\mathbf{R}_K$, in place of the density or the basis functions:

$$Q_{lm}^{IJ}(\mathbf{R}_K) = \int_{\Omega_s} R_{lm}(\mathbf{r} - \mathbf{R}_K) O_{\mu\nu}^{IJ}(\mathbf{r}) \, d\mathbf{r}. \qquad (9.114)$$

The number of such terms required for the evaluation of the full electrostatic energy scales like N^2, because there is one term per pair of centers. If the number of atoms is large, then there are algorithms that cut down the cost to order-N scaling by using the fact that the magnitude of the electrostatic interaction decreases like $1/R$, with R the distance between centers. More distant pairs contribute less to the multi-pole expansion, and then are not required to be calculated as accurately as close pairs.

The idea of the fast multi-pole method (FMM) (Greengard and Rokhlin, 1987) is to keep the accuracy of all terms approximately constant. This is achieved by dividing the system into boxes that contain a reasonably small number of particles (between 10 and 100), and then constructing a hierarchy of boxes by a coarse-graining procedure. At every level the *parent* box contains eight *child* boxes (in three dimensions). At the highest level there is only one box that contains all the particles in the system. At the lowest level, for every child there is a *near-field* region that contains a first shell of neighboring boxes, where the interactions are non-classical (there is overlap of the basis functions). These have to be treated explicitly, but the number of such interactions scales like N. The size of the boxes is chosen such that beyond the first shell, the interactions can effectively be treated as classical. Next, there is a *local far-field* region corresponding to the near-field boxes of its parent. These interactions can be treated as classical and calculated as a multi-pole expansion. Finally, there is a *remote far-field* region. If this region was treated as the local far-field, then the scaling would still be N^2. The strategy is then to calculate its contribution as the far-field of the parent. In order to do this it is necessary to express the multi-pole moments of a child in terms of those of the parent, which correspond to a different (but nearby) center. This is done by using the translation relation

$$Q_{lm}(\mathbf{R}_P) = \sum_{l'm'} R_{l-l',m-m'}(\mathbf{R}_C - \mathbf{R}_P) Q_{l'm'}(\mathbf{R}_C). \tag{9.115}$$

The far-field of the parent is also divided into a local and a remote far-field, which are treated in the same way but at a higher level in the hierarchy. The resulting algorithm scales like N. In the case of basis functions that decay with different rates (exponents), this procedure can be generalized by grouping together orbitals of the same extent (similar exponents) into *branches* (White and Head-Gordon, 1994).

9.4.7 Generalized forces

The Hellmann–Feynman theorem states that, if the total energy has been variationally optimized with respect to arbitrary variations in the one-electron wave functions (or the electronic density in DFT methods), then the change in the expectation value of the Hamiltonian (the total energy) with respect to changes in some external parameter λ, e.g. the nuclear coordinates **R**, can be calculated as the expectation value of the derivative of the Hamiltonian

$$F_\lambda(\text{HF}) = \left\langle \Phi(\lambda) \left| \frac{\partial \hat{\mathcal{H}}(\lambda)}{\partial \lambda} \right| \Phi(\lambda) \right\rangle. \tag{9.116}$$

This is valid if there is full variational freedom. However, if the wave functions are expanded in a truncated basis set, the hypotheses of the theorem are violated, and additional contributions to the force appear due to the dependence of the wave function on the external parameter.

$$F_\lambda = \frac{\partial E(\lambda)}{\partial \lambda} = \left\langle \Phi(\lambda) \left| \frac{\partial \hat{\mathcal{H}}(\lambda)}{\partial \lambda} \right| \Phi(\lambda) \right\rangle + 2 \left\langle \frac{\partial \Phi(\lambda)}{\partial \lambda} \left| \hat{\mathcal{H}} \right| \Phi(\lambda) \right\rangle. \tag{9.117}$$

The second term contains contributions from variations of the basis functions and of the expansion coefficients, although these latter disappear in approaches of variational nature. Therefore, retaining only the dependence on the basis functions, the contribution can be calculated using a procedure devised by Pulay (1969), which consists of augmenting the standard force with terms proportional to the first derivatives of the basis functions. Starting from the Hartree–Fock energy expression (9.91), we obtain the following expression for the generalized force arising from the variation of an external parameter λ, which could be an atomic coordinate:

$$\begin{aligned}
F_\lambda = & \sum_{\mu,\nu=1}^{M} \sum_{I,J=1}^{P} \rho_{\nu\mu}^{JI} \frac{\partial h_{1\mu\nu}^{IJ}}{\partial \lambda} + \frac{\partial V_{nn}}{\partial \lambda} \\
& + \frac{1}{2} \sum_{\mu\gamma\nu\delta=1}^{M} \sum_{IKJL=1}^{P} \left(\rho_{\nu\mu}^{JI} \rho_{\delta\gamma}^{LK} - \rho_{\delta\mu}^{LI} \rho_{\nu\gamma}^{JK} \right) \frac{\partial \langle \mu\gamma | \nu\delta \rangle}{\partial \lambda} \\
& - \sum_{\mu,\nu=1}^{M} \sum_{I,J=1}^{P} W_{\nu\mu}^{JI} \frac{\partial S_{\mu\nu}^{IJ}}{\partial \lambda},
\end{aligned} \tag{9.118}$$

where

$$W_{\nu\mu}^{JI} = \sum_{i=1}^{N} \varepsilon_i c_{i\nu}^{J} c_{i\mu}^{I*} \tag{9.119}$$

is the energy-weighted density matrix. The first two terms are the pure Hellmann–Feynman force, while the last two arise due to the parameter-dependence of the basis functions. The details of the derivation can be found in Jensen (1999). Here we just mention that, to obtain this expression, the generalized eigenvalue equation (9.81) was used, together with the orthonormality of the wave functions, which for a non-orthogonal basis set is written

$$\sum_{\mu=1}^{M} \sum_{\nu=1}^{M} c_{i\mu}^{I*} S_{\mu\nu}^{IJ} c_{j\nu}^{J} = \delta_{ij}. \tag{9.120}$$

By deriving this expression with respect to the parameter λ, and using the fact that the derivative of the whole expression is zero because δ_{ij} is a constant, the derivatives of the expansion coefficients $c_{j\nu}^{J}$ can be written in terms of the derivatives of the overlap matrix. As usual, similar expressions are obtained for the Kohn–Sham forces, using Expression (9.94).

While the last two terms vanish for floating basis sets such as PW, because these do not depend on the atomic positions, for atom-centered basis sets such as LCAOs, STOs or GTOs, the dependence of the basis functions on the atomic positions takes its toll. Both of these terms require the calculation of the derivatives of the basis functions, and integrals that involve these derivatives. For GTO, the derivatives are still GTO but corresponding to an angular momentum decreased in one unit. Therefore, these integrals can be calculated analytically in the same way as all other integrals (Amos *et al.*, 1995). For other basis sets, such as numerical AOs, the derivatives of the basis functions can be obtained numerically. The computational effort, however, is not negligible, as is the case for PW. For atom-centered basis sets, the cost of calculating the force on the atoms can be of the same order of magnitude of a self-consistency iteration. Notice that the last term in (9.118) vanishes for orthogonal basis sets.

A similar situation arises in methods based on augmentation spheres such as APWs or LMTOs, because the spheres move with the nuclei. Force theorems for arbitrary basis sets, which avoid the computation of Pulay forces, have been devised by Methfessel and van Schilfgaarde (1993) following the ideas behind the Harris functional (Harris, 1985) (see Section 10.1.2). These methods are based on displacing the density together with the spheres.

It is also possible to calculate second derivatives with respect to external parameters, thus allowing for the calculation of linear response functions such as vibrational frequencies, infrared activity, dipole polarizability, and even higher-order response functions such as hyperpolarizabilities and Raman activity, amongst others. Magnetic and nuclear spin responses such as NMR spectra are also accessible, but they require the calculation of matrix elements of the angular momentum operator (see Jensen, 1999, Chapter 10).

9.5 Gaussian basis sets

Due to the wide use of Gaussian basis sets, it is worthwhile to describe in some depth how the Hamiltonian and overlap matrix elements are calculated.

The complete and detailed expressions can be found in computational chemistry books (see e.g. Helgaker *et al.*, 2000). As mentioned before, the great advantage of GTOs is that all the integrals can be calculated either analytically, or with a limited computational effort (one-dimensional integrals), thus saving precious computer time. Moreover, not only matrix elements and energy, but also derivatives of the energy at any order can be calculated analytically, thus allowing for the calculation of forces, vibrational frequencies, polarizabilities, and many other properties with high accuracy and a reasonable computational effort. There are many codes that work with Gaussian basis sets, including GAUSSIAN (Frisch *et al.*, 2004), GAMESS (Schmidt *et al.*, 1993), CADPAC (Amos *et al.*, 1995), DGAUSS (Andzelm and Wimmer, 1992), and NWChem (Kendall *et al.*, 2000), amongst others.

More rare are Gaussian codes that incorporate periodic boundary conditions, by constructing basis functions in the form of Bloch sums. The most advanced code of this class is CRYSTAL (Pisani and Dovesi, 1980). The critical technical issue here is the calculation of exchange integrals, which has been addressed by Causà *et al.* (1988). Therefore, this code allows for Hartree–Fock calculations in solids, and naturally permits extensions to hybrid HF-KS schemes such as B3LYP.

9.5.1 Overlap integrals

The overlap integrals for primitive s functions centered on different atomic sites are:

$$S_{ss}^{IJ} = \int e^{-\alpha(\mathbf{r}-\mathbf{R}_I)^2} e^{-\beta(\mathbf{r}-\mathbf{R}_J)^2}\, d\mathbf{r} = e^{-\gamma(\mathbf{R}_I-\mathbf{R}_J)^2} \left(\frac{\pi}{\kappa}\right)^{3/2}, \qquad (9.121)$$

with $\kappa = \alpha + \beta$ and $\gamma = \alpha\beta/(\alpha+\beta)$. This can be readily extended to contracted Gaussians:

$$S_{\mu\nu}^{IJ} = \sum_{i=1}^{k} \sum_{j=1}^{n} a_i b_j e^{-\gamma_{ij}(\mathbf{R}_I-\mathbf{R}_J)^2} \left(\frac{\pi}{\kappa_{ij}}\right)^{3/2}, \qquad (9.122)$$

where $\kappa_{ij} = \alpha_i + \beta_j$ and $\gamma_{ij} = \alpha_i\beta_j/(\alpha_i+\beta_j)$, and a_i, b_j are the contraction coefficients. The overlap matrix elements for any pair of primitive Gaussians requires the use of the notation and expressions developed in Section 8.4.5. Using the Gaussian product formula (8.59) and integrating, we obtain

$$S_{\mu\nu}^{IJ} = e^{-\gamma(\mathbf{R}_I-\mathbf{R}_J)^2} I_x I_y I_z, \qquad (9.123)$$

with

$$I_x = \sum_{i=0}^{(i_1+i_2)/2} C_{2ix}^{i1,i2}(IJK)\frac{(2i-1)!!}{(2\kappa)^i}\left(\frac{\pi}{\kappa}\right)^{1/2}, \qquad (9.124)$$

where the fact has been used that the integrals containing odd powers of x vanish, and κ, γ, and $\mathbf{R}_K$ are defined in Formula (8.54). Similar expressions apply to I_y and I_z. It can be seen that the overlap between any two GTOs decays exponentially fast with the distance between the two centers, although how fast depends on whether these are core, valence, or diffuse orbitals.

9.5.2 Kinetic integrals

The first Hamiltonian matrix element to consider is the kinetic contribution, which is given by Expression (9.101). For a pair of uncontracted GTOs this expression becomes:

$$T^{IJ}_{\mu\nu} = \frac{\hbar^2}{2m}(T_x + T_y + T_z), \qquad (9.125)$$

where T_x, T_y, and T_z are obtained using the symmetric form

$$T_x = \int \int \int \left[i_1(x-x_I)^{i_1-1} - 2\alpha(x-x_I)^{i_1+1} \right] (y-y_I)^{j_1}(z-z_I)^{k_1}$$
$$\times \left[i_2(x-x_J)^{i_2-1} - 2\beta(x-x_J)^{i_2+1} \right] (y-y_J)^{j_2}(z-z_J)^{k_2}$$
$$\times e^{-\alpha(\mathbf{r}-\mathbf{R}_I)^2} e^{-\beta(\mathbf{r}-\mathbf{R}_J)^2} dx\, dy\, dz. \qquad (9.126)$$

This triple integral factorizes into the product of three Cartesian integrals that are straightforward to calculate. For a pair of s-type primitive GTOs, the kinetic integrals are:

$$T^{IJ}_{ss} = \frac{\hbar^2}{m}\left[(\mathbf{R}_K - \mathbf{R}_I) \cdot (\mathbf{R}_K - \mathbf{R}_J) + \frac{1}{2\kappa} \right] S^{IJ}_{ss}. \qquad (9.127)$$

9.5.3 Electron–nuclear interaction

The nuclear attraction terms are more complicated because the Coulomb interaction does not factorize in Cartesian coordinates. There are several possible strategies to compute these integrals, but a particularly elegant one is via integral transforms of the Coulomb potential. Boys proposed to use the Gaussian transform identity (Boys, 1950),

$$\frac{1}{|\mathbf{r}-\mathbf{r}_C|} = \frac{1}{\sqrt{\pi}} \int_{-\infty}^{\infty} e^{-s^2(\mathbf{r}-\mathbf{r}_C)^2} ds, \qquad (9.128)$$

which transforms the Coulomb potential into a Cartesian separable form that looks like an s-type Gaussian of exponent s^2, centered at $\mathbf{R}_C$. Equally useful expressions can be obtained by using Laplace or Fourier integral transforms. The resulting nuclear attraction integral is then re-written as

$$U_{\mu\nu}^{IJ} = \sum_{C=1}^{P} \int_{-\infty}^{\infty} \left(\int G_{i_1 j_1 k_1}(\mathbf{r}, \alpha, \mathbf{R}_I) e^{-s^2(\mathbf{r}-\mathbf{r}_C)^2} G_{i_2 j_2 k_2}(\mathbf{r}, \beta, \mathbf{R}_J) \, d\mathbf{r} \right) \frac{ds}{\sqrt{\pi}},$$
(9.129)

which can be calculated by applying twice the Gaussian product formula and then integrating over the auxiliary variable s.

For two s-type primitive Gaussian orbitals the above matrix element assumes the following elegant expression:

$$U_{ss}^{IJ} = e^{-\gamma(\mathbf{R}_I - \mathbf{R}_J)^2} \left(\frac{\pi}{\kappa} \right)^{3/2} \sum_{C=1}^{P} \frac{\text{erf}(\sqrt{\kappa}|\mathbf{R}_K - \mathbf{R}_C|)}{|\mathbf{R}_K - \mathbf{R}_C|},$$
(9.130)

where γ, κ, and $\mathbf{R}_K$ are given by (8.54), and "erf" is the error function defined as

$$\text{erf}(x) = \frac{2}{\sqrt{\pi}} \int_0^x e^{-t^2} dt = \frac{2x}{\sqrt{\pi}} \int_0^1 e^{-x^2 t^2} dt.$$
(9.131)

The integral transform approach can be readily extended to primitive GTOs of arbitrary angular momentum, where we obtain expressions similar to (9.124), but now involving the so-called *Boys function* (related to the *incomplete Gamma function*):

$$F_n(x) = \int_0^1 e^{-xt^2} t^{2n} dt.$$
(9.132)

The Boys function can be calculated numerically and tabulated at the outset of a calculation. Then, interpolation can be used to obtain its value at the required value of x. $F_n(x)$ is a smooth, monotonically decaying, exponential-like function (Helgaker et al., 2000).

9.5.4 Two-electron integrals

The electron–electron interaction terms for an atom-centered basis set are given by Expression (9.89), where the basis functions are primitive Cartesian GTOs. Usually these terms are indicated by the following shorthand notation:

$$V_{\mu\gamma\nu\delta}^{IKJL} = \langle \mu\gamma | \nu\delta \rangle,$$
(9.133)

where the centers of the basis functions are not indicated explicitly.

The first operation in calculating these integrals is to combine the primitive GTO ϕ_μ with ϕ_γ, and ϕ_ν with ϕ_δ, using the Gaussian product formula. The result of each of the two products is a linear combination of Gaussians as in

(8.59). Next, the repulsive Coulomb potential $1/|\mathbf{r}-\mathbf{r}'|$ in (9.89) is converted into a third Gaussian using any of the integral transforms (Gaussian, Laplace, Fourier), exactly as for the electron–nuclear interaction. The integrals over the two sets of spatial coordinates can be performed analytically, and the remaining integral over the auxiliary variable can be expressed in terms of some function that can be easily evaluated numerically.

The explicit expression for the Coulomb repulsion matrix element between four s-type primitive GTOs is:

$$\langle ss|ss\rangle = \frac{2\pi^{5/2}}{\kappa_1\kappa_2\sqrt{\kappa_1+\kappa_2}} e^{-\gamma_1(\mathbf{R}_I-\mathbf{R}_K)^2} e^{-\gamma_2(\mathbf{R}_J-\mathbf{R}_L)^2} F_0(\sqrt{\Gamma}|\mathbf{R}_A-\mathbf{R}_B|), \quad (9.134)$$

where

$$\gamma_1 = \frac{\alpha_1\alpha_2}{\alpha_1+\alpha_2}, \qquad \gamma_2 = \frac{\alpha_3\alpha_4}{\alpha_3+\alpha_4}, \qquad \Gamma = \frac{\gamma_1\gamma_2}{\gamma_1+\gamma_2}, \quad (9.135)$$

$$\mathbf{R}_A = \frac{\alpha_1\mathbf{R}_I + \alpha_2\mathbf{R}_K}{\alpha_1+\alpha_2}, \qquad \mathbf{R}_B = \frac{\alpha_3\mathbf{R}_J + \alpha_4\mathbf{R}_L}{\alpha_3+\alpha_4}, \quad (9.136)$$

and $F_0(x)$ is the Boys function of order 0 given by (9.132), which is directly related to the error function as in (9.130).

Calculating these matrix elements for anything different from $\langle ss|ss\rangle$ is not complicated, but it proved not to be a very efficient strategy for evaluating this type of integral. The problem is that terms involving primitive GTOs of higher order imply the calculation of summations of up to 15 different indices. Other methods have been proposed that are more efficient, and some of them will be briefly mentioned below. For an exhaustive description, the reader is referred to specific books on the subject (Helgaker et al., 2000). The important issue is that these integrals can be written as a sum of K terms, where $K = \sum_{l=1}^{4} i_l + j_l + k_l$, and each term carries a Boys function $F_m(\sqrt{\Gamma}|\mathbf{R}_A-\mathbf{R}_B|)$, with $m = 0,\ldots,K$. This requires the calculation of one-dimensional integrals of the form

$$I = \int_0^1 Q_K(s) e^{-\Gamma(\mathbf{R}_A-\mathbf{R}_B)^2 s^2} ds, \quad (9.137)$$

where $Q_K(s)$ is a $2K$-degree polynomial in s, which depends only on the order of the four Cartesian GTOs involved (the indices i_1, j_1, k_1, etc.). Having to calculate one such integral per each possible combination of four basis orbitals is an operation that scales like M^4. Even if one-dimensional integrals are not very demanding computationally, their number grows quite rapidly with the number of atoms. Therefore, it is important to calculate these integrals with the least possible effort so that they do not become the bottleneck of the calculation (eventually they will, but the larger the number of atoms that can be reached, the better).

Numerical integration methods

The integrals above can be calculated using a variety of numerical techniques. The most efficient ones are based on *quadratures*, i.e. using the theory of orthogonal polynomials (Abramowitz and Stegun, 1965). The main idea is that one-dimensional integrals can be calculated exactly as a sum over a set of special points s_i,

$$\int_a^b f(s)w(s)\mathrm{d}s = \sum_{i=1}^n f(s_i)w_i, \tag{9.138}$$

where the weights w_i are computed as $w_i = \int_a^b L_i(s)w(s)\mathrm{d}s$, and $L_i(s)$ are the Lagrange interpolation polynomials of degree n, which pass through all the points s_i. This requires the existence of a set of polynomials $P_i(s)$ that are orthogonal with respect to the positive measure $w(s)$. The special points s_i are the n roots of the polynomial $P_n(s)$ of degree n. The equivalence between integral and sum is exact if $f(s)$ is a polynomial of degree smaller than the number of special points n. In the case of the Coulomb integrals (9.137), since the degree of $Q_K(s)$ is $2K$, then $n \geq 2K$, but making use of the fact that $Q_K(s)$ is a polynomial of degree K in s^2, the number of quadrature points can be reduced by a half, so that effectively $n \geq K$.

Expression (9.137) is suggestive because the measure $w(s) = \mathrm{e}(-s^2)$, defined in the $(-\infty, \infty)$ interval, is consistent with Hermite's orthogonal polynomials. This would be Gauss Hermite integration. However, since the integration interval is $(0, 1)$ the use of quadratures requires a modification of Hermite polynomials to verify the correct orthogonality condition. The modified Hermite polynomials are called Rys's polynomials, and are orthogonal with respect to the measure $w(s) = \mathrm{e}(-\alpha s^2)$ in the $(0, 1)$ interval (Dupuis *et al.*, 1976). These are called Gauss–Rys quadratures. The advantage of using this approach is that the coefficients in (9.137) for *shells* of different angular momenta can be calculated efficiently using either numerical or recursive methods (Rys *et al.*, 1983; Ishida, 1998), and bypassing the summations over 15 indices.

There are several other useful schemes to evaluate Coulomb integrals over non-spherical Gaussians, such as the one proposed by McMurchie and Davidson (1978) (see also Helgaker *et al.*, 2000, Chapter 8).

9.5.5 Obara–Saika recurrence relations

In 1986, Obara and Saika proposed an alternative method for calculating Cartesian Gaussian integrals exploiting the translational invariance property of GTOs (Obara and Saika, 1986). Since Gaussian integrals depend only on the distance between different centers, they are invariant if a rigid translation of all the centers is applied. This idea translates into a scheme where integrals of higher order can be obtained from the lower order ones by a recursive algorithm.

For Cartesian primitive GTOs the overlap integrals can be written as the product of three Cartesian integrals, $S_{\mu\nu}^{IJ} = S_{i_1,i_2}^{IJ} S_{j_1,j_2}^{IJ} S_{k_1,k_2}^{IJ}$. Obara and Saika obtained the following simple recurrence relations for the Cartesian integrals:

$$S_{n+1,m}^{IJ} = (x_K - x_I)S_{n,m}^{IJ} + \frac{1}{2\kappa}\left(nS_{n-1,m}^{IJ} + mS_{n,m-1}^{IJ}\right), \qquad (9.139)$$

$$S_{n,m+1}^{IJ} = (x_K - x_J)S_{n,m}^{IJ} + \frac{1}{2\kappa}\left(nS_{n-1,m}^{IJ} + mS_{n,m-1}^{IJ}\right), \qquad (9.140)$$

where the subindices n and m indicate the power to which the quantities $(x - x_I)$ and $(x - x_J)$ appear in the two basis functions. Therefore, all the overlap matrix elements can be obtained recursively from the overlap integral for two s-type Gaussians:

$$S_{00}^{IJ} = e^{-\gamma(x_I - x_J)^2}\left(\frac{\pi}{\kappa}\right)^{1/2}. \qquad (9.141)$$

The Obara–Saika method can also be used to calculate the kinetic, one-electron, and two-electron Coulomb integrals recursively. The resulting expressions can be quite complicated and will not be reproduced here. The interested reader is referred to the original paper (Obara and Saika, 1986) or to specialized books (Helgaker *et al.*, 2000). Improvements to the Obara–Saika scheme for uncontracted basis functions have been proposed under the name of vertical and horizontal recurrence relations by Head-Gordon and Pople (1988), and recurrence relations have also been derived for contracted GTOs (Gill and Pople, 1991).

9.5.6 Multi-pole expansion

For sufficiently large systems, a significant fraction of the two-electron integrals corresponds to the classical interaction between two non-overlapping charge distributions. While the methods previously described can be used for this, a more efficient way of evaluating them is by performing a multi-pole expansion of the Coulomb interaction, as explained in Section 9.4.6.

In the case of GTOs, the Gaussian product formula indicates that the overlap distributions are linear combinations of Gaussians, but centered at $\mathbf{R}_K = (\alpha\mathbf{R}_I + \beta\mathbf{R}_J)/(\alpha + \beta)$, as given by Expression (8.59). Therefore, each two-electron integral gives rise to a number of three-center integrals that can be calculated via multi-polar expansions. The fact that the overlap distributions are still one-center functions is very important, because then the multi-pole moments vanish beyond a certain value for the angular momentum. Therefore, the multi-pole expansion can be carried out exactly by adding up a finite number of terms. The case of four s-type primitive GTOs is particularly simple, because the only non-zero multi-pole moment is $Q_{00}(\mathbf{R}_K) = S_{00}^{IJ}$, so that the matrix elements between non-overlapping s-type Gaussians is $\langle ss|ss \rangle = S_{00}^{IJ} S_{00}^{LM}/R_{KN}$, where $\mathbf{R}_K$ and $\mathbf{R}_N$ are the centers of the overlap distributions between atoms I and J, and L and M. Fast multi-pole methods become important only when dealing with very large molecules such as proteins; they are implemented in codes like GAUSSIAN (Frisch *et al.*, 2004) and GAMESS (Schmidt *et al.*, 1993).

References

Abramowitz, M. and Stegun, I. A. (1965). *Handbook of Mathematical Functions*. New York, Dover.
Allen, M. P. and Tildesley, D. J. (1987). *Computer Simulation of Liquids*. Oxford, Clarendon Press.
Amos, R. D., Alberts, I. L., Andrews, J. S. *et al.* (1995). *CADPAC: The Cambridge Analytic Derivatives Package Issue 6. A Suite of Quantum Chemistry Programs*. Cambridge, University of Cambridge. www-theor.ch.cam.ac.uk/software/cadpac.html.
Andersen, O. K. (1973). Simple approach to the band-structure problem. *Solid State Comm.* **13**, 133–136.
  (1975). Linear methods in band theory. *Phys. Rev. B* **12**, 3060–3083.
  (1984). Linear methods in band theory. In *The Electronic Structure of Complex Systems*, P. Phariseau and W. M. Temmerman, eds. New York, Plenum Press, 11–66.
Andersen, O. K. and Wooley, R. G. (1973). Muffin-tin orbitals and molecular calculations: general formalism. *Mol. Phys.* **26**, 905–927.
Andersen, O. K., Postnikov, A. V., and Savrasov, S. Y. (1992). The muffin-tin-orbital point of view. In *Applications of Multiple Scattering Theory to Materials Science*, W. H. Butler, P. H. Dederichs, A. Gonis, and R. L. Weaver, eds. MRS Symposium Proceedings, vol. 253. Pittsburgh, Materials Research Society, 37–70.
Andersen, O. K., Saha-Dasgupta, T., Tank, R. W., Arcangeli, C., Jepsen, O., and Krier, G. (2000). Developing the MTO formalism. In *Electronic Structure and the Physical Properties of Solids: the Uses of the LMTO Method*, H. Dreyssé, ed. Lecture Notes in Physics. Berlin, Springer Verlag, 1–84.
Andzelm, J. and Wimmer, E. (1992). Density functional Gaussian-type-orbital approach to molecular geometries, vibrations, and reaction energies. *J. Chem. Phys.* **96**, 1280–1303.
Baroni, S., de Gironcoli, S., Dal Corso, A., and Giannozzi, P. (2001). Phonons and related crystal properties from density-functional perturbation theory. *Rev. Mod. Phys.* **73**, 515–562.
Blaha, P., Schwarz, K., Sorantini, P., and Trickey, S. B. (1990). Full-potential, linearized augmented plane wave programs for crystalline systems. *Comput. Phys. Commun.* **59**, 399–415.
Blöchl, P. (1994). Projector augmented-wave method. *Phys. Rev. B* **50**, 17953–17979.
Boys, S. F. (1950). Electron wave functions I. A general method for calculation for the stationary states of any molecular system. *Proc. Roy. Soc. London* **200**, 542–554.
Causà, M., Dovesi, R., Orlando, R., Pisani, C., and Saunders, V. R. (1988). Treatment of the exchange interactions in Hartree–Fock LCAO calculation of periodic systems. *J. Phys. Chem.* **92**, 909–913.

Dupuis, M., Rys, J., and King, H. F. (1976). Evaluation of molecular integrals over Gaussian basis functions. *J. Chem. Phys.* **65**, 111–116.

Estrín, D. A., Corongiu, G., and Clementi, E. (1993). Structure and dynamics of molecular systems using density functional theory. In *METECC, Methods and Techniques in Computational Chemistry*. Cagliari, Stef, Chapter 12, 541–567.

Ewald, P. P. (1921). Die Berechnung optischer und elektrostatischer Gitterpotentiale. *Ann. Phys.* **64**, 253–287.

Finnis, M. W. (2003). *Interatomic Forces in Condensed Matter*. Oxford Series on Materials Modelling, Oxford, Oxford University Press.

Frisch, M. J., Trucks, G. W., Schlegel, H. B. *et al.* (2004). *Gaussian 03, Revision C.02*. Wallingford, CT, Gaussian, Inc. www.gaussian.com.

Gill, P. M. W. and Pople, J. A. (1991). The prism algorithm for 2-electron integrals. *Int. J. Quantum Chem.* **40**, 753–772.

Godbout, N., Salahub, D. R., Andzelm, J., and Wimmer, E. (1992). Optimization of Gaussian-type basis sets for local spin density functional calculations. Part I. Boron through Neon, optimization technique and validation. *Can. J. Chem.* **70**, 560–571.

Gonis, A. (1992). *Green functions for ordered and disordered systems*. Amsterdam, North-Holland.

Greengard, L. and Rokhlin, V. (1987). A fast algorithm for particle simulations. *J. Comput. Phys.* **73**, 325–348.

Ham, F. S. and Segall, B. (1961). Energy bands in periodic lattices – Green's function method. *Phys. Rev.* **124**, 1786–1796.

Harris, J. (1985). Simplified method for calculating the energy of weakly interacting fragments. *Phys. Rev. B* **31**, 1770–1779.

Head-Gordon, M. and Pople, J. A. (1988). A method for two-electron Gaussian integral and integral derivative evaluation using recurrence relations. *J. Chem. Phys.* **89**, 5777–5786.

Helgaker, T., Jorgensen, P., and Olsen, J. (2000). *Molecular Electronic Structure Theory*. Chichester, Wiley.

Herring, C. (1940). A new method for calculating wave functions in crystals. *Phys. Rev.* **57**, 1169–1177.

Holtzwarth, N. A. W., Matthews, G. E., Tackett, A. R., and Dunning, R. B. (1997). Comparison of the projector augmented-wave, pseudopotential and linearized augmented-plane-wave formalism for density-functional calculations of solids. *Phys. Rev. B* **55**, 2005–2017.

Ihm, J., Zunger, A., and Cohen, M. L. (1979). Momentum-space formalism for the total energy of solids. *J. Phys. C: Solid State Phys.* **12**, 4409–4422.

Ishida, K. (1998). Rapid algorithm for computing the electron repulsion integral over higher order Gaussian-type orbitals: accompanying coordinate expansion method. *J. Comput. Chem.* **19**, 923–934.

Jackson, J. D. (1975). *Classical Electrodynamics*, 2nd edn. New York, Wiley.

Jensen, F. (1999). *Introduction to Computational Chemistry*. Chichester, Wiley.

Jepsen, O., Madsen, J., and Andersen, O. K. (1982). Spin-polarized electronic structure of the Ni (001) surface and thin films. *Phys. Rev. B* **26**, 2790–2809.

Kendall, R. A., Apra, E., Bernholdt, D. E. *et al.* (2000). High performance computational chemistry: an overview of NWChem a distributed parallel application. *Comput. Phys. Commun.* **128**, 260–283. www.emsl.pnl.gov/docs/nwchem/nwchem.html.

King-Smith, R. D., Payne, M. C., and Lin, J. S. (1991). Real-space implementation of nonlocal pseudopotentials for first-principles total-energy calculations. *Phys. Rev. B* **44**, 13063–13066.

Kleinman, L. and Bylander, D. M. (1982). Efficacious form for model pseudopotentials. *Phys. Rev. Lett.* **48**, 1425–1428.

Koelling, D. D. and Arbman, G. O. (1975). Use of energy derivative of the radial solution in an augmented plane wave method: application to Cu. *J. Phys. F: Metal Phys.* **5**, 2041–2054.

Koelling, D. D., Freeman, A. J., and Mueller, F. M. (1970). Shifts in the electronic band structure of metals due to non-muffin-tin potentials. *Phys. Rev. B* **1**, 1318–1324.

Kohn, W. and Rostoker, N. (1954). Solution of the Schrödinger equation in periodic lattices with an application to metallic lithium. *Phys. Rev.* **94**, 1111–1120.

Korringa, J. (1947). On the calculation of the energy of a Bloch wave in a metal. *Physica* **13**, 392–400.

Kresse, G. and Joubert, D. (1999). From ultrasoft pseudopotentials to the projector augmented-wave method. *Phys. Rev. B* **59**, 1758–1775.

Laasonen, K., Pasquarello, A., Car, R., Lee, C., and Vanderbilt, D. (1993). Car–Parrinello molecular dynamics with Vanderbilt ultrasoft pseudopotentials. *Phys. Rev. B* **47**, 10142–10153.

Loucks, T. L. (1965). Fermi surfaces of Cr, Mo and W by the augmented-plane-wave method. *Phys. Rev.* **139**, A1181–A1188.

  (1967). *The Augmented Plane Wave Method.* New York, Benjamin.

McMurchie, L. E. and Davidson, E. R. (1978). One- and two-electron integrals over cartesian Gaussian functions. *J. Comput. Phys.* **26**, 218–231.

Methfessel, M. and van Schilfgaarde, M. (1993). Derivation of force theorems in density-functional theory: application to the full-potential LMTO method. *Phys. Rev. B* **48**, 4937–4940.

Methfessel, M., Rodriguez, C. O., and Andersen, O. K. (1989). Fast full-potential calculations with a converged basis of atom-centered linear muffin-tin orbitals: structural and dynamic properties of silicon. *Phys. Rev. B* **40**, 2009–2012.

Methfessel, M., van Schilfgaarde, M., and Casali, R. A. (2000). A full-potential LMTO method based on smooth Hankel functions. In *Electronic Structure and the Physical Properties of Solids: the Uses of the LMTO Method*, H. Dreyssé, ed. Lecture Notes in Physics. Berlin, Springer Verlag, 114–147.

Nielsen, O. H. and Martin, R. M. (1983). First-principles calculation of stress. *Phys. Rev. Lett.* **50**, 697–700.

  (1985a). Quantum-mechanical theory of stress and force. *Phys. Rev. B* **32**, 3780–3791.

  (1985b). Stresses in semiconductors: *ab initio* calculations on Si, Ge, and GaAs. *Phys. Rev. B* **32**, 3792–3805.

Obara, S. and Saika, A. (1986). Efficient recursive computation of molecular integrals over Cartesian Gaussian functions. *J. Chem. Phys.* **84**, 3963–3974.

Parrinello, M., Hutter, J., Marx, D. *et al.* (2004). CPMD V3.9. IBM Corp. 1990–2004, MPI für Festkörperforschung Stuttgart 1997–2001. www.cpmd.org.

Payne, M. C., Teter, M. P., Allan, D. C., Arias, T. A., and Joannopoulos, J. D. (1992). Iterative minimization techniques for *ab initio* total-energy calculations: molecular dynamics and conjugate gradients. *Rev. Mod. Phys.* **64**, 1045–1097.

Pisani, C. and Dovesi, R. (1980). Exact exchange Hartree–Fock calculations for periodic systems. I. Illustration of the method. *Int. J. Quantum Chem.* **17**, 501–516. www.crystal.unito.it.

Pople, J. A. and Beveridge, D. L. (1970). *Approximate Molecular Orbital Theory.* Advanced Chemistry, London, McGraw-Hill.

Press, W. H., Teukolsky, S. A., Vetterling, W. T., and Flannery, B. P. (1992). *Numerical Recipes. The Art of Scientific Computing.* Cambridge, Cambridge University Press.

Pulay, P. (1969). *Ab initio* calculation of force constants and equilibrium geometries in polyatomic molecules. I. Theory. *Mol. Phys.* **17**, 197–204.

Rys, J., Dupuis, M., and King, H. F. (1983). Computation of electron repulsion integrals using the Rys quadrature method. *J. Comput. Chem.* **4**, 154–157.

Savrasov, S. Y. and Savrasov, D. Y. (1992). Full-potential linear-muffin-tin-orbital method for calculating total energies and forces. *Phys. Rev. B* **46**, 12181–12195.

Schmidt, M. W., Baldridge, K. K., Boatz, J. A. *et al.* (1993). General atomic and molecular electronic structure system. *J. Comput. Chem.* **14**, 1347–1363. www.msg.ameslab.gov/GAMESS/GAMESS.html.

Segall, M. D., Lindan, P. J. D., Pickard, C. J. *et al.* (2002). First-principles simulation: ideas, illustrations and the CASTEP code. *J. Phys. Condens. Matter* **14**, 2717–2744.

Skriver, H. (1984). *The LMTO Method*. New York, Springer.

Slater, J. C. (1937). Wave functions in a periodic potential. *Phys. Rev.* **51**, 846–851.

Soler, J. M. and Williams, A. R. (1989). Simple formula for the atomic forces in the augmented-plane-wave method. *Phys. Rev. B* **40**, 1560–1564.

  (1990). Augmented-plane-wave forces. *Phys. Rev. B* **42**, 9728–9731.

Soler, J. M., Artacho, E., Gale, J. D. *et al.* (2002). The SIESTA method for *ab initio* order-N materials simulation. *J. Phys. Condens. Matter* **14**, 2745–2780. www.siesta.org.

VandeVondele, J., Krack, M., Mohamed, F., Parrinello, M., Chassaing, T., and Hutter, J. (2005). QUICKSTEP: fast and accurate density functional calculations using a mixed gaussian and plane waves approach. *Comput. Phys. Commun.* **167**, 103–128.

White, C. A. and Head-Gordon, M. (1994). Derivation and efficient implementation of the fast multipole method. *J. Chem. Phys.* **101**, 6593–6605.

Williams, A. R. (1970). Non-muffin-tin energy bands for silicon by the Korringa–Kohn–Rostoker method. *Phys. Rev. B* **1**, 3417–3426.

Wimmer, E., Krakauer, H., Weinert, M., and Freeman, A. J. (1981). Full-potential self-consistent linearized-augmented-plane-wave method for calculating the electronic structure of molecules and surfaces: O_2 molecule. *Phys. Rev. B* **24**, 864–875.

Yu, R., Singh, D., and Krakauer, H. (1991). All-electron and pseudopotential force calculations using the linearized-augmented-plane-wave method. *Phys. Rev. B* **93**, 6411–6422.

10
Simplified approaches to the electronic problem

When the electronic many-body problem is solved at any level of theory, its solution provides energies and atomic forces that allow for the determination of structural and thermodynamic properties. Forces also open the road to molecular dynamics simulation and the computation of dynamical properties. In addition, and at variance with classical force fields, any such theory would give access to electronic properties such as electronic energy levels or bands, electronic density, charge transfer, and, in principle, also optical and electronic transport properties.

It is clear that the simpler the treatment of the electronic variables, the more efficient the calculation. Therefore, larger systems can be studied, and longer MD simulations permit us to accumulate more reliable statistics and give access to dynamical phenomena occurring on longer time scales. If the information arising from the electronic degrees of freedom is not strictly necessary, then a strategy that removes the electrons altogether out of the picture by means of effective interatomic potentials (classical force fields) is a winning strategy. However, as long as electrons are treated explicitly, the determination of interatomic potentials requires the solution of the Schrödinger equation, and this is far more expensive than just replacing distances and angles in an explicit formula.

Linear scaling with the system size in the solution of Hartree–Fock or Kohn–Sham equations can only be achieved for very large systems. But even in that limit, when using an atom-centered basis set, the calculation of the matrix elements and the diagonalization of the Hamiltonian matrix require a major computational effort; the latter scales as N^3. Therefore, unless order-N diagonalization algorithms are used (Ordejón *et al.*, 1993; Kim *et al.*, 1995), eventually this operation becomes the bottleneck of the calculation. A reasonable goal is to try and generate a range of simplified schemes that can provide semiquantitative answers such as reproducing correct trends when comparing related systems, e.g. solvation energies for a family of molecules or for different conformations of a molecule in solution (Murdock *et al.*, 2002), with much smaller computational effort. These methods should be flexible enough that, by increasing the level of complexity, they can be made into more and more accurate schemes. Having identified the bottlenecks of a Hartree–Fock or Kohn–Sham calculation, the simplification proceeds by approximating in the operations where the reduction of the computational cost is most effective. A simplified scheme, however, should be constructed having in mind a number of general conditions that keep it under control. Elaborating on the conditions suggested by Pople and Beveridge (1970), the following appears to constitute a reasonable set of features that simplified models should fulfill.

(i) The method should be simple enough to allow for its application to large systems with a reasonable computational effort. The measure of what constitutes a *large* system is a concept that evolves in time, according to the progress in the speed of CPUs. Nowadays, large may be considered to apply to a biomolecule such as a protein, perhaps including a piece of cellular membrane, or a piece of material sufficiently large to accommodate point-like and extended defects such as dislocations, i.e. several thousands, or even millions of particles, as in metallurgy simulations.

(ii) The approximations introduced should not be that severe that they modify the physical forces that determine structural and dynamical properties. Directionality of chemical bonds, ordering of energy levels, electrostatic forces, and energetic ordering of different isomers or bulk phases are major features that are desirable to retain.

(iii) The approximate wave function should be as unbiased as possible, avoiding the explicit introduction of pre-existent qualitative ideas such as electron pairing in a chemical bond.

(iv) The method should account for all the chemically active electrons, i.e. those in the valence shells or bands. This is in contrast to π-bonding models for organic planar molecules such as benzene, which take into account explicitly only the π electronic states.

(v) The approach should be sufficiently general to allow for systematic improvements, from the simplest possible up to the *ab initio* level. It should be possible to derive the methodology from a precise starting point (Hartree–Fock or Kohn–Sham), by means of a series of more or less controlled approximations. When the approximations are not controlled, then they should be carefully validated, and the limit of validity should be clearly determined.

The first step in order to reduce the computational cost without compromising the above conditions is to eliminate the *core electrons* from the description, seeing that these assume a really minor role in chemical bonding. Core electrons can be removed by replacing the ionic core (nucleus plus core electrons) with a pseudopotential. Another possible way is to replace the nuclear charge by an effective value Z_{eff}, which would typically be the valence charge Z_V, while retaining the coulombic form of the nuclear attraction. A second natural step is to consider only a *single-ζ* or *minimal basis set*, i.e. the minimum number of basis functions necessary to describe the neutral atom of each of the participating species (see Section 8.4.6). These two approximations help to reduce the size of the Hamiltonian matrix, and thus the computational cost of the diagonalization operation.

In Section 10.1 we shall review a first class of methods that is based on simplifying the calculation of Hamiltonian matrix elements in atom-centered basis sets. Physicists have used this kind of approach at the level of empirical models for a long time under the name of *tight-binding* methods (TB). More recently, it was shown that there is a clear connection between tight-binding and density functional theory. In parallel, chemists have followed a similar road by simplifying the calculation of the matrix elements in Hartree–Fock theory, thus giving rise to a variety of *semiempirical* methods. These approaches are based on neglecting the most expensive matrix elements, while the remaining ones are calculated using STO minimal basis sets. The exponents of the STO and other parameters are fitted to atomic and molecular data. Semiempirical methods will be described in Section 10.2, and a connection between these two worlds will be attempted in Section 10.3.

We shall then briefly mention many-body approaches like embedded atom models and bond order potentials, methods that take into account the electronic polarizability in a simplified way like the shell model, and some common empirical force fields. In Section 10.6 we shall focus on hybrid methods that treat a portion of the system quantum-mechanically and the rest at a classical level. These are quite useful to study molecules in solution, and also to isolate the chemically active region of a system, thus avoiding the use of valuable computational resources in describing at a quantum level a portion of the system that can be dealt with accurately enough using classical potentials. Finally, we shall devote the last section to orbital-free density functionals of the Thomas–Fermi family, which have recently received renewed attention due to new developments in the field.

10.1 Tight-binding methods

The tight-binding (TB) approach to the electronic structure of solids is complementary to the nearly free electron picture. While the latter is a reasonably good representation of the electronic structure of simple metals, TB provides a more faithful representation of systems where the electrons are localized in chemical bonds of different degrees of covalency. The starting point of this model is to consider that, in a first approximation, electrons are localized in a single atom, but they have the possibility to jump to neighboring atoms. TB methods range from very basic empirical models to the most sophisticated *ab initio* schemes, where the Hamiltonian matrix is derived from density functional theory and an orthogonal basis set of localized orbitals is constructed from atom-centered basis

10.1 Tight-binding methods

functions by means of an appropriate linear transformation (see e.g. Andersen *et al.*, 2000). In this section, we first introduce the basic ideas of the empirical TB method, and then we discuss more elaborate approaches based on simplifications of the Kohn–Sham formalism.

10.1.1 Empirical tight-binding

Let us first introduce, for pedagogical reasons, the simplest possible form of the TB model. The state of an electron in atom I is indicated by the ket $|I\rangle$. In a real-space representation, this would be the atomic eigenstate $\phi_I(\mathbf{r} - \mathbf{R}_I)$. If the atoms are sufficiently far apart, then, in the crudest approximation, electrons in every atom will have the same *on-site* energy value ε_0. This situation can be represented by a model Hamiltonian of the form $\hat{\mathcal{H}}_0 = \varepsilon_0 \sum_I |I\rangle\langle I|$.

When the atoms are brought together, we need to take into account the possibility of the electrons jumping from one atom to any of its neighbors. This is achieved by including off-diagonal terms in the Hamiltonian, which now reads

$$\hat{\mathcal{H}}_{\text{TB}} = \varepsilon_0 \sum_I |I\rangle\langle I| + t \sum_I \sum_J |I\rangle\langle J|. \tag{10.1}$$

In a model situation, the sum on J runs only over the nearest neighbors of I. The parameter t is usually called *hopping* or *hopping integral*.

In order to understand the meaning of the hopping integrals, let us consider the real-space version of the TB Hamiltonian for a single electron in a molecule,

$$\hat{\mathcal{H}}_{\text{TB}} = -\frac{\hbar^2}{2m}\nabla^2 + \sum_K v_K(\mathbf{r} - \mathbf{R}_K), \tag{10.2}$$

and propose a linear combination of atomic orbitals for the TB wave function, i.e. $\varphi(\mathbf{r}) = \sum_J C_J \phi_J(\mathbf{r} - \mathbf{R}_J)$, where

$$\left[-\frac{\hbar^2}{2m}\nabla^2 + v_J(\mathbf{r})\right]\phi_J(\mathbf{r}) = \varepsilon_0 \phi_J(\mathbf{r}). \tag{10.3}$$

By replacing this expression into Schrödinger's equation, we obtain

$$\hat{\mathcal{H}}_{\text{TB}}\varphi(\mathbf{r}) = \sum_J C_J \left[-\frac{\hbar^2}{2m}\nabla^2 + v_J(\mathbf{r} - \mathbf{R}_J) + \sum_{K \neq J} v_K(\mathbf{r} - \mathbf{R}_K)\right]\phi_J(\mathbf{r})$$

$$= \varepsilon_0 \sum_J C_J \phi_J(\mathbf{r}), \tag{10.4}$$

which, multiplying by $\phi_K^*(\mathbf{r})$ and integrating over $\mathbf{r}$, gives rise to diagonal terms of the form

$$H_{\text{TB}}^{JJ} = \int \phi_J^*(\mathbf{r})\left[-\frac{\hbar^2}{2m}\nabla^2 + v_J(\mathbf{r} - \mathbf{R}_J)\right]\phi_J(\mathbf{r})d\mathbf{r} = \varepsilon_0, \tag{10.5}$$

where ε_0 is the on-site energy, and off-diagonal terms

$$H_{TB}^{IJ} = -\frac{\hbar^2}{2m}\int \phi_I^*(\mathbf{r})\nabla^2\phi_J(\mathbf{r})d\mathbf{r} + \sum_K \int \phi_I^*(\mathbf{r})v_K(\mathbf{r}-\mathbf{R}_K)\phi_J(\mathbf{r})d\mathbf{r}. \quad (10.6)$$

These are precisely the hopping terms, which involve two- and three-center integrals. The two-center integrals are the most important contribution. If three-center integrals ($K \neq I \neq J$) are neglected, what remains is

$$H_{TB}^{IJ} \approx \int \phi_I^*(\mathbf{r})\left[-\frac{\hbar^2}{2m}\nabla^2 + v_I(\mathbf{r}-\mathbf{R}_I) + v_J(\mathbf{r}-\mathbf{R}_J)\right]\phi_J(\mathbf{r})d\mathbf{r}, \quad (10.7)$$

which corresponds to the electron being shared between two atoms I and J. There is a kinetic energy contribution to hopping, but more important are the potentials attracting the electron to the two atoms. The most substantial part of the kinetic energy is actually associated with the on-site terms.

As an illustration, we can use the model Hamiltonian (10.1) to study the energy bands of a a crystalline solid. We start by constructing translationally invariant electronic states of the form

$$|\mathbf{k}\rangle = \frac{1}{\sqrt{N}}\sum_J e^{i\mathbf{k}\cdot\mathbf{R}_J}|J\rangle, \quad (10.8)$$

where N is the number of atomic sites. It can be easily shown that these wave functions satisfy Bloch's theorem, and are thus eigenstates of the tight-binding Hamiltonian. By operating with the above Hamiltonian on $|\mathbf{k}\rangle$, and assuming the orthogonality of the atomic eigenstates $|J\rangle$, which in general is not true, but in the present model situation is justified, we obtain

$$\hat{\mathcal{H}}_{TB}|\mathbf{k}\rangle = \frac{1}{\sqrt{N}}\sum_J e^{i\mathbf{k}\cdot\mathbf{R}_J}\left(\varepsilon_0|J\rangle + t\sum_\mathbf{a}|J+\mathbf{a}\rangle\right), \quad (10.9)$$

where $\mathbf{a}$ are the lattice vectors connecting atom J with its nearest neighbors. Since $\sum_J \exp(i\mathbf{k}\cdot\mathbf{R}_J)|J+\mathbf{a}\rangle = \exp(-i\mathbf{k}\cdot\mathbf{a})\sum_J \exp(i\mathbf{k}\cdot\mathbf{R}_J)|J\rangle$,

$$\hat{\mathcal{H}}_{TB}|\mathbf{k}\rangle = \left[\varepsilon_0 + t\sum_\mathbf{a} e^{-i\mathbf{k}\cdot\mathbf{a}}\right]|\mathbf{k}\rangle. \quad (10.10)$$

Therefore, the energy bands are given by the dispersion relation

$$\varepsilon(\mathbf{k}) = \varepsilon_0 + t\sum_\mathbf{a} \exp(-i\mathbf{k}\cdot\mathbf{a}), \quad (10.11)$$

10.1 Tight-binding methods

which for a simple cubic lattice becomes

$$\varepsilon(\mathbf{k}) = \varepsilon_0 + 2t\left[\cos(k_x a) + \cos(k_y a) + \cos(k_z a)\right] \quad (10.12)$$

with a band width $W = 12t$. Therefore, the hopping parameter can be interpreted in terms of the band width, and fitted to experimental or *ab initio* calculated values. At variance with the nearly-free electron model, the spectrum of the tight-binding Hamiltonian is bound, and gives rise to an electronic density of states $g(\varepsilon)$ that begins at energy $\varepsilon_0 - W/2$, rising sharply to peak at the center of the band ε_0, and dropping again to end at $\varepsilon_0 + W/2$. For the above oversimplified model the density of states is symmetric, but for more realistic models it may also be asymmetric with respect to the center of the band.

The electronic energy in the empirical tight-binding model, as in *ab initio* tight-binding methods, is given simply by the sum of the eigenvalues of the Hamiltonian up to the highest occupied energy level, which in the case of an infinite solid becomes an integral up to the Fermi energy:

$$E_{\text{TB}} = \int_{-\infty}^{\epsilon_F} \varepsilon g(\varepsilon) \mathrm{d}\varepsilon, \quad (10.13)$$

where, effectively, the lower limit of the integral is the bottom of the band, i.e. $\varepsilon_0 - W/2$. The shape of the density of states can be modeled, e.g. as $g(\varepsilon) \approx 2[(W/2)^2 - (\varepsilon - \varepsilon_0)^2]^{1/2}$, where the factor 2 stands for spin degeneracy, or can also be expanded in terms of the moments of the eigenvalue distribution, thus giving rise to a family of classical, many-body force fields that have been successfully used to model metallic systems (Finnis and Sinclair, 1984).

The above empirical tight-binding model can be made more realistic by taking into account the identity of the atoms. This is achieved by considering that the electrons in the isolated atom occupy their corresponding atomic orbitals, and these orbitals hybridize to give rise to energy bands. Therefore, for each atomic species we consider a minimal set of atomic valence orbitals, i.e. one orbital for each valence state occupied in the isolated atom. Within this picture first- and second-row elements are represented by one s and three p orbitals, while transition metal atoms require one s and five d orbitals. This more elaborate tight-binding Hamiltonian can be expressed as:

$$\hat{\mathcal{H}}_{\text{TB}} = \sum_I \sum_\alpha \varepsilon_\alpha^I |\phi_\alpha^I\rangle\langle\phi_\alpha^I| + \sum_{\alpha\beta} \sum_{I \neq J} t_{\alpha\beta}^{IJ} |\phi_\alpha^I\rangle\langle\phi_\beta^J|, \quad (10.14)$$

where $|\phi_\alpha^I\rangle$ represents an atomic orbital of symmetry α, e.g. an s-type orbital, centered on atom I. Now, ε_α^I are the *on-site* atomic energies associated with these orbitals, and depend on the type of orbital. The hopping parameter t also depends on the type of orbitals between which the electron jumps, and on the distance between the atoms. The most general form is to consider all the hopping integrals

$t_{\alpha\beta}^{IJ}$, which represent the jump of an electron from state α centered on atom I to state β centered on atom J. Later, the hopping matrix can be simplified to take into account only jumps between nearest neighboring atoms.

This type of Hamiltonian can accommodate several bands of different symmetry, e.g. s and d bands for transition metals, or s and p bands for systems involving first- and second-row atoms such as Al and Si. It can also be used to describe compounds such as oxides (SiO_2 or ZrO_2), or inter-metallic alloys. In principle the approach is completely general, and the diagonalization of the TB Hamiltonian is straightforward. The parameters of the empirical tight-binding model, i.e. the on-site energies and hopping integrals, can be fitted either to calculated *ab initio* band structures, or to experimental data. The *ab initio* approach to the determination of the model parameters will be discussed in Section 10.1.2. The success of this type of model depends on how well the empirical TB Hamiltonian can reproduce *ab initio* or experimental band structures, which is not always guaranteed for minimal basis sets. In general, TB models tend to be more successful when the underlying atomic-like picture is a faithful representation of interatomic bonding. This is the case of electrons more tightly bound to the atom, which give rise to relatively narrow bands. Such is the case of $3d$ bands in transition metals.

It is interesting to notice the similarity between the tight-binding Hamiltonian and one-electron *ab initio* Hamiltonians such as Kohn–Sham and Hartree–Fock, when expressed in an atom-centered basis set (see Section 9.4). Formally, the expression is the same, but in the empirical tight-binding scheme two-electron Coulomb contributions are not calculated explicitly. They are empirically incorporated into the model parameters, which can then be interpreted as renormalized one-electron contributions. In particular, Expression (9.87) for the band energy is the same, only the matrix elements $h_{I\mu\nu}^{IJ}$ are replaced by the on-site energies and hopping integrals. Another common element is that, normally, atom-centered basis sets are non-orthogonal. Therefore, the wave functions are the solutions of the generalized eigenvalue problem

$$(\mathbf{H}_{\text{TB}} - \varepsilon_n \mathbf{S})\varphi_n = 0, \tag{10.15}$$

where ε_n are the energies of the electronic levels and $\mathbf{S}$ is the overlap matrix, whose elements are given by (9.82). The band energy is given by

$$E_{\text{TB}} = \sum_{n=1}^{N} \varepsilon_n = \text{Tr}\left(\hat{\rho}\hat{\mathcal{H}}_{\text{TB}}\right), \tag{10.16}$$

where N is the highest occupied energy level. The density matrix $\hat{\rho}$ is defined in terms of the expansion coefficients of the wave functions φ_n in the atom-centered basis set, as discussed in Section 9.4.2.

The interatomic forces arising from the TB Hamiltonian are purely attractive. Therefore, a TB model useful for geometry optimization or molecular dynamics simulation requires an additional energy term that takes into account the repulsive part of the potential. This term arises from a combination of inter-nuclear Coulomb and electron–electron repulsive interactions that have been neglected in the TB Hamiltonian. The simplest and most efficient way to introduce this aspect is by means of an additional energy term

$$E_{\text{rep}} = \frac{1}{2} \sum_{I \neq J} V_{\text{rep}}(|\mathbf{R}_I - \mathbf{R}_J|), \qquad (10.17)$$

with $V_{\text{rep}}(R)$ an effective two-body potential that depends on the species of the two atoms involved, and is fitted to reproduce bulk properties such as lattice constant, bulk modulus, and elastic constants of the solid.

In general, empirical tight-binding schemes are non-self-consistent, in the sense that the Hamiltonian does not depend on the solutions of the Schrödinger equation. There are, however, self-consistent versions.

10.1.2 Ab initio *tight-binding*

As suggested by the similarity in the form of the Hamiltonians, it is possible to make a connection between the Kohn–Sham scheme for an atom-centered basis set and the tight-binding approach. In this section we follow the formulation and the ideas developed in Chapter 7 of Finnis (2003). To start with, we write down the Kohn–Sham Hamiltonian matrix elements:

$$H_{\text{KS}\mu\nu}^{IJ} = \langle \phi_\mu^I | \left[-\frac{\hbar^2}{2m} \nabla^2 + \hat{v}_{\text{KS}}[\rho] \right] | \phi_\nu^J \rangle. \qquad (10.18)$$

The matrix elements of the Kohn–Sham potential can be written as

$$v_{\text{KS}\mu\nu}^{IJ}[\rho] = U_{\mu\nu}^{IJ} + \mu_{\text{XC}\mu\nu}^{IJ}[\rho] + \sum_{\gamma\delta=1}^{M} \sum_{KL=1}^{P} \rho_{\delta\gamma}^{LK} \langle \mu^I \gamma^K | \nu^J \delta^L \rangle, \qquad (10.19)$$

with $U_{\mu\nu}^{IJ}$ the matrix element of the nuclear attraction given by (9.102), and $\mu_{\text{XC}\mu\nu}^{IJ}$ the matrix element of the exchange-correlation potential. The last term is the Hartree potential written in terms of the density matrix and the Coulomb two-electron matrix elements given by (9.89). This form shows clearly the dependence of the Kohn–Sham Hamiltonian on the density, which is also implicitly present in the exchange-correlation potential.

Hamiltonian matrix elements

The calculation of the Kohn–Sham matrix elements formally involves the calculation of integrals ranging from one up to four centers, although the latter are reduced to two-center integrals if the density is expressed in an atom-centered basis set, or if the Hartree potential is expressed as a sum of atom-centered potentials (see Section 9.4.5). Otherwise, four-center integrals of the form $\langle \mu^I \gamma^K | \nu^J \delta^L \rangle$ and three-center integrals where two of the basis functions are centered on the same atom are the most expensive of them all, and are almost invariably neglected in tight-binding schemes. There is another type of three-center integral involving the nuclear attraction, which can be written $U^{IJ}_{\mu\nu} = \sum_K \langle \mu^I | V_K | \nu^J \rangle$. These are also neglected, so that the remaining terms involve either two basis functions centered on the same atom, $\langle \mu^I | V_K | \nu^I \rangle$, or one of the basis functions and the nuclear potential centered on the same atom, $\langle \mu^I | V_J | \nu^J \rangle$. It is expected that the neglected terms can be absorbed in a parameterization of the remaining one- and two-center integrals, and in a redefinition of the inter-nuclear repulsion term. This is a rather uncontrolled approximation, but can be approximately justified by actually calculating these terms and showing that they assume small values.

One-center integrals correspond to the on-site matrix elements ε^I_μ, while some of the two-center integrals correspond to the hopping terms $t^{IJ}_{\mu\nu}$ in (10.14). There are, however, two-center integrals that cannot be written as hopping terms. These will be eliminated by making another somewhat uncontrolled approximation, namely that the Kohn–Sham potential can be written as a sum of atom-centered potentials, $\hat{v}_{KS} = \sum_K \hat{v}^{(K)}_{KS}$. Expression (10.19) shows that the only contribution to the Kohn–Sham potential that can be strictly written in this way is the external potential. The Hartree contribution is linear in the density, but in general the total density cannot be written as a superposition of atom-centered partial densities. The exchange-correlation term is not even linear in the total density. Having said this, the atomic sphere approximation (ASA) in electronic structure calculations based on augmentation spheres consists precisely of approximating the Kohn–Sham potential as a superposition of atom-centered, overlapping spherical potentials (see Section 9.2). The ASA proved quite successful in describing close-packed structures. Within this assumption the Kohn–Sham matrix elements conveniently take the same form as the nuclear attraction terms, including three-center integrals where the two basis functions and the potential are centered on different atoms. To be consistent all three-center integrals are neglected. Within these approximations, the only remaining terms are *on-site* one-center and *hopping* two-center integrals, which are the basic ingredients of the tight-binding model. Notice that there are two-center integrals that involve basis functions centered on the same atom, while the potential is centered on another atom. These represent the influence of the crystal field on the on-site energies and, in principle, they can be absorbed in the parameterization of the model.

To avoid having to calculate the integrals every time they are required, a standard procedure in TB schemes is to parameterize them as a function of the type of basis function (angular momentum) and the distance between the centers. The seminal work of Slater and Koster has been invaluable in this respect (Slater and Koster, 1954). Two-center integrals involve one orbital at each center, and assume the form

$$t_{\mu\nu}^{IJ} = \int \phi_\mu^*(\mathbf{r}-\mathbf{R}_I) \hat{\mathcal{H}}_{KS}^{TB} \phi_\nu(\mathbf{r}-\mathbf{R}_J) d\mathbf{r}, \quad (10.20)$$

with

$$\hat{\mathcal{H}}_{KS}^{TB} = -\frac{\hbar^2}{2m}\nabla^2 + v_{KS}^{(I)}(\mathbf{r}-\mathbf{R}_I) + v_{KS}^{(J)}(\mathbf{r}-\mathbf{R}_J). \quad (10.21)$$

If the two orbitals are of the s-type, then the integral is called $ss\sigma$. If one of the orbitals is of s-type and the other is of p-type, then in principle there are two possibilities: either the p orbital is oriented along the $\mathbf{R}_{IJ}$ direction or it is lying in the perpendicular plane. The former are different from zero, and are called $sp\sigma$ or $ps\sigma$, according to whether $\hat{\mathcal{H}}_{KS}^{TB}$ operates on the p or on the s orbital. The latter are zero by symmetry; the negative and positive lobes of the p orbitals produce contributions that are equal and of opposite sign. A similar situation arises when the two orbitals are of p-type. If the two orbitals are aligned with the $\mathbf{R}_{IJ}$ direction, then the matrix element is called $pp\sigma$, and if they are both perpendicular to that direction, it receives the name of $pp\pi$. The other combinations are ruled out by symmetry reasons, as above. When the orbitals are not aligned or perpendicular to the $\mathbf{R}_{IJ}$ direction, the integrals can be obtained as a linear combination of the above after rotating the coordinate axes. The same analysis has been carried out by Slater and Koster for d-type orbitals and their matrix elements with s- and p-type ones.

For an sp system there are five independent integrals: $ss\sigma$, $sp\sigma$, $ps\sigma$, $pp\sigma$, and $pp\pi$ per pair of different atomic species, and four different ones ($sp\sigma = -ps\sigma$) for atoms belonging to the same species. Therefore, in a binary sp system such as SiO_2 there are 13 independent hopping integrals. If d-type basis functions are included, then the following nine combinations are non-vanishing: $sd\sigma$, $ds\sigma$, $pd\sigma$, $dp\sigma$, $pd\pi$, $dp\pi$, $dd\sigma$, $dd\pi$, and $dd\delta$. Therefore, an sd transition metal requires five hopping integrals ($sd\sigma = ds\sigma$), and a transition metal oxide like TiO_2 requires 14 of them, four for the O-O pairs, five for the Ti-Ti pairs, and five for the Ti-O pairs. The overlap matrix elements $S_{\mu\nu}^{IJ}$ can be calculated following the same rules as the hopping integrals.

The overlap and hopping integrals not only depend on the type and orientation of the orbitals, but are also functions of the distance between centers. The exact functional form depends on the form of the basis functions. For example, overlap integrals between GTOs decay as $\exp(-\gamma R_{IJ}^2)$, but they turn into exponentials for STOs. Although early parameterizations of the hopping integrals tended to use exponential forms (Ducastelle, 1970), their behavior with distance is better described by an inverse power-law decay of the form

$$t_{\mu\nu}^{IJ}(R_{IJ}) = t_0^{IJ} R_{IJ}^{-n(\mu,\nu)}, \tag{10.22}$$

where t_0^{IJ} is a constant that depends on the two atomic species, and $n(\mu, \nu)$ is an exponent that depends on the type of orbitals. For example, for sp-bonded systems it can be argued that an exponent of $n = 2$ is appropriate (Harrison, 1980).

The inverse power-law behavior is computationally inconvenient because the values of the integrals decay too slowly, and matrix elements have to be calculated for atoms that are quite far apart. It also reduces the sparsity of the TB matrix. However, these matrix elements are probably of the order of or even smaller than the three- and four-center integrals already neglected. Therefore, there is little additional harm caused if the functional form of the hopping integrals with respect to distance is chosen to approach zero smoothly at some cutoff radius R_0 (Goodwin *et al.*, 1989). Functional forms along these lines contain a number of parameters that are fitted to experimental or calculated band structures. It is this fitting, together with the repulsive two-body potential E_{rep}, that is expected to take care of the most severe approximations made along the way.

The Harris functional

In principle, the solution of Kohn–Sham equations requires the self-consistent determination of the density matrix. It is possible, however, to devise a non-self-consistent approach akin to the empirical tight-binding approach. This is based on the approximation to the energy proposed by Harris and Foulkes (Harris, 1985; Foulkes, 1987). This energy functional was originally devised as an approximation to the energy of weakly interacting fragments of the system. The idea is to construct the electronic density as a superposition of densities associated with different parts of the system, $\rho_{\text{in}}(\mathbf{r}) = \sum_i \tilde{\rho}_i(\mathbf{r})$, and then evaluate the energy *only* in terms of this *input* density, according to:

$$E_{\text{Harris}}[\rho_{\text{in}}] = \sum_{n=1}^{N} \varepsilon_n[\rho_{\text{in}}] - \frac{1}{2} \int \frac{\rho_{\text{in}}(\mathbf{r})\rho_{\text{in}}(\mathbf{r}')}{|\mathbf{r}-\mathbf{r}'|} d\mathbf{r} d\mathbf{r}'$$

$$+ \int (\epsilon_{\text{XC}}[\rho_{\text{in}}] - \mu_{\text{XC}}[\rho_{\text{in}}]) d\mathbf{r} + V_{\text{nn}}. \tag{10.23}$$

The one-electron eigenvalues ε_n are obtained by solving Kohn–Sham equations *non-self-consistently* using the Kohn–Sham potential evaluated for the *input* density. Notice that this energy functional coincides with the Kohn–Sham functional when $\rho_{\rm in}$ is the self-consistent charge density. However, for a general density this is not the case because the Hamiltonian is constructed with the input density, but the energy uses the output density obtained from the solutions of the eigenvalue problem, i.e.

$$\rho_{\rm out}({\bf r}) = \sum_{n=1}^{N} |\varphi_n[\rho_{\rm in}]({\bf r})|^2. \qquad (10.24)$$

This inconsistency introduces an error in the energy, which prevents us from using Kohn–Sham's functional as an estimator for the energy unless the density is close to self-consistency. In the case of the Harris functional, it can be shown that the error in the energy is of *second order* in the difference between the input and the self-consistent density, $(\rho_{\rm in} - \rho_{\rm sc})$. A good account of the variational properties and errors of the exact Kohn–Sham functional and approximations such as the Harris functional has been given in Finnis (2003) in Chapters 3 and 4, in terms of perturbations and linear response theory. Therefore, the self-consistent density is still a stationary point of the Harris functional. However, at variance with the Kohn–Sham functional, it is not a minimum but a saddle point (Zaremba, 1990).

Therefore, the behavior of the Harris energy as an indicator of convergence in a self-consistent procedure is not necessarily monotonically decreasing. It can be oscillatory, but in general it converges from below. This, however, depends heavily on the basis set. For plane waves the curvature of the energy functional in terms of the wave vectors ${\bf G}$, given by the kernel

$$K({\bf G}, {\bf G}') = \frac{4\pi}{G^2}\delta_{{\bf G},{\bf G}'} + \frac{\delta^2 E_{\rm XC}}{\delta\rho({\bf G})\delta\rho({\bf G}')}, \qquad (10.25)$$

assumes positive values for small ${\bf G}$, but crosses over to negative values at a wave vector ${\bf G}_{\rm c}$ because the exchange-correlation contribution is negative. Therefore, for large wave vectors the Harris energy is a maximum, but for small ${\bf G}$ is a minimum. This makes impracticable approaches based on the optimization of the Harris functional using PW basis sets, such as the Car–Parrinello approach (see Chapter 12). Interestingly, this saddle point behavior does not seem to manifest itself when using atom-centered basis sets. Apparently, in the space of coefficients of the density matrix, the Harris functional corresponds to a maximum and not to a saddle point.

Ab initio *non-self-consistent tight-binding*

The Harris functional just described can then be used as a starting point for a tight-binding scheme. The form of the energy (10.23) is the same as for the empirical tight-binding scheme (10.16), except for the double-counting terms. One possibility is to absorb these terms into the repulsive energy

$$E_{\text{rep}} = \int \frac{\rho_{\text{in}}(\mathbf{r})\rho_{\text{in}}(\mathbf{r}')}{|\mathbf{r}-\mathbf{r}'|}\,d\mathbf{r}\,d\mathbf{r}' + \int (\epsilon_{\text{XC}}[\rho_{\text{in}}] - \mu_{\text{XC}}[\rho_{\text{in}}])\,d\mathbf{r} + V_{\text{nn}}, \quad (10.26)$$

which is usually written as a sum of pairwise empirical potentials fitted to bulk properties as in Expression (10.17). This approach is called the *tight-binding band model* (Finnis, 2003), and the energy is given by:

$$E_{\text{band}}[\rho_{\text{in}}] = \text{Tr}\,(\hat{\rho}(\text{out})\,\hat{\mathcal{H}}_{\text{TB}}\,[\rho_{\text{in}}]) + E_{\text{rep}}, \quad (10.27)$$

where $\hat{\rho}$ is the density matrix obtained after solving the eigenvalue problem, i.e. the output density matrix. This shows the connection between density functional theory and empirical tight-binding models; the tight-binding Hamiltonian corresponds to the Kohn–Sham Hamiltonian evaluated at some appropriate input density, which normally takes the form of a superposition of atom-centered densities (Chetty et al., 1991). In that case, the first and last term in (10.26) are in fact sums of pairwise potentials. The second term is not necessarily so, although Finnis has given a justification in the case of smoothly varying input densities (Finnis, 2003, Chapter 7).

The TB band model has been widely used. However, it has some limitations. The most severe one is that on-site elements of the density matrix, which are related to the charge associated to the different atoms in the system, are only meaningful if the charges are adjusted self-consistently. Otherwise, charge transfers tend to be overestimated and unrealistic. For that reason some models impose the condition of local charge neutrality (Sankey and Niklewski, 1989), but this is not too satisfactory either, because then charge transfer is inhibited.

An alternative route is to make use of the concept of *bond energy* as opposed to *band energy*. The bond energy is defined as the part of the band energy that corresponds to matrix elements centered on different atoms (hence the name *bond*):

$$E_{\text{bond}}[\rho_{\text{in}}] = \sum_{\nu=1}^{M}\sum_{J\neq I}^{P}\left(\sum_{\mu=1}^{M}\sum_{I=1}^{P}\rho_{\nu\mu}^{JI}(\text{out})H_{\mu\nu}^{IJ}[\rho_{\text{in}}]\right). \quad (10.28)$$

The density matrix elements appearing in this definition are the *bond orders*, and the Hamiltonian matrix elements $H_{\text{TB}\mu\nu}^{IJ}$ are precisely the tight-binding *hopping integrals* $t_{\mu\nu}^{IJ}$.

The bond energy accounts for the cohesion (in a crystal) or chemical bonding (in a molecule). Therefore, it makes sense to use E_{bond} to approximate the *cohesive* or *binding* energy in place of the total energy. The binding energy is the difference between the total energy of the solid or molecule and the Kohn–Sham energies of the isolated components, i.e. $E_{\text{bind}} = E_{\text{KS}}^{\text{tot}} - \sum_I E_{\text{KS}}^I$. By using the Harris approximation to Kohn–Sham energies and the further approximation that the effective Kohn–Sham potentials of the isolated atoms are additive, the binding energy can be written as

$$E_{\text{bind}} = E_{\text{bond}} + \tilde{E}_{\text{prom}} + \tilde{E}_{\text{rep}}, \quad (10.29)$$

where $\tilde{E}_{\text{prom}}$ is associated with the promotion of electrons from the ground state configuration of the isolated atoms to the configuration polarized by the field of the neighboring atoms (Bester and Fähnle, 2001). To avoid an uncontrolled charge transfer in non-self-consistent models, the promotion energy is made into a constant (Finnis, 2003). The remaining term $\tilde{E}_{\text{rep}}$ is the repulsive energy, which is again modeled via effective pairwise additive potentials. It is a modified form of the repulsive energy of the TB band model E_{rep} in that the contributions of the isolated atoms are subtracted. This is called the *tight-binding bond model*, and it is discussed in detail by Pettifor (1995).

Therefore, in the TB bond model the on-site elements of the Hamiltonian do not contribute to the binding energy, but they are part of the Hamiltonian and are necessary to compute the bond orders $\rho_{\mu\nu}^{IJ}$ through the non-self-consistent solution of the eigenvalue problem. They are normally fitted to *ab initio* calculated band structures and bulk properties, but they can also be adjusted to ensure local charge neutrality, i.e. so that the charge in each atom $q_I = \sum_\mu \rho_{\mu\mu}^{II}$ remains constant. This latter, however, is a form of self-consistency requirement. The two schemes described above approximate the overlap matrix with the unit matrix. The non-orthogonality of the basis can be retained in the picture by redefining appropriately the bond energy (Pettifor, 1995; Børnsen *et al.*, 1999; Finnis, 2003).

10.1.3 Introducing self-consistency

In principle, self-consistency is expected to play an important role whenever there are charge transfers between different atoms in the system. If the extent of these transfers is roughly known, as in the case of ionic systems (e.g. salts like NaCl), then they can be built into the input density to be used in a non-self-consistent calculation. However, when charge transfer and electronic polarization effects are the result of complex rearrangements, then self-consistency in the charge density becomes an important ingredient.

To fix ideas, let us consider the case of a water molecule. The purely ionic picture would be that the hydrogen atoms donate their electron to the more electronegative oxygen, thus completing its electronic shell. This is actually not the case, because oxygen and hydrogen form covalent bonds, where the electrons are shared. In fact, the dipole moment of the water molecule can only be reproduced by a point-charge model if the charges are significantly smaller than +1 and -2. Asking for local charge neutrality would be a disaster in this case. Moreover, the positive charges in the vicinity of the oxygen atom polarize the electronic density quite strongly; so much so that the geometry and dynamics of the water molecule cannot be satisfactorily reproduced with purely point charges, but an additional polarization potential that depends on the geometry of the molecule is required, as wisely pointed out by Stillinger and David (1978). The polarized charge distribution gives rise to a non-spherical electric field, which acts back on the charge distribution that originated the polarization, and so on in a self-consistent cycle. Therefore, charge transfer and electronic polarization can only be accounted for satisfactorily within a self-consistent scheme.

A straightforward self-consistent extension of the tight-binding method is to re-calculate the Kohn–Sham potential using the output density – or a suitable mix of the output and input densities – and repeat the operation until the two coincide within the desired accuracy. This scheme is not too different from *ab initio* approaches using atom-centered basis sets, except for the use of a minimal basis and the parameterization of the matrix elements in terms of the distance between centers and the type of basis functions, instead of calculating the integrals on-the-fly.

Another approach based on a second order functional that incorporates charge density fluctuations, i.e. quadratic in $(\rho - \rho_{in})$, was proposed in the late nineties independently by several groups (Finnis *et al.*, 1998; Elstner *et al.*, 1998; Esfarjani and Kawazoe, 1998; Schelling *et al.*, 1998). The first level of self-consistency that can be introduced in the tight-binding scheme allows only for charge transfer between atoms. In this model the band energy is supplemented with a term proportional to the variation of the atomic charges from the input charge distribution. This term contains on-site and inter-site contributions of the form:

$$E_{\text{SCCT}} = E_{\text{band}} + \frac{1}{2} \sum_{I=1}^{P} U_I \delta q_I^2 + \frac{1}{2} \sum_{I=1}^{P} \sum_{J \neq I}^{P} U_{IJ} \delta q_I \delta q_J, \qquad (10.30)$$

where $\delta q_I = (q_I^{\text{out}} - q_I^{\text{in}})$ and $q_I = \sum_\mu \rho_{\mu\mu}^{II}$ are the Mulliken charges. This model introduces an on-site Coulomb repulsion whose magnitude is determined by the species-dependent parameter U_I, and inter-site Coulomb interactions that are normally assumed to be of the form $U_{IJ} = 1/|\mathbf{R}_I - \mathbf{R}_J|$.

The matrix elements for this simple version of the self-consistent tight-binding model can be readily derived from (10.30). In the case of orthogonal tight-binding, only the diagonal matrix elements are modified in the following way:

$$H_{\mu\nu}^{IJ}(\text{SCCT}) = H_{\mu\nu}^{IJ}[\rho_{\text{in}}] + \left[U_I \delta q_I + \sum_{K \neq I}^{P} U_K \delta q_K \right] \delta_{\mu\nu} \delta_{IJ}, \qquad (10.31)$$

but if the non-orthogonality of the basis set is retained, then there is an additional term that involves the overlap matrix elements (Finnis, 2003). The TB bond model described in the previous section can be thought of as a self-consistent charge transfer model where the on-site repulsion parameters U_I are so large that local charge neutrality ($\delta q_I = 0$) is verified.

Once the eigenvalue problem for an initial input density has been solved, then the electronic density is re-calculated and new charges are derived. These enter Expressions (10.30) and (10.31), thus modifying the Hamiltonian so that the eigenvalue equation has to be solved again to obtain a new set of atomic charges, and so on until the input and output charges coincide.

A successive level of sophistication has been introduced by Finnis et al. (1998), who generalized the idea of self-consistent point charges to a multi-site, multi-polar expansion where not only are charges adjusted, but also higher multi-pole moments are assigned to the atomic sites, typically dipoles and quadrupoles. The computational procedure remains the same as for point charges, but the energy and Hamiltonian have to be generalized. Using the multi-pole expansion of the electrostatic energy given in (9.112), and retaining the on-site Coulomb repulsion separately to account for local exchange and correlation effects, the energy can be written as

$$E_{\text{SCTB}} = E_{\text{band}} + \frac{1}{2}\sum_{I=1}^{P} U_I \delta q_I^2$$

$$+ \frac{1}{2}\sum_{I=1}^{P}\sum_{J\neq I}^{P}\sum_{l,m}\sum_{l',m'} \delta q_{lm}(\mathbf{R}_I) B_{lm}^{l'm'}(\mathbf{R}_I - \mathbf{R}_J)\delta q_{l'm'}(\mathbf{R}_J), \quad (10.32)$$

where the spherically symmetric terms ($l = l' = 0$) correspond to the inter-site charge transfer terms already discussed above. The quantities δq_{lm} are the multi-pole moments of the deformation of the charge distribution with respect to atomic densities, which in terms of the density matrix are written:

$$\delta q_{lm}(\mathbf{R}_I) = \sum_{\mu=1}^{M}\sum_{\nu=1}^{M} \rho_{\nu\mu}^{II} D_{\mu\nu\, lm}(\mathbf{R}_I), \quad (10.33)$$

with

$$D_{\mu\nu\, lm}(\mathbf{R}_I) = \int \phi_\nu^*(\mathbf{r}-\mathbf{R}_I) R_{lm}(r)\phi_\mu(\mathbf{r}-\mathbf{R}_I)d\mathbf{r}. \quad (10.34)$$

With the above definitions, the Hamiltonian matrix elements in an orthogonal basis set become:

$$H_{\mu\nu}^{IJ}(\text{SCTB}) = H_{\mu\nu}^{IJ}[\rho_{\text{in}}] + U_I \delta q_I \delta_{\mu\nu}\delta_{IJ}$$

$$+ \sum_{K\neq I}^{P}\sum_{l,m}\sum_{l',m'} D_{\mu\nu\, lm}^{I}(\mathbf{R}_I) B_{lm}^{l'm'}(\mathbf{R}_I - \mathbf{R}_K)\delta q_{l'm'}(\mathbf{R}_K)\delta_{IJ}. \quad (10.35)$$

This extension of the self-consistent charge transfer method has been designed with the goal of describing oxide materials, where the large electronic polarizability of the oxygen atoms requires the self-consistent determination of, at least, the dipolar term. It is sometimes called the polarizable tight-binding model, and has been successfully applied to describe the phase diagram and the behavior of oxygen defects in zirconia (Fabris *et al.*, 2001, 2002).

Tight-binding methods are efficient because the calculation of the Hamiltonian matrix elements is an inexpensive operation, these being given by explicit, distance-dependent expressions. These methods, however, still require a matrix diagonalization, and hence are significantly more expensive computationally than empirical force fields made of classical potentials. It is of interest, then, to try and simplify the tight-binding description in such a way as to eliminate the quantum-mechanical step, but keeping its spirit as much as possible. Several approaches have been proposed to achieve this goal, mostly during the mid-eighties. These are described in a forthcoming section, but first we concentrate on quantum chemical methods that are similar in spirit to the tight-binding scheme.

10.2 Semiempirical approaches in quantum chemistry

Semiempirical approaches in quantum chemistry start from the Hartree–Fock equation, adopt a minimal basis set, and replace the atomic nuclei plus core electrons by ionic cores of charge Z_V (the valence charge). There is a range of such methods, which differ basically on the treatment of the matrix elements of the Fock operator.

10.2.1 Extended Hückel theory

A very simple model that fulfills the requirements for a semiempirical method was developed by R. Hoffmann and received the name of *extended Hückel theory* (Hoffmann, 1963). The diagonal elements of the Fock matrix are approximated by atomic ionization potentials, and the off-diagonal elements are taken proportional to the overlap matrix elements between STO basis functions, where the proportionality constant involves the average of the diagonal elements:

$$F_{\mu\nu}^{IJ}(\text{EHT}) = \begin{cases} -I_\mu^I, & \mu = \nu, \ I = J, \\ -K\left(I_\mu^I + I_\nu^J\right) S_{\mu\nu}^{IJ}, & \mu \neq \nu. \end{cases} \quad (10.36)$$

Usually $K = 0.875$, and the overlaps are calculated for STOs with the exponents given by Slater's rules. The great advantage of this model is that the matrix elements are parameterized directly, in such a way that they do not depend on the electronic density matrix as in the Roothaan–Hall equations. Therefore, the extended Hückel method is non-self-consistent, but it provides the density matrix as a result of the diagonalization of the Fock matrix. The overlap matrix is fully retained, thus constituting a non-orthogonal method. This approach is not suitable for quantitative calculations, but it is extremely useful for a quick qualitative determination of the wave functions. Also, it often predicts qualitatively correct geometric and energetic trends.

10.2.2 Neglect of differential overlap

If, however, we wish to retain the structure of the Fock matrix in terms of one- and two-electron matrix elements, then the extended Hückel method is too drastic an approximation. A systematic strategy for simplifying the Fock matrix is to construct a hierarchy of models where different matrix elements are discarded under certain well-founded assumptions.

The discussion starts with the introduction of the concept of *differential overlap*, which was already defined in Expression (9.113) of Section 9.4 under the name of overlap distribution.

$$O_{\mu\nu}^{IJ}(\mathbf{r}) = \phi_\mu^*(\mathbf{r} - \mathbf{R}_I)\phi_\nu(\mathbf{r} - \mathbf{R}_J). \tag{10.37}$$

If $I \neq J$ the differential overlaps are non-negligible only in a region between the two atoms, but they assume small values that decay very fast with the distance between atoms. In the case of GTOs this decay is Gaussian, and is given by the product formula (8.59). By actually calculating Coulomb repulsion integrals that involve $O_{\mu\nu}^{IJ}(\mathbf{r})$ for $I \neq J$, it can be observed that they tend to assume very small values, especially for $\mu \neq \nu$. These functions are called *diatomic differential overlaps*. The central approximation of semiempirical methods in quantum chemistry is the *zero differential overlap* approximation (ZDO). Here, all the diatomic differential overlaps are set to zero, except when they are involved in kinetic matrix elements or in matrix elements of the nuclear attraction where one of the orbitals is centered on the same atomic site of the potential. The latter, especially if the two centers are close, are extremely important because they are responsible for chemical bonding. Notice that it is the overlap distributions as functions of the electronic coordinate, and not only the overlap integrals $S_{\mu\nu}^{IJ}$, that are set to zero. If $I = J$ and $\mu \neq \nu$ the *monoatomic* differential overlap $O_{\mu\nu}^{II}(\mathbf{r})$ is not a small quantity, and cannot be happily set to zero. However, what is strictly zero is the overlap matrix element $S_{\mu\nu}^{II}$ due to the orthogonality of basis functions centered on the same atom.

Immediate consequences of the ZDO approximation are the following.

(i) $S_{\mu\nu}^{IJ} = \delta_{\mu\nu}\delta_{IJ}$, and thus the overlap matrix $\mathbf{S}$ reduces to the identity matrix. As a consequence, the Roothaan–Hall equations become a standard diagonalization problem, $\mathbf{Fc} = \Lambda\mathbf{c}$.

(ii) In terms of differential overlaps, nuclear attraction one-electron integrals given by (9.102) are written as $U_{\mu\nu}^{IJ} = \int O_{\mu\nu}^{IJ}(\mathbf{r})u_K(\mathbf{r} - \mathbf{R}_K)d\mathbf{r}$. Consequently, three-center integrals where $I \neq J$ are automatically set to zero. Two-center integrals are retained, including those involving diatomic differential overlaps.

(iii) One-center and two-center kinetic energy integrals, including those for basis functions centered on different atoms, are retained. The latter form part of the hopping integrals in tight-binding methods, and together with nuclear attraction integrals take into account the formation of chemical bonds.

(iv) According to (9.89), two-electron matrix elements can be written as $\langle \mu^I \lambda^K | \nu^J \sigma^L \rangle = \int\int O_{\mu\nu}^{IJ}(\mathbf{r})v_2(\mathbf{r} - \mathbf{r}')O_{\lambda\sigma}^{KL}(\mathbf{r}')d\mathbf{r}\,d\mathbf{r}'$. Therefore, under the ZDO approximation, only those integrals with $I = J$ and $K = L$ survive, i.e. $\langle \mu^I \lambda^K | \nu^J \sigma^L \rangle = \langle \mu^I \lambda^K | \nu^J \sigma^K \rangle \delta_{IJ}\delta_{KL}$. This implies that three- and four-center two-electron integrals are set to zero.

If the only approximation made is the ZDO, the resulting scheme receives the name of *neglect of diatomic differential overlap* (NDDO). The overlap matrix becomes the identity matrix, and the one-electron matrix elements are approximated by

$$\bar{h}_{1\mu\nu}^{IJ} = \begin{cases} T_{\mu\nu}^{IJ} + U_{\mu\nu}^{IJ}(I) + U_{\mu\nu}^{IJ}(J), & I \neq J, \\ T_{\mu\nu}^{II} + U_{\mu\nu}^{II}(I) + \sum_{K \neq I}^{P} U_{\mu\nu}^{II}(K), & I = J, \end{cases} \quad (10.38)$$

with

$$U_{\mu\nu}^{IJ}(K) = \int \phi_\mu^*(\mathbf{r} - \mathbf{R}_I) u_K^{\text{ECP}}(\mathbf{r} - \mathbf{R}_K) \phi_\nu(\mathbf{r} - \mathbf{R}_J) d\mathbf{r} \quad (10.39)$$

and $u_K^{\text{ECP}}(r)$ a species-dependent effective core potential. A particularly simple choice is $u_K^{\text{ECP}}(r) = -Z_K/r$, where the valence charge Z_K replaces the nuclear charge. Notice that only two-center integrals $U_{\mu\nu}^{IJ}(I)$, $U_{\mu\nu}^{IJ}(J)$, and $U_{\mu\nu}^{II}(K)$ are retained. By adding the non-zero two-electron integrals, we eventually obtain the expression for the Fock matrix in the NDDO approximation:

$$\bar{F}_{\mu\nu}^{II} = \bar{h}_{1\mu\nu}^{II} + \sum_{\lambda,\sigma=1}^{M}\sum_{K=1}^{P} \rho_{\sigma\lambda}^{KK}\langle \mu^I \lambda^K | \nu^I \sigma^K \rangle - \sum_{\lambda,\sigma=1}^{M} \rho_{\sigma\lambda}^{II}\langle \mu^I \lambda^I | \nu^I \sigma^I \rangle,$$

$$\bar{F}_{\mu\nu}^{IJ} = \bar{h}_{1\mu\nu}^{IJ} - \sum_{\lambda,\sigma=1}^{M} \rho_{\nu\mu}^{JI}\langle \mu^I \lambda^J | \sigma^I \nu^J \rangle. \quad (10.40)$$

The first line corresponds to basis functions μ and ν both centered on atom I (on-site). In the second line, the functions μ and ν are centered on different atoms

I and J, respectively (hopping). As in Hartree–Fock theory, the total energy is given by

$$E = \frac{1}{2}\text{Tr}\left[\hat{\rho}\left(\hat{h}_1 + \hat{\mathcal{F}}\right)\right] + V_{nn}, \tag{10.41}$$

where the core–core repulsion V_{nn}, in principle, is kept in its original form, i.e. $V_{nn} = 1/2 \sum_{I \neq J} Z_I Z_J / |\mathbf{R}_I - \mathbf{R}_K|$.

This extraordinary simplification comes at the price that the remaining integrals cannot retain their original meaning. The *damage* caused by neglecting a large number of terms must be absorbed in a redefinition of the integrals retained in the model, which then become fitting parameters. Obviously, there is no universal criterion to do this parameterization. Fitting to Hartree–Fock calculations with a minimal basis set is not adequate because these results are normally of quite a poor quality. One possible road is to calculate some of the integrals directly from the basis functions, and make some others into parameters that are obtained from atomic experimental data. Another way is to make all the integrals into parameters and fit them to large sets of experimental molecular data on many different molecules of the same class, e.g. organic molecules.

An important observation is that the matrix elements (10.40) require the knowledge of the density matrix $\rho_{\mu\nu}^{IJ}$. This latter is constructed with the coefficients of the wave functions that are obtained by solving the modified Roothaan–Hall equations in the NDDO approximation. Therefore, this problem has to be solved self-consistently, exactly in the same way of Hartree–Fock. Initially, a density matrix is guessed from an extended Hückel (non-self-consistent) calculation, and is used to construct the Fock matrix. The eigenvalue problem is solved, and the procedure cycled to convergence.

An important aspect to keep in mind when approximating the matrix elements is the *invariance* under unitary transformations of the basis set. The Hartree–Fock wave function is such an invariant, and it is desirable that approximate wave functions retain this property. This aspect has been studied by Pople, Santry, and Segal, who showed that the NDDO approximation is indeed invariant under unitary transformations, but further approximations must be done very carefully (Pople *et al.*, 1965).

The NDDO scheme can be further simplified by neglecting partially or totally the monoatomic differential overlaps, i.e. those for which the basis functions are centered on the same atom. Two such schemes will be reviewed below: CNDO and INDO, together with their most important modifications (MINDO). Then, we will concentrate on the semiempirical models most widely used at present, namely AM1 and PM3, which belong to the class of modified NDDO models. All these models require the self-consistent solution of the approximate Roothaan–Hall equations.

10.2.3 CNDO

The simplest level of approximation is to totally neglect monoatomic differential overlaps, i.e. $O^{IJ}_{\mu\nu}(\mathbf{r}) = \delta_{\mu\nu}\delta_{IJ}$. This scheme, which retains only the gross features of the electron–electron interaction, is known as *complete neglect of differential overlap* (CNDO) (Pople et al., 1965). Although this approach is rarely used nowadays, it is instructive because the meaning of the different terms is particularly clear. The justification of the additional approximation is not as natural as in NDDO because, even if the monoatomic overlap integrals for $\mu \neq \nu$ are strictly zero, it is not generally true that the product of two functions centered on the same atom assumes small values everywhere in space.

Two-electron matrix elements are $\langle \mu^I \lambda^K | \nu^J \sigma^L \rangle = \langle \mu\lambda | \mu\lambda \rangle \delta_{\mu\nu}\delta_{\lambda\sigma}\delta_{IJ}\delta_{KL}$. This, however, is not sufficient to make the approximation invariant against transformations of the basis functions. It is also required that these integrals are made independent of the type of basis function, i.e.

$$\langle \mu^I \lambda^K | \nu^J \sigma^L \rangle = \gamma_{IK}\delta_{\mu\nu}\delta_{\lambda\sigma}\delta_{IJ}\delta_{KL}. \tag{10.42}$$

This represents an additional simplification, which is unavoidable if the scheme is required to be invariant. A positive aspect is that the number of parameters in the model is reduced. However, this also entails a negative aspect, namely that specific features that depend on the character (s, p, d) of the individual orbitals and on the spin orientation (exchange) are lost. The surviving terms depend only on the type of atoms and on the distance between them, and represent an average electrostatic repulsion between electrons. At long distances $\gamma_{IK} = Z_I Z_K / |\mathbf{R}_I - \mathbf{R}_K|$, with Z_I and Z_K the valence charges of atoms I and K, respectively.

One-electron matrix elements can be divided into one-center and two-center contributions. Two-center integrals are of the type $U^{IJ}_{\mu\nu}(I)$ and $U^{IJ}_{\mu\nu}(K)$. The first type involves diatomic overlaps on different atoms, and cannot be neglected without compromising chemical bonding. The matrix elements for $I \neq J$ are given by

$$h^{IJ}_{1\mu\nu} = \langle \mu^I | (-\frac{\hbar^2}{2m}\nabla^2 + u_I^{\text{ECP}} + u_J^{\text{ECP}}) | \nu^J \rangle = \beta^{IJ}_{\mu\nu} \tag{10.43}$$

and are called *resonance integrals*. They can be thought of as the energy gain when an electron is shared by two atomic centers, and are then clearly related to the concept of chemical bond. Notice that these integrals are nothing but the hopping integrals (10.7) in tight-binding methods.

The second type of two-center integral involves products of monoatomic overlaps with the nuclear attraction centered on a different atom. In the CNDO scheme these are set to zero, unless $\mu = \nu$. In addition, the remainder of the one-electron integrals when the basis functions are centered on the same atom are the matrix elements of the atomic Hamiltonian $\langle \mu^I | -\hbar^2 \nabla^2 / 2m + u_I^{\text{ECP}} | \nu^I \rangle = \varepsilon_\mu^I \delta_{\mu\nu}$, with ε_μ^I the atomic eigenvalues. Therefore, the one-electron matrix elements are simplified to:

$$\tilde{h}_{1\mu 1'}^{IJ} = \begin{cases} \beta_{\mu\nu}^{IJ}, & I \neq J, \\ \delta_{\mu\nu}\left[\varepsilon_\mu^I + \sum_{K \neq I}^P U_{\mu\mu}^{II}(K)\right], & I = J. \end{cases} \quad (10.44)$$

This, however, is not sufficient to preserve rotational invariance. The diagonal elements $U_{\mu\mu}^{II}(K)$ have to be made independent of the type of basis function so that $U_{\mu\mu}^{II}(K) = V_{IK}$ for all basis functions. Notice that, in general, $V_{IK} \neq V_{KI}$. This term represents the interaction of an electron in atom I with the ionic core of atom K, which depends on the type of atom at K. In addition, the resonance integrals are made proportional to the overlap matrix elements $\beta_{\mu\nu}^{IJ} = \beta_{IJ} S_{\mu\nu}^{IJ}$, as in the extended Hückel approach, where β_{IJ} is taken to be independent of the basis functions. The overlap integral, which is neglected in the normalization of the wave functions, is included here to take into account the fact that chemical bonding should be stronger when the overlap of the orbitals is more substantial.

Summarizing, the Fock matrix in the CNDO approximation is written:

$$\tilde{F}_{\mu\nu}^{IJ} = \beta_{IJ} S_{\mu\nu}^{IJ} - \rho_{\mu\nu}^{IJ} \gamma_{IJ} \quad (10.45)$$

for $I \neq J$ and $\mu \neq \nu$, and

$$\tilde{F}_{\mu\nu}^{II} = \delta_{\mu\nu}\left[\varepsilon_\mu^I + \sum_{K \neq I}^P \left(\rho^{KK} \gamma_{IK} + V_{IK}\right) + \rho^{II} \gamma_{II}\right] - \rho_{\mu\nu}^{II} \gamma_{II} \quad (10.46)$$

for $I = J$, where $\rho^{KK} = \sum_\gamma \rho_{\gamma\gamma}^{KK}$ is the partial trace of the density matrix with respect to the orbitals centered on atom K. ρ^{KK} is simply the total electronic charge associated with atom K, i.e. the Mulliken charge.

A big advantage of the CNDO approximation is that the total energy can be decomposed into monoatomic and diatomic contributions, $E_{\text{CNDO}} = \sum_I \mathcal{E}_I + \frac{1}{2} \sum_{I \neq J} \mathcal{E}_{IJ}$. This facilitates the interpretation of the different energy terms, which is not straightforward in Hartree–Fock or Kohn–Sham methods. A CNDO parameterization requires the specification of the atomic energies ε_μ^I, nuclear attraction integrals V_{IJ}, Coulomb repulsion integrals γ_{IJ}, resonance integrals β_{IJ}, and overlap matrix elements $S_{\mu\nu}^{IJ}$. The quantities $S_{\mu\nu}^{IJ}$, V_{IJ}, and γ_{IJ} are calculated analytically using s-type STOs, while the other terms are fitted to experimental atomic data. In the original parameterization due to Pople and Segal (1965) (CNDO/1) the atomic energies were obtained from experimental ionization potentials. The resonance integrals were taken as $\beta_{IJ} = (\beta_I + \beta_J)/2$, and the parameters β_I were chosen to reproduce Hartree–Fock calculations with a minimal basis set. Since the CNDO/1 model was soon observed to over-bind diatomic molecules, Pople and

Segal (1966) immediately proposed a second parameterization (CNDO/2) where the atomic energies were calculated using also experimental data on electron affinities. In addition, the nuclear attraction integrals, instead of being calculated, were made proportional to the electron repulsion $V_{IJ} = -Z_J \gamma_{IJ}$, thus canceling errors made by neglecting overlap integrals.

10.2.4 INDO and MINDO

One of the limitations of the CNDO approximation is that it does not make any distinction between parallel and antiparallel spins in the retained exchange integrals. This problem is manifested in systems with spin polarization, like triplet states or open-shell molecules. A possible solution is the *intermediate neglect of differential overlap* approach (INDO) (Pople et al., 1967), where exchange integrals that involve monoatomic differential overlaps – neglected in the CNDO scheme – are now included. This is achieved by neglecting monoatomic differential overlaps in two-center integrals, while retaining them in one-center integrals. The only difference in the matrix elements with respect to CNDO is that the term $\rho^{II}\gamma_{II}$ appears not only for the same type of basis function, but also for $\mu \neq \nu$.

$$\tilde{F}^{II}_{\mu\nu} = \delta_{\mu\nu}\left[\varepsilon^{I}_{\mu} + \sum_{K \neq I}^{P}\left(\rho^{KK}\gamma_{IK} + V_{IK}\right)\right] + \left(\rho^{II} - \rho^{II}_{\mu\nu}\right)\gamma_{II}, \qquad (10.47)$$

where the parameters are calculated according to the CNDO/2 philosophy.

For a molecule containing only two species, e.g. a hydrocarbon, both the CNDO/2 and the INDO methods require the parameterization of three two-electron integrals, γ_{HH}, γ_{CC}, and γ_{CH}, two resonance integrals β_H and β_C, the atomic energy ε^{H}_{s} of the hydrogen s orbital, and the atomic energies ε^{C}_{s} and ε^{C}_{p} of the carbon s and p orbitals. That is five parameters and three functions of the distance between centers (the two-electron integrals). These numbers rise significantly in the NDDO method, where there are no invariance constraints on the integrals, and 27 different one- and two-center integrals must be parameterized for a single atomic species with s and p orbitals.

A useful modification of the INDO method, in the spirit of extended Hückel theory, consists of retaining the two-center character of the resonance integrals by writing $h^{IJ}_{1\mu\nu} = \beta_{IJ}S^{IJ}_{\mu\nu}(I^{I}_{\mu} + I^{J}_{\nu})$, where $I \neq J$, I^{I}_{μ} is the ionization potential for a μ-type electron in atom I, and similarly for I^{J}_{ν}. This approximation, called *modified intermediate neglect of differential overlap* (MINDO) improves significantly over the original INDO approach, especially the MINDO/3 version (Bingham et al., 1975).

10.2.5 MNDO, AM1, and PM3

With the advent of more powerful computers, it ceased to be important to neglect monoatomic differential overlaps for the sake of computational convenience. Therefore, NDDO semiempirical methods that neglect three- and four-center integrals, but nothing else, came into the foreground, in the form of different parameterizations. In all the NDDO methods the electronic integrals are treated in the same way, by calculating them analytically for STOs. Mostly s and p electrons have been considered, because the introduction of d orbitals increases the number of parameters to several hundreds. Therefore, these methods are useful mostly for organic molecules, but have also been parameterized for inorganic elements such as Si, Al, Zn, Sn, and Pb. The introduction of d basis functions is a rather recent addition (Thiel, 1981). Below are summarized the most relevant aspects of NDDO calculations.

(i) The one-electron integrals are parameterized in terms of atomic energies, two-electron integrals, and resonance parameters in the following way:

$$h_{1\mu\nu}^{IJ} = \begin{cases} \delta_{\mu\nu}\varepsilon_\mu^I - \sum_{K \neq I} Z_K \langle \mu^I s^K | \nu^I s^K \rangle, & I = J, \\ \frac{1}{2} S_{\mu\nu}^{IJ}(\beta_I + \beta_J), & I \neq J, \end{cases} \quad (10.48)$$

where the two-electron integrals are computed for s-type orbitals centered on the K atoms, while retaining the character of the orbitals in atom I. This is basically the same type of approximation as the CNDO/2 method.

(ii) Two-electron one-center integrals are calculated for s and p orbitals in different orientations, thus giving rise to four different direct Coulomb integrals, and a single exchange integral.

(iii) The different type and orientation of the orbitals in each atom originates 22 different two-electron two-center integrals. For historical reasons, these are written only in terms of the exponents of the STOs, ζ_s and ζ_p, and the distance between atoms.

(iv) The exact core–core repulsion term $V_{nn} = 1/2 \sum_{I \neq J} Z_I Z_J / |\mathbf{R}_I - \mathbf{R}_K|$ is not canceled exactly by the electron–electron and electron–core interaction at long distances, due to the neglected three- and four-center integrals. Therefore, it is necessary to modify this interaction so that the correct limiting behavior is recovered.

This last aspect is the central difference between the various flavors of the *modified neglect of differential overlap* (MNDO) method, which has been mostly developed in the group of M. Dewar. These require the determination of 11 parameters for the electron integrals, i.e. $\zeta_s, \zeta_p, \beta_s, \beta_p, \varepsilon_s, \varepsilon_p$, and the five two-electron two-center integrals. In addition, they involve some extra parameters related to the modified core–core repulsion.

The original MNDO method (Dewar and Thiel, 1977) uses

$$V_{nn}^{MNDO}(I, J) = Z_I Z_J \langle s^I s^J | s^I s^J \rangle \left(1 + e^{-\alpha_I R_{IJ}} + e^{-\alpha_J R_{IJ}}\right), \qquad (10.49)$$

except for O-H and N-H bonds, where the second term in the parentheses is replaced by $e^{-\alpha_I R_{IH}}/R_{IH}$. In this model the exponents α_I are parameters that, together with the STO exponents, the resonance parameters, and the orbital energies, are fitted to experimental molecular data containing heats of formation, ionization potentials, geometries, and dipole moments. The five two-center two-electron integrals are taken from atomic spectra.

The MNDO method suffers from systematic problems due to a core–core repulsion that is too strong. This was corrected by Dewar et al. (1985) by adding an extra term to the MNDO repulsion of the form

$$V_{nn}^{AM1}(I, J) = V_{nn}^{MNDO}(I, J) \qquad (10.50)$$

$$+ \frac{Z_I Z_J}{R_{IJ}} \left(\sum_n e^{-b_{nI}(R_{IJ}-c_{nI})^2} + \sum_n e^{-b_{nJ}(R_{IJ}-c_{nJ})^2} \right),$$

where the number of Gaussians depends on the atom, and the coefficients b_{nI} and c_{nI} are additional fitting parameters. This variant of the MNDO approach is called the *Austin model 1* (AM1), and is one of the semiempirical methods most widely used at present (Dewar et al., 1985). The fitting procedure for the MNDO and AM1 models was carried out by varying the parameters and looking for a good agreement in a few molecules.

A more robust fitting procedure based on the minimization of a penalty function over a large molecular data set was introduced by Stewart (1989). The functional form of the AM1 scheme was retained, the number of Gaussians was set to two, and the whole set of parameters was re-optimized. This is called the *parametric model 3* (PM3), and is another widely used model. A good account of the successes and failures of MNDO models is given in Jensen (1999) Chapter 3. This is still an active research area, and new semiempirical models and parameterizations are frequently proposed. It is also common to parameterize the AM1 model for specific systems.

10.3 Relation between tight-binding and semiempirical methods

A first observation is that the extended Hückel approach is just a non-orthogonal empirical tight-binding method where the on-site energies are identified with the ionization potential, $\varepsilon_\alpha^I = -I_\alpha^I$, and the hopping integrals are taken to be $t_{\alpha\beta}^{IJ} = -K(I_\alpha^I + I_\beta^J) S_{\alpha\beta}^{IJ}$.

Apart from the extended Hückel method, semiempirical approaches in quantum chemistry require the self-consistent solution of the simplified Roothaan–Hall equations. With little effort the Fock Hamiltonian can be converted into the Kohn–Sham Hamiltonian, by modifying the treatment of the exchange interaction. Therefore, semiempirical methods have to be compared to self-consistent tight-binding models. They share the common features of using a minimal basis set and eliminating the core electrons.

The main approximations made in the construction of semiempirical models are at the level of discarding certain types of matrix elements and approximating others, being careful not to violate invariance properties of the Hamiltonian with respect to rotations of the reference frame. The main approximation is to neglect two-electron Coulomb repulsion integrals and one-electron nuclear attraction integrals that involve diatomic differential overlaps, i.e. products of basis functions on the same electronic coordinate, but centered on different atoms. This takes care of three- and four-center integrals, and turns the overlap matrix into the unit matrix. Therefore, except for the extended Hückel approach, semiempirical methods assume an orthogonal basis set. Tight-binding models also neglect three- and four-center integrals, so that up to this point semiempirical methods correspond to self-consistent orthogonal tight-binding models.

By looking at Expression (10.40) it can be observed that the on-site elements (first line) contain a contribution arising from the density in all the other atoms. This term takes care of charge transfer and polarization when the density is adjusted self-consistently. In the CNDO/2 approximation the Coulomb integrals are all taken to be independent of the orbitals, with the consequence that the direct Coulomb interaction assumes the form $\sum_{K \neq I}(\rho^{KK} - Z_K)\gamma_{IK}$, which is the same as the inter-site contribution in the TB self-consistent charge transfer model (10.31) once it is realized that $\delta q_K = \rho^{KK} - Z_K$. Therefore, the integrals γ_{IK} play the role of the inter-site Coulomb interactions U_{IK}. The on-site Coulomb repulsion is also present in the term $\varepsilon_\mu^I(\text{CNDO}) + \rho^{II}\gamma_{II}$ once the on-site energies are written as $\varepsilon_\mu^I(\text{CNDO}) = -(I_\mu + A_\mu)/2 - Z_I\gamma_{II}$. In that case, the diagonal elements of the TB Hamiltonian are $H_{\mu\mu}^{II} = \varepsilon_\mu^I(\text{TB}) = -(I_\mu + A_\mu)/2$, and the other term, $(\rho^{II} - Z_I)\gamma_{II}$, is identified with the on-site Hubbard term with $U_I = \gamma_{II}$.

The exchange term is treated differently in the two approaches due to the different starting points, Kohn–Sham vs Hartree–Fock. In the former, exchange is approximated by a local expression given by the LDA, and a local correlation term is also added. Correlation is absent in Hartree–Fock, though, but this is not a crucial difference. Tight-binding models can also be constructed using the non-local exchange of Hartree–Fock, and semiempirical methods can surely deal with the LDA.

The two-site Fock matrix elements in the second line of (10.40) look very much like the hopping integrals in TB, except for the difference in the exchange term. The Hartree contribution between two different centers disappears in the

NDDO approach because it involves diatomic differential overlaps. This includes two-center integrals of the form $\langle \mu^I \gamma^I | \nu^J \delta^J \rangle$ and $\langle \mu^I \gamma^J | \nu^J \delta^I \rangle$, which are not obviously neglected in TB schemes. This is not particularly important, however, because in the end the inter-site matrix elements, or resonance integrals, are parameterized in the same way as TB hopping integrals. In the case of semiempirical methods, these are written as $\tilde{F}^{IJ}_{\mu\nu} = \beta_{IJ} S^{IJ}_{\mu\nu}$, where usually $\beta_{IJ} = (\beta_I + \beta_J)/2$. The hopping term in TB is written in a similar way, but it is not assumed to be proportional to the overlap matrix element. The functional dependence of resonance and hopping integrals with the distance between centers is normally different, unless the exponential form of Ducastelle is used within the TB scheme. This difference is probably taken care of by different parameterizations of the repulsive potential.

In summary, the tight-binding charge transfer model appears to be a close relative of the CNDO/2 semiempirical method, while modified NDDO methods such as AM1 and PM3 take care of more general polarization effects beyond charge transfer.

10.4 Many-body classical potentials

Tight-binding and semiempirical methods have two important advantages over *ab initio* approaches: (1) they use minimal basis sets, so that the Hamiltonian matrix is the smallest possible that is still meaningful; (2) the Hamiltonian matrix elements are parameterized to explicit analytical forms, and thus their calculation is reduced to the evaluation of a formula in terms of interatomic geometrical quantities. The sum of the Hamiltonian eigenvalues represents a substantial part of the energy. This quantity can be expressed in terms of the density matrix and the matrix elements, as for the band energy (10.27) or, equivalently, the bond energy (10.28). Both the eigenvalues and the density matrix are the result of the diagonalization of the Hamiltonian, which is the next computational bottleneck of the calculation. Therefore, in order further to simplify the approach, the next target is to devise models for the density matrix. Another possible, related, alternative is to develop models for the density of states and then calculate the band energy using Expression (10.13).

The first class of models acquires a more clear meaning if the energy is written in terms of bond orders rather than the full density matrix, as these are direct indicators of bond strength. Models that approximate the bond orders avoiding the diagonalization of the Hamiltonian are called *bond order potentials* (BOP). There are two types of BOP: those where the bond orders are obtained from a rigorous theoretical approach, by means of a series of controlled approximations (Pettifor, 1989), and those where the bond orders (and in fact the potential itself) are given some physically motivated empirical form (Tersoff, 1986). It is beyond the scope

10.4 Many-body classical potentials

of this book to describe these approaches in detail. A very good account is given in Finnis (2003) Chapter 7. Here, it suffices to mention that analytic models for the bond orders can be obtained via the recursion method (Haydock et al., 1972), where the Green's function $\hat{G}(\varepsilon) = (\hat{H} - \varepsilon)^{-1}$ is written as a truncated continued fraction expansion. The bond orders are related to the matrix elements of the Green's function in an atom-centered representation by (see Finnis, 2003, Chapter 1):

$$\rho_{\mu\nu}^{IJ} = -\frac{2}{\pi} \text{Im} \int_{-\infty}^{\epsilon_F} \langle \phi_{\mu}^{I} | \hat{G}(\varepsilon + i0^{+}) | \phi_{\nu}^{J} \rangle \, d\varepsilon. \tag{10.51}$$

While this approach is theoretically quite robust and controlled, the approximations required to describe the correct physics, e.g. bonds of the appropriate valence, are far from obvious and have been worked out case by case. Nevertheless, there are high hopes that general BOPs will be developed for classes of materials such as semiconductors, transition metals, organic molecules, etc.

The alternative road of developing models for the electronic density of states is also related to the recursion method, and has been pioneered by Finnis and Sinclair (1984). The bond energy is given in terms of the local density of states $g_{\nu}^{I}(\varepsilon)$ as

$$E_{\text{bond}} = \sum_{I=1}^{P} \sum_{\nu=1}^{M} \int_{-\infty}^{\epsilon_F} (\varepsilon - H_{\nu\nu}^{II}) g_{\nu}^{I}(\varepsilon) \, d\varepsilon, \tag{10.52}$$

and the LDOS is further expanded in terms of its moments given by $\mu_{i}^{I} = \sum_{\nu=1}^{M} \int_{-\infty}^{\infty} \varepsilon^{i} g_{\nu}^{I}(\varepsilon) \, d\varepsilon$. These can also be expressed as a product of Hamiltonian matrix elements along closed paths that start at atom I in orbital ν and return to the same place, visiting i other atoms in between. Hence, the second moment of the LDOS is given by $\mu_{2}^{I} = \sum_{J \neq I} \sum_{\nu\sigma} (H_{\nu\sigma}^{IJ})^{2}$. Furthermore, it can be shown that for any model of the LDOS whose shape is determined only by its second moment, the bond energy is given by $E_{\text{bond}} = \sum_{I} \sqrt{\mu_{2}^{I}}$. This is called the *second moments* approximation (SMA) (Finnis and Sinclair, 1984), and the total energy is given by the bond energy plus a repulsive pair potential

$$E_{\text{SMA}} = -a \sum_{I=1}^{P} \sqrt{\sum_{J \neq I} \sum_{\nu\sigma} (H_{\nu\sigma}^{IJ})^{2}} + \frac{1}{2} \sum_{I \neq J} V_{\text{rep}}(|\mathbf{R}_{I} - \mathbf{R}_{J}|), \tag{10.53}$$

where a is some appropriate constant. The SMA can be extended to cases where the density of states is bimodal by introducing fourth moments. These can be obtained from the recursion method or, directly, via path counting.

A more flexible approximation is obtained by proposing an electronic energy of the form $E_{\text{EAM}} = \sum_{I} G_{I} \left[\sum_{J \neq I} \rho_{J}^{a}(|\mathbf{R}_{I} - \mathbf{R}_{J}|) \right]$, where $G_{I}[\rho_{h}]$ is an atomic *embedding* function that represents the change in total energy when an atom I is introduced into a uniform electron gas of density ρ_h, and ρ_h is the density of the host at position R_I, but excluding atom I. In practice, these are classical potentials that include many-body interactions in a form that depends on the local environment of the atoms, especially on the coordination. In fact, they share this common feature with bond order potentials. They receive the generic names of *embedded atom model* (Daw and Baskes, 1984), *effective medium theory* (Jacobsen et al., 1987), or *glue potentials* (Ercolessi et al., 1986). For metallic systems, they represent a significant improvement over simple pair potentials.

Another road that has been extensively exploited in the context of oxides and, in particular, ferroelectric perovskite materials is that of the so-called *shell models* (Dick and Overhauser, 1958). Here, atoms are represented by a core (the ionic core) and a shell (the electronic valence shell). The core–shell interaction can be tailored to represent the electronic polarization, either linear or non-linear. The non-linear oxygen shell polarization proved to be extremely relevant, especially in describing the ferroelectric instability of perovskite oxides (Migoni et al., 1976).

10.5 Classical force fields

Classical force fields have been proposed for many specific applications. The simplest possibility is to consider only two-body potentials. These are useful for systems where the interactions are not directional, such as rare gases and metallic systems. For metals, however, pair potentials describe cohesion rather poorly, and effective many-body approaches such as those in the previous section are required. For ionic systems such as salts, pair potentials should include electrostatic interactions. Semiconducting systems require also angular forces in the form of three-body contributions. Examples of this are the bond-order-based Brenner and Tersoff potentials (Tersoff, 1986; Brenner, 1990), which have been very successful in describing the different phases of carbon and silicon. Another interesting approach to three-body potentials, or angular-dependent forces is that of Stillinger. A successful, all-atom potential was proposed for water, which includes the oxygen electronic response through a polarization potential (Stillinger and David, 1978). Another very useful empirical force field, the Stillinger–Weber potential, which includes angular forces, was devised for silicon and used extensively due to its extremely good quality (Stillinger and Weber, 1985).

These force fields are tailored for specific applications, and hence have excellent quality. This approach, however, is somewhat restrictive because it implies that for every different system, the force field has to be parameterized and tested thoroughly. This issue becomes particularly relevant in computational chemistry, especially in the context of organic and biomolecular systems or molecular liquids,

10.5 Classical force fields

due to the large variety of systems of interest. For this reason, empirical force-field packages have been developed along the years by carrying out extensive and careful parameterizations for a wide spectrum of systems of interest, but mostly involving first-row atoms, and a few other biochemically relevant interactions such as disulphide bridges (S—S) or some metal atoms like Fe, Cu, Mn, or Pt that are found in the active sites of proteins. Some popular packages are CHARMM (Brooks *et al.*, 1983), UFF (Rappe *et al.*, 1992), AMBER (Cornell *et al.*, 1995), OPLS (Jorgensen, 1998), and GROMOS (van Gunsteren and Berendsen, 1990), amongst many other force-field packages in use.

These force fields include a number for different interactions. For a start, the intra-molecular interactions are expanded into two-body distance-dependent forces, three-body angular-dependent forces, and four-body torsional forces. Stretching potentials are usually taken to be harmonic:

$$V_2(R_{IJ}) = k_{IJ}(R_{IJ} - R_{IJ}^0)^2, \tag{10.54}$$

with R_{IJ}^0 the equilibrium distance between atoms I and J. A useful alternative is the Morse potential $V(r) = D(1 - e^{ar})^2$, which describes properly the non-linear attractive regime at long distances. At short distance, repulsion can be modeled via power-law or exponential potentials (Hill, 1948). Bending potentials are also usually taken to be harmonic in the bending angle:

$$V_3(\theta_{IJK}) = k_{IJK}(\theta_{IJK} - \theta_{IJK}^0)^2, \tag{10.55}$$

with θ_{IJK}^0 the equilibrium angle between atoms I, J, and K. An out-of-plane bending quadratic potential that describes pyramidalization of tri-atomic bonds is also included. Torsional potentials must be periodic in the angle, and are thus written as a Fourier series:

$$V_4(\omega) = \sum_{n=1} V_n[1 + (-1)^{n-1}\cos(n\omega)], \tag{10.56}$$

with ω the torsional angle. The terms included in the expansion depend on the particular case. The parameters can be determined by fitting either to electronic structure calculations or to experimental data.

Inter-molecular interactions are mostly electrostatic and dispersive. Electrostatics is modeled by assigning partial charges to the atomic sites. These charges are not a physically measurable quantity, and they have to be determined from electronic structure calculations. There are several possible strategies to do this. The ESP approach requests that the partial charges reproduce the electrostatic potential of a molecule (or set of molecules) in a grid of points. It can also be restricted so as to reproduce dipole and quadrupole moments (RESP). Other strategies are based on reproducing the moments of the charge distribution, and can be extended to distributed multi-pole models, where multi-pole moments are assigned to every atomic site or to bonds (Stone, 1981). The partial charges interact electrostatically with the partial charges in other molecules in the system according to

$$V_{es}(R_{IJ}) = \frac{Q_I Q_J}{\varepsilon R_{IJ}}, \tag{10.57}$$

with ε a dielectric constant. To take into account steric (shape-related) constraints, electrostatic forces are supplemented with short-range van der Waals-type forces acting between atoms in different molecules or parts of the molecule. These are typically represented by Lennard-Jones 12-6 potentials (Lennard-Jones, 1924)

$$V_{\text{rep-disp}}(R_{IJ}) = \epsilon_{IJ} \left[\left(\frac{\sigma_{IJ}}{R_{IJ}} \right)^{12} - \left(\frac{\sigma_{IJ}}{R_{IJ}} \right)^{6} \right], \tag{10.58}$$

or variants such as the Buckingham exponential potential for the repulsive part. This type of potential is also used to describe hydrogen-bonds, e.g. Lennard-Jones 12-10.

The main goal of these packages is to be as universal and quick as possible. Therefore, the electronic information, which could be included via polarization functions or electronic shells, is discarded because it would represent an additional, significant computational cost.

10.6 Hybrid QM-MM methods

It is often the case that the chemically relevant part of a large system, i.e. the portion requiring a quantum-mechanical description, is confined to a small region. This is the case of solvated molecules and chemical reactions in solution, which are extremely ubiquitous phenomena in physical and organic chemistry, especially chemical synthesis. Here, the solute and perhaps a few surrounding solvent molecules require quantum mechanics, while the rest of the solvent can be described via classical potentials. It is also the situation encountered in enzyme catalysis, where the chemically important region is confined to the *active site* of the enzyme. Such an active site can be modeled by the substrate, a number of surrounding residues of the enzyme, and perhaps a few water molecules from the solution. The rest of the protein acts basically as a scaffolding, to create the favorable geometry for selecting a specific substrate, and catalyzing a specific chemical reaction such as bond cleavage, hydrogenation, oxidation, etc. The solvent surrounding the enzyme acts as a medium that favors the appropriate folding. The molecular scaffold can be efficiently described at the level of empirical potentials, provided that these are able to reproduce the geometrical structure of the protein.

10.6 Hybrid QM-MM methods

This class of system can be studied by combining a quantum-mechanical description of the chemically active region with a lower level, classical description of the surrounding medium, as shown schematically in Fig. 10.1. This idea has been proposed and implemented in the context of Hartree–Fock methods (Field *et al.*, 1990), and then extended to DFT approaches (Estrín *et al.*, 1997; Eichinger *et al.*, 1999). Approaches of this type are usually called QM-MM, where the acronym MM stands for *molecular mechanics*, which is the name given to classical mechanical methods in the computational chemistry community.

There are two main types of interface between QM and MM regions. The simplest case is when there are no chemical bonds between the two regions. This is mostly the case of molecules in solution. A more complex situation arises when the division of the system in QM and MM regions requires us to section a number of chemical bonds. This is the situation for enzymes.

In the first case, the total energy is written as the sum of three terms, one for the QM region, another for the MM region, and a third one for the interaction between the two:

$$E_{\text{QM-MM}}[\rho](\mathbf{R}, Q, \tau) = E_{\text{QM}}[\rho](\mathbf{R}) + E_{\text{MM}}(Q, \tau)$$
$$+ E_{\text{int}}[\rho](\mathbf{R}, Q, \tau), \quad (10.59)$$

where τ and Q are the coordinates and partial charges of the particles in the MM region, $\mathbf{R}$ the coordinates of the nuclei in the QM region, and $\rho(\mathbf{r})$ the electronic charge density. This expression applies irrespectively of the choice of quantum-mechanical description.

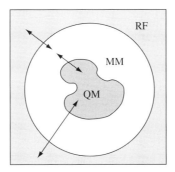

Fig. 10.1 Schematic diagram of a QM-MM calculation. The inner region (QM) is treated quantum-mechanically, and is surrounded by a region (MM) that is described using classical potentials. The interaction between the two regions is electrostatic supplemented with a short-range potential. The outer region (RF) represents a reaction field that incorporates the dielectric response of the infinite medium. It interacts electrostatically with both the QM and MM regions, as indicated by the arrows.

The MM region is usually described by a classical force field of the kind presented in Section 10.5, consisting of partial charges in the nuclear sites supplemented with short-range forces acting between atoms in different molecules. Intra-molecular forces in the MM region typically include stretching, bending, and torsional components, but also rigid molecular models can be used, like SPC/E or TIP4P for water. The QM region is treated within any of the schemes described in this book, i.e. DFT, Hartree–Fock, or semiempirical methods. The new ingredient is the interaction between the QM and MM regions, which is represented by a combination of electrostatic and short-range forces. Since the QM region includes a distributed electronic charge and point-like nuclear charges, the interaction energy between the QM and MM regions is given by

$$E_{\text{int}}[\rho(\mathbf{r})](\mathbf{R}, \mathbf{Q}, \tau) = -\sum_{K=1}^{S} Q_K \int \frac{\rho(\mathbf{r})}{|\mathbf{r} - \tau_K|} d\mathbf{r} + \sum_{K=1}^{S}\sum_{I=1}^{P} \frac{Q_K Z_I}{|\mathbf{R}_I - \tau_K|}$$
$$+ \sum_{K=1}^{S}\sum_{I=1}^{P} V_{\text{SR}}(|\mathbf{R}_I - \tau_K|), \qquad (10.60)$$

where S and P are the number of particles in the MM and QM regions, respectively, Z_I are the nuclear charges of the particles in the QM region, and $V_{\text{SR}}(R)$ is a short-range potential required to enforce steric constraints, and also to avoid the collapse of negative partial charges in the MM region onto the nuclei in the QM region. Typically, $V_{\text{SR}}(R)$ is chosen to be of the Lennard-Jones 12-6 form, but some other form such as a Buckingham potential can also be used.

Within a DFT scheme, the QM-MM energy functional gives rise to the following modified Kohn–Sham Hamiltonian

$$\hat{\mathcal{H}}_{\text{QM-MM}}[\rho] = \hat{\mathcal{H}}_{\text{KS}}[\rho] - \sum_{K=1}^{S} \frac{Q_K}{|\mathbf{r} - \tau_K|}, \qquad (10.61)$$

showing that the partial charges of the MM particles act as an external electrostatic potential for what concerns the quantum-mechanical electronic problem in the QM region. In addition, the nuclear coordinates in the QM region feel not only the Hellmann–Feynman quantum-mechanical forces, but also the forces arising from the MM region:

$$\mathbf{F}_I^{\text{QM}} = -\frac{\partial E_{\text{KS}}[\rho](\mathbf{R})}{\partial \mathbf{R}_I} - Z_I \sum_{K=1}^{S} Q_K \frac{\mathbf{R}_I - \tau_K}{|\mathbf{R}_I - \tau_K|^3}$$
$$- \sum_{K=1}^{S} \frac{\partial V_{\text{SR}}(|\mathbf{R}_I - \tau_K|)}{\partial \mathbf{R}_I}. \qquad (10.62)$$

Conversely, the electronic density and nuclear coordinates in the QM region also exert forces onto the particles in the MM region:

$$\mathbf{F}_K^{\mathrm{MM}} = -\frac{\partial E_{\mathrm{MM}}(\mathbf{Q}, \tau)}{\partial \tau_K} + Q_K \int \rho(\mathbf{r}) \frac{\tau_K - \mathbf{r}}{|\mathbf{r} - \tau_K|^3} \, d\mathbf{r}$$

$$- Q_K \sum_{I=1}^{P} Z_I \frac{\tau_K - \mathbf{R}_I}{|\mathbf{R}_I - \tau_K|^3} - \sum_{I=1}^{P} \frac{\partial V_{\mathrm{SR}}(|\mathbf{R}_I - \tau_K|)}{\partial \tau_K}. \quad (10.63)$$

When the QM-MM approach requires the sectioning of chemical bonds, the interfacing procedure is more delicate because the electrons in the QM part of the bond should behave as if on the other side there were electronic orbitals to make the bond. In particular, they should be orthogonalized to other orbitals that have been removed from the description. Moreover, care must be taken with charge transfer effects across the interface. As a rule of thumb, charge transfer problems are expected to be minimized by sectioning bonds between atoms of the same type, e.g. C-C bonds in proteins. The first issue has been attacked in a number of ways. One possibility is to retain a set of frozen ghost (unoccupied) orbitals in the MM part of the interface, so that the orbitals in the QM part can orthogonalize to them (Monard et al., 1996). A more robust and successful approach proposed by Laio et al. (2002) consists of replacing the potential due to the nucleus in the MM region with an appropriate pseudopotential that mimics the removed electrons much in the spirit of atomic pseudopotentials.

An issue related to the statistical sampling of the hybrid system that has rarely been explored up to now is that the force fields in the MM region and at the interface should be constructed in such a way that they are compatible with the first-principles description used for the QM region. Otherwise, there is a big risk of corrupting the thermodynamic and dynamical properties of the system. For example, if a solvated molecule is modeled by a first solvation shell of QM solvent molecules embedded in a large cluster of classical solvent molecules, then the two solvent models must be as close as possible to each other. Otherwise the situation is equivalent to a two-fluid system, where spurious energy transfers can occur. In this sense, QM-MM schemes are expected to behave better if the MM and interface potentials are obtained by fitting the parameters to QM calculations at the same level that will be used in the hybrid scheme.

In addition, the QM-MM system can be embedded into a continuum that mimics the dielectric response of the infinite medium, as shown in Fig. 10.1. This is called a *reaction field* (RF), and different flavors are possible. The simplest one is the Onsager RF, which considers the QM-MM system as a sphere with a dipole moment interacting with the dielectric medium as in continuum electrostatics (Onsager, 1936). The electric field due to the dielectric medium is given by

$$\mathbf{E}_{\rm rf} = \frac{(2\varepsilon - 1)\mathbf{p}}{(2\varepsilon + 1)R_{\rm c}^3}, \qquad (10.64)$$

with $\mathbf{p}$ the dipole moment of the cluster and $R_{\rm c}$ the radius of the cavity, which in the present context is the radius of the QM-MM cluster which is to be embedded into the dielectric medium. The quantity ε is the dielectric constant of the medium. Other more sophisticated approaches take into account more accurately the shape of the QM-MM region (Miertus et al., 1981). A pending assignment in this area is to combine QM-MM methods with periodic boundary conditions. The problem here is how to approach the issue of the replication of the QM part. A generalization of this approach to a hierarchy of layers that are treated at different levels has been realized in the ONIOM model (Svensson et al., 1996).

10.7 Orbital-free density functional approaches

The general ideas of orbital-free functionals were already introduced in Section 4.1. In the Thomas–Fermi–Dirac (TFD) approach, the kinetic and exchange terms in the energy functional, which are known in terms of single-particle orbitals but not as explicit functionals of the density, are replaced by local averages of the homogeneous electron gas expressions (LDA). While Kohn–Sham calculations that treat the kinetic term exactly have shown that the LDA for exchange, in general, is not such a bad approximation, the local approximation for the kinetic energy is extremely poor for systems that depart from the uniform electron gas, such as atoms and molecules. The most severe limitations are the lack of shell structure and the incorrect behavior of the density at short and long distances. Therefore, more accurate explicit expressions for the energy in terms of the density are required. The TFD approximation (4.7) is the best that can be done at the *local* level. Leaving aside exchange, which has already been discussed in Chapter 5, we now concentrate on non-local orbital-free approximations to the kinetic energy.

It is important to remark that such an approach is really worthwhile because orbital-free schemes require only the solution of the inverse problem for the density. Therefore, they are computationally much more efficient than solving Schrödinger's equation as in the Kohn–Sham or Hartree–Fock theories. Unfortunately, this road proved quite difficult, and the approximations proposed up to date are not yet accurate enough to become the way of choice. In this section we briefly describe these approaches.

A computationally convenient way of introducing non-locality is to perform a semi-local gradient expansion of the kinetic energy in the spirit of GGA and meta-GGA for exchange and correlation. This approach takes into account local inhomogeneities of the density, and it was first proposed by Hohenberg and Kohn in their seminal paper, by analyzing the linear response of the electron gas in the limit of long-wavelength perturbations to the homogeneous density (Hohenberg and Kohn, 1964). The expansion of the kinetic energy to fourth order in the density gradient was derived earlier by Kirzhnits (1957), and can be written in the following way:

$$T_{GEA}[\rho] = T_{TF}[\rho] + T_2[\rho] + T_4[\rho], \tag{10.65}$$

where $T_{TF}[\rho]$ is the Thomas–Fermi expression, and T_2 and T_4 are the second- and fourth-order terms given by Wang et al. (2001):

$$T_2[\rho] = \frac{1}{72} \int \frac{|\nabla \rho|^2}{\rho} d\mathbf{r} \tag{10.66}$$

and

$$T_4[\rho] = \frac{(3\pi^2)^{-2/3}}{540} \int \rho^{1/3} \left[\left(\frac{\nabla^2 \rho}{\rho} \right)^2 - \frac{9}{8} \frac{\nabla^2 \rho}{\rho} \left(\frac{|\nabla \rho|^2}{\rho} \right)^2 + \frac{1}{3} \left(\frac{|\nabla \rho|^2}{\rho} \right)^4 \right] d\mathbf{r}. \tag{10.67}$$

As usual with perturbative expansions, the convergence with increasing number of terms is not guaranteed. In this case, the second order is known to improve over Thomas–Fermi, while the inclusion of T_4 does not represent a significant advance over T_2. In addition, this expansion cannot account for rapidly varying densities such as those occurring in vacuum regions (outside molecules or surfaces). In fact, for atoms the density calculated with the truncated gradient expansion still diverges at long distances (Chacón et al., 1985).

The correct second-order correction for one-electron systems, and also for two-electron systems where the two electrons occupy the same spatial orbital (opposite spins), was derived by von Weizsäcker (1935). The functional form is the same as (10.66), but the coefficient in front is different:

$$T_{vW}[\rho] = \frac{1}{8} \int \frac{|\nabla \rho|^2}{\rho} d\mathbf{r} = 9 T_2[\rho]. \tag{10.68}$$

Any approximate functional should be able to retrieve von Weizsäcker's expression in those cases, but should also reproduce the gradient expansion (10.66) in the limit of slowly varying densities. Several advances in this direction have been proposed along the years, and have become more reliable in recent times.

One possibility is to propose effective re-summation schemes similar to GGA for exchange (DePristo and Kress, 1987; Lee et al., 1991; Wang et al., 2001; King and Handy, 2001):

$$T_s[\rho] = c_0 \int \rho^{5/3}(\mathbf{r}) P(s[\rho, \nabla \rho]) d\mathbf{r}, \tag{10.69}$$

with $s = (c_2/c_0)|\nabla\rho/\rho^{4/3}|^2$. The function $P(s)$ should observe the two limits $P(s) \to 1+s$ for small s, and $P(s) \to 9s$ for $s \to \infty$, in order to respect the behavior of the energy in the two known limits. Besides this, there is much freedom on how to do this. Lee et al. (1991) showed that the gradient contributions from the kinetic and the exchange terms are closely related, and proposed a functional form consistent with Becke's exchange. This improves the energetics of atomic and molecular systems, but does not lead to the correct ground state density by minimization of the energy functional. This issue was addressed by Wang et al. (2001), who derived a differential equation for $P(s)$ in terms of the Kohn–Sham potential, for spherically symmetric systems. They concluded that in order for a kinetic functional to achieve this goal, it must be at least of the meta-GGA type, i.e. including the Laplacian of the density. King and Handy (2001) explored a similar road, but using

$$T_s[\rho] = c_0 \int \rho^{5/3}(\mathbf{r}) \left(\Phi(\mathbf{r}) + \frac{1}{8} \frac{|\nabla\rho|^2}{\rho} \right) d\mathbf{r}, \qquad (10.70)$$

where $\Phi(\mathbf{r})$ is a modulating function that goes to zero at long distances. Interestingly, using this approach they obtained the correct atomic shell structure.

A cleaner, although computationally heavier, route to the shell structure is to abandon semi-local approximations in favor of a fully non-local approach to the kinetic term. This has been done in the spirit of the weighted density approximation for exchange (see Section 5.3) (Chacón et al., 1985; García-González et al., 1996). The idea here is to divide the kinetic energy into a Weizsäcker term T_{vW} and a non-local contribution T_{nl} written as

$$T_{nl}[\rho] = \frac{8}{5} \int \rho(\mathbf{r}) t[\tilde{\rho}(\mathbf{r})] d\mathbf{r} - \frac{3}{5} T_{TF}[\rho], \qquad (10.71)$$

where $t[\rho]$ is the kinetic energy density in the LDA (Thomas–Fermi), and

$$\tilde{\rho}(\mathbf{r}) = \int \rho(\mathbf{r}') w(\zeta(\mathbf{r},\mathbf{r}'), |\mathbf{r}-\mathbf{r}'|) d\mathbf{r}' \qquad (10.72)$$

is a non-spherical average density defined by the weight function w. This latter depends on the local density through a function $\zeta(\mathbf{r},\mathbf{r}')$ which, importantly, is symmetric in $\mathbf{r}$ and $\mathbf{r}'$. A universal weight function is determined by imposing that the linear response of the homogeneous electron gas is reproduced.

This idea of enforcing the linear response has also been pursued by other authors (Wang and Teter, 1992; Perrot, 1994; Smargiassi and Madden, 1994; Wang et al., 1999), but in a slightly different form. The kinetic functional is written as $T_s = T_{TF} + T_{vW} + T_{lr}$, with

$$T_{lr}[\rho] = \int \int \rho^\alpha(\mathbf{r}) \omega_{\alpha\beta}(\zeta(\mathbf{r},\mathbf{r}'), |\mathbf{r}-\mathbf{r}'|) \rho^\beta(\mathbf{r}') d\mathbf{r} d\mathbf{r}', \qquad (10.73)$$

and $\omega_{\alpha\beta}$ is again determined by imposing the correct linear response. There have also been proposals of incorporating the second order response into this approach (Foley and Madden, 1996). Unfortunately, the feasibility of this scheme relies on a Taylor expansion around some uniform density, and is thus of little applicability to molecular systems. A good account of this latter methodology has been given by Watson and Carter (2000).

References

Andersen, O. K., Saha-Dasgupta, T., Tank, R. W., Arcangeli, C., Jepsen, O., and Krier, G. (2000). Developing the MTO formalism. In *Electronic Structure and the Physical Properties of Solids: the Uses of the LMTO method*, H. Dreyssé, ed. Lecture Notes in Physics. Berlin, Springer Verlag.

Bester, G. and Fähnle, M. (2001). Interpretation of *ab initio* total energy results in a chemical language: I. Formalism and implementation into a mixed-basis pseudopotential code. *J. Phys. Condens. Matter* **13**, 11541–11550.

Bingham, R. C., Dewar, M. J. S., and Lo, D. H. (1975). Ground states of molecules. XXVI. MINDO/3 calculations for hydrocarbons. *J. Am. Chem. Soc.* **97**, 1294–1301.

Børnsen, N., Meyer, B., Grotheer, O., and Fähnle, M. (1999). E_{cov} – a new tool for the analysis of electronic structure in a chemical language. *J. Phys. Condens. Matter* **11**, L287–L293.

Brenner, D. W. (1990). Empirical potential for hydrocarbons for use in simulating the chemical vapor deposition of diamond films. *Phys. Rev. B* **42**, 9458–9471.

Brooks, R., Bruccoleri, R. E., Olafson, B. D., States, D. J., Swaminatham, S., and Karplus, M. (1983). CHARMM: a program for macromolecular energy, minimization, and dynamics calculations. *J. Comput. Chem.* **4**, 187–217.

Chacón, E., Alvarellos, J. E., and Tarazona, P. (1985). Nonlocal kinetic energy functional form nonhomogeneous electron systems. *Phys. Rev. B* **32**, 7868–7877.

Chetty, N., Jacobsen, K. W., and Nørskov, J. K. (1991). Optimized and transferable densities from first principles local density calculations. *J. Phys. Condens. Matter* **3**, 5437–5443.

Cornell, W. D., Cieplak, P., Bayly, C. I. *et al.* (1995). A second generation force field for the simulation of proteins, nucleic acids, and organic molecules. *J. Am. Chem. Soc.* **117**, 5179–5197.

Daw, M. S. and Baskes, M. I. (1984). Embedded atom method: derivation and application to impurities, surfaces and other defects in metals. *Phys. Rev. B* **29**, 6443–6453.

DePristo, A. E. and Kress, J. D. (1987). Rational function representation for accurate exchange energy functionals. *J. Chem. Phys.* **86**, 1425–1428.

Dewar, M. J. S. and Thiel, W. (1977). The MNDO method. Approximations and parameters. *J. Am. Chem. Soc.* **99**, 4899–4907.

Dewar, M. J. S., Zoebisch, E. G., Healy, E. F., and Stewart, J. J. P. (1985). AM1: a new general purpose quantum mechanical molecular model. *J. Am. Chem. Soc.* **107**, 3902–3909.

Dick, B. G. and Overhauser, J. A. W. (1958). Theory of the dielectric constants of alkali halide crystals. *Phys. Rev.* **112**, 90–103.

Ducastelle, F. (1970). Elastic modulus of transition metals. *J. de Phys.* **31**, 1055–1062.

Eichinger, M., Tavan, P., Hutter, J., and Parrinello, M. (1999). A hybrid method for solutes in complex solvents: density functional theory combined with empirical force fields. *J. Chem. Phys.* **110**, 10452–10467.

Elstner, M., Porezag, D., Jungnickel, G., Elsner, J., Haugk, M., Frauenheim, T., Suhai, S., and Seifert, G. (1998). Self-consistent-charge density-functional tight-binding method for simulations of complex materials properties. *Phys. Rev. B* **58**, 7260–7268.

Ercolessi, F., Tosatti, E., and Parrinello, M. (1986). Au(100) surface reconstruction. *Phys. Rev. Lett.* **57**, 719–722.

Esfarjani, K. and Kawazoe, Y. (1998). Self-consistent tight-binding formalism for charged systems. *J. Phys. Condens. Matter* **10**, 8257–8267.

Estrín, D. A., Kohanoff, J., Laría, D. H., and Weht, R. O. (1997). A hybrid quantum and classical mechanical Monte Carlo simulation of the interaction of hydrogen chloride with solid water clusters. *Chem. Phys. Lett.* **280**, 280–286.

Fabris, S., Paxton, A. T., and Finnis, M. W. (2001). Free energy and molecular dynamics calculations for the cubic-tetragonal phase transition in zirconia. *Phys. Rev. B* **63**, 094101.

  (2002). A stabilisation mechanism of zirconia based on oxygen vacancies only. *Acta Materialia* **50**, 5171–5178.

Field, M. J., Bash, P. A., and Karplus, M. J. (1990). A combined quantum mechanical and molecular mechanical potential for molecular dynamics simulations. *J. Comput. Chem.* **11**, 700–733.

Finnis, M. W. (2003). *Interatomic Forces in Condensed Matter*. Oxford Series on Materials Modelling, Oxford, Oxford University Press.

Finnis, M. W. and Sinclair, J. E. (1984). A simple empirical N-body potential for transition metals. *Phil. Mag.* **50**, 45–55.

Finnis, M. W., Paxton, A. T., Methfessel, M., and van Schilfgaarde, M. (1998). The crystal structure of zirconia from first principles and self-consistent tight-binding. *Phys. Rev. Lett.* **81**, 5149–5152.

Foley, M. and Madden, P. A. (1996). Further orbital-free kinetic-energy functionals for *ab initio* molecular dynamics. *Phys. Rev. B* **53**, 10589–10598.

Foulkes, W. M. C. (1987). Ph.D. thesis, University of Cambridge.

García-González, P., Alvarellos, J. E., and Chacón, E. (1996). Kinetic-energy density functional: atoms and shell structure. *Phys. Rev. A* **54**, 1897–1905.

Goodwin, L., Skinner, A. J., and Pettifor, D. G. (1989). Generating transferable tight-binding parameters – application to silicon. *Europhys. Lett.* **9**, 701–706.

Harris, J. (1985). Simplified method for calculating the energy of weakly interacting fragments. *Phys. Rev. B* **31**, 1770–1779.

Harrison, W. A. (1980). *Electronic Structure and the Properties of Solids*. San Francisco, Freeman.

Haydock, R., Heine, V., and Kelly, M. J. (1972). Electronic structure based on the local atomic environment for tight-binding bands. *J. Phys. C* **5**, 2845–2858.

Hill, T. L. (1948). Steric effects. I. Van der Waals potential energy curves. *J. Chem. Phys.* **16**, 399–404.

Hoffmann, R. (1963). An extended Hückel theory. I. Hydrocarbons. *J. Chem. Phys.* **39**, 1397–1412.

Hohenberg, P. and Kohn, W. (1964). Inhomogeneous electron gas. *Phys. Rev.* **136**, B864–867.

Jacobsen, K. W., Nørskov, J. K., and Puska, M. J. (1987). Interatomic interaction in the effective medium theory. *Phys. Rev. B* **35**, 7423–7442.

Jensen, F. (1999). *Introduction to Computational Chemistry*. Chichester, Wiley.

Jorgensen, W. L. (1998). OPLS force fields. In *Encyclopedia of Computational Chemistry*, P. v. R. Schleyer, ed. New York, Wiley, 1986–1989.

Kim, J., Mauri, F., and Galli, G. (1995). Total-energy global optimizations using nonorthogonal localized orbitals. *Phys. Rev. B* **52**, 1640–1648.

King, R. A. and Handy, N. C. (2001). Kinetic energy functionals for molecular calculations. *Mol. Phys.* **99**, 1005–1009.

Kirzhnits, D. A. (1957). Quantum corrections to the Thomas–Fermi equation. *Zh. Eksp. Teor. Fiz.* **32**, 115. *Sov. Phys. JETP* **5**, 64–72 (1957).

Laio, A., VandeVondele, J., and Röthlisberger, U. (2002). A Hamiltonian electrostatic coupling scheme for hybrid Car–Parrinello molecular dynamics simulations. *J. Chem. Phys.* **116**, 6941–6947.

Lee, H., Lee, C., and Parr, R. G. (1991). Cojoint gradient correction to the Hartree–Fock kinetic- and exchange-energy density functionals. *Phys. Rev. A* **44**, 768–771.

Lennard-Jones, J. E. (1924). The determination of molecular fields. II. From the equation of state of a gas. *Proc. Roy. Soc. London* **106**, 463–477.

Miertus, S., Scrocco, E., and Tomasi, J. (1981). Electrostatic interaction of a solute with a continuum. A direct utilization of *ab initio* molecular potentials for the prevision of solvent effects. *Chem. Phys.* **55**, 117–129.

Migoni, R., Bilz, H., and Bauerle, D. (1976). Origin of Raman scattering and ferroelectricity in oxidic perovskites. *Phys. Rev. Lett.* **37**, 1155–1158.

Monard, G., Loos, M., Théry, V., Baka, K., and Rivail, J.-L. (1996). Hybrid classical quantum force field for modeling very large molecules. *Int. J. Quant. Chem.* **58**, 153–159.

Murdock, S. E., Lynden-Bell, R. M., Kohanoff, J., and Sexton, G. J. (2002). Determining the electronic structure and chemical potentials of molecules in solution. *Phys. Chem. Chem. Phys.* **4**, 3016–3021.

Onsager, L. (1936). Electric moments of molecules in liquids. *J. Am. Chem. Soc.* **58**, 1486–1493.

Ordejón, P., Drabold, D. A., Grumbach, M. P., and Martin, R. M. (1993). Unconstrained minimization approach for electronic computations that scales linearly with system size. *Phys. Rev. B* **48**, 14646–14649.

Perrot, F. (1994). Hydrogen–hydrogen interaction in an electron gas. *J. Phys. Condens. Matter* **6**, 431–446.

Pettifor, D. (1989). New many-body potential for the bond order. *Phys. Rev. Lett.* **63**, 2480–2483.

Pettifor, D. G. (1995). *Bonding and Structure in Molecules and Solids*. Oxford, Clarendon Press.

Pople, J. A. and Beveridge, D. L. (1970). *Approximate Molecular Orbital Theory*. Advanced Chemistry, London, McGraw-Hill.

Pople, J. A. and Segal, G. A. (1965). Approximate self-consistent molecular orbital theory. II. Calculations with complete neglect of differential overlap. *J. Chem. Phys.* **43**, S136–S151.

(1966). Approximate self-consistent molecular orbital theory. III. CNDO results for AB_2 and AB_3 systems. *J. Chem. Phys.* **44**, 3289–3296.

Pople, J. A., Santry, D. P., and Segal, G. A. (1965). Approximate self-consistent molecular orbital theory. I. Invariant procedures. *J. Chem. Phys.* **43**, S129–S135.

Pople, J. A., Beveridge, D. L., and Dobosh, P. A. (1967). Approximate self-consistent molecular-orbital theory. V. Intermediate neglect of differential overlap. *J. Chem. Phys.* **47**, 2026–2033.

Rappe, A. K., Casewit, C. J., Colwell, K. S., Goddard III, W. A., and Skiff, W. M. (1992). UFF, a full periodic table force field for molecular mechanics and molecular dynamics simulations. *J. Am. Chem. Soc.* **114**, 10024–10035.

Sankey, O. F. and Niklewski, D. J. (1989). *Ab initio* multicenter tight-binding model for molecular dynamics simulations and other applications in covalent systems. *Phys. Rev. B* **40**, 3979–3995.

Schelling, P. K., Yu, N., and Halley, J. W. (1998). Self-consistent tight-binding atomic-relaxation model of titanium dioxide. *Phys. Rev. B* **58**, 1279–1293.

Slater, J. C. and Koster, G. F. (1954). Simplified LCAO method for the periodic potential problem. *Phys. Rev.* **94**, 1498–1524.

Smargiassi, E. and Madden, P. A. (1994). Orbital-free kinetic-energy functionals for first-principles molecular dynamics. *Phys. Rev. B* **49**, 5220–5226.

Stewart, J. J. P. (1989). Optimization of parameters for semiempirical methods. I. Method. *J. Comput. Chem.* **10**, 209–220.

Stillinger, F. H. and David, C. W. (1978). Polarization model for water and its ionic dissociation products. *J. Chem. Phys.* **69**, 1473–1484.

Stillinger, F. H. and Weber, T. A. (1985). Computer simulation of local order in condensed phases of silicon. *Phys. Rev. B* **31**, 5262–5271.

Stone, A. J. (1981). Distributed multipole analysis; or how to describe a molecular charge distribution. *Chem. Phys. Lett.* **83**, 233–239.

Svensson, M., Humbel, S., Froesse, R. D. J., Matsubara, T., Sieber, S., and Morokuma, K. (1996). ONIOM: a multilayered integrated MO + MM method for geometry optimizations and single point energy predictions. A test for Diels–Alder reactions and Pt(P(t-Bu)$_3$)$_2$ + H$_2$ oxidative addition. *J. Phys. Chem.* **100**, 19357–19363.

Tersoff, J. (1986). New empirical model for the structural properties of silicon. *Phys. Rev. Lett.* **23**, 542–548.

Thiel, W. (1981). The MNDOC method, a correlated version of the MNDO model. *J. Am. Chem. Soc.* **103**, 1413–1420.

Van Gunsteren, W. F. and Berendsen, H. J. C. (1990). *Computer Simulation of Molecular Dynamics: Methodology, Applications and Perspectives in Chemistry.* www.igc.ethz.ch/gromos.

Von Weizsäcker, C. F. (1935). Zur Theorie der Kernmassen. *Z. Phys.* **96**, 431–458.

Wang, B., Stott, M. J., and von Barth, U. (2001). Approximate electron kinetic-energy functionals. *Phys. Rev. A* **63**, 052501-1–052501-8.

Wang, L.-W. and Teter, M. P. (1992). Kinetic-energy functional of the electron density. *Phys. Rev. B* **45**, 13196–13220.

Wang, Y. A., Govind, N., and Carter, E. A. (1999). Orbital-free kinetic-energy density functionals with a density-dependent kernel. *Phys. Rev. B* **60**, 16350–16358.

Watson, S. C. and Carter, E. A. (2000). Linear-scaling parallel algorithms for the first principles treatment of metals. *Comput. Phys. Commun.* **128**, 67–92.

Zaremba, E. (1990). Extremal properties of the Harris functional. *J. Phys. Condens. Matter* **2**, 2479–2486.

11

Diagonalization and electronic self-consistency

Two important practical aspects of any electronic structure calculation are the diagonalization of the Hamiltonian matrix, and the convergence towards the self-consistent solution, where the input density (or potential) equals the output one. In this chapter we shall discuss a number of algorithms that are normally used in computer codes. To remind the reader, the problem at hand is to solve the Kohn–Sham eigenvalue problem:

$$\hat{\mathcal{H}}_{KS}[\rho_{in}]\,\varphi_i(\mathbf{r}) = \varepsilon_i\,\varphi_i(\mathbf{r}), \qquad (11.1)$$

where the Kohn–Sham potential is evaluated at some input electronic density $\rho_{in}(\mathbf{r})$, i.e.

$$v_{KS}[\rho_{in}](\mathbf{r}) = v_{ext}(\mathbf{r}) + \int \frac{\rho_{in}(\mathbf{r}')}{|\mathbf{r}-\mathbf{r}'|}\,d\mathbf{r}' + \mu_{XC}[\rho_{in}](\mathbf{r}), \qquad (11.2)$$

and the output density is constructed with the solutions of the above eigenvalue equation:

$$\rho_{out}(\mathbf{r}) = \sum_{i=1}^{N} |\varphi_i[\rho_{in}](\mathbf{r})|^2. \qquad (11.3)$$

The self-consistency condition implies that $\rho_{out}(\mathbf{r})$ equals $\rho_{in}(\mathbf{r})$ within a certain numerical tolerance. Since potential and density are univocally connected (apart from an additive constant in the potential) via the Hohenberg–Kohn theorem, the above self-consistency condition can also be stated in terms of the Kohn–Sham potential, i.e. that $v_{KS}[\rho_{in}]$ and $v_{KS}[\rho_{out}]$ coincide within some requested numerical tolerance.

This is the traditional approach in electronic structure calculations, and it consists of a nested procedure of matrix diagonalizations at fixed input density, and the construction of the input density for the next iteration via mixing and extrapolation procedures. This methodology can be applied to any self-consistent diagonalization problem such as Hartree–Fock.

Diagonalization is the subject of the first section of this chapter, and mixing schemes for self-consistency are discussed in Section 11.2. The self-consistent diagonalization problem can also be attacked from the minimization angle, where the Kohn–Sham energy functional is directly minimized by varying the Kohn–Sham orbitals. Here we enter the vast terrain of multi-dimensional optimization procedures like steepest descent or conjugated gradients. Such methods will be discussed in Section 11.3.

11.1 Diagonalization

According to the size of the basis set and the sparsity of the Hamiltonian matrix, several different algorithms can be used to solve the diagonalization problem. If the matrix is reasonably small, then perhaps standard routines as implemented in linear algebra packages like LAPACK (Anderson *et al.*, 1999) are the most convenient option. These routines have also been adapted to parallel platforms, and showed favorable scaling properties (Blackford *et al.*, 1997).

If the matrix is too large, then it becomes more convenient to use iterative algorithms that filter out the lowest-lying eigenstates, up to half the number of electrons in the case of insulating (closed-shell) systems, or a few more states in the case of metallic (open-shell) systems to allow for fractional occupation. This is usually the case when using a PW basis set. While atom-centered basis sets typically involve between 10 and 20 basis functions per atom, the number of PWs per atom is of the order of 100–200, i.e. one order of magnitude larger, and it grows linearly with the volume of the supercell. Another case where huge matrices have to be diagonalized is in configuration interaction methods. These are, then, cases for iterative diagonalization. Here we shall illustrate the idea with a simple example, and then we shall describe two widespread iterative algorithms, namely those proposed by Davidson (1975) and Lanczos (1950).

Iterative diagonalization methods do not require the knowledge of the full matrix, but only the action of the matrix on an arbitrary vector, i.e.

$$\Psi^{(n+1)} = \mathbf{H}\,\Psi^{(n)}. \tag{11.4}$$

The vectors $\Psi^{(n)}$ form a sequence from which the lowest-lying eigenstates can be extracted. To visualize how these methods work, let us consider the following simple procedure. Starting from a trial vector $\Psi^{(0)}$, we perform the following operation:

$$\Psi^{(1)} = e^{-\tau \mathbf{H}}\,\Psi^{(0)}, \tag{11.5}$$

11.1 Diagonalization

with τ a positive quantity. If we expand the trial vector in the basis of eigenstates of matrix **H**, the action of the exponential can be written as

$$\Psi^{(1)} = e^{-\tau \mathbf{H}} \left(\sum_{i=1}^{M} c_i^{(0)} \varphi_i \right) = \sum_{i=1}^{M} e^{-\tau \varepsilon_i} c_i^{(0)} \varphi_i = \sum_{i=1}^{M} c_i^{(1)} \varphi_i, \qquad (11.6)$$

where M is the dimension of the matrix, given by the size of the basis set. $c_i^{(1)} = e^{-\tau \varepsilon_i} c_i^{(0)}$ are the coefficients of the iterated vector and, since the ground state has the lowest energy, it is easy to see that this operation enhances the relative weight of the ground state with respect to higher energy levels. Therefore, if this operation is repeated many times, say N_{iter}, the result will be

$$\Psi^{(N_{\text{iter}})} = \sum_{i=1}^{M} e^{-N_{\text{iter}} \tau \varepsilon_i} c_i^{(0)} \varphi_i = \sum_{i=1}^{M} c_i^{(N_{\text{iter}})} \varphi_i, \qquad (11.7)$$

so that the relative weights of the higher eigenstates with respect to the lowest-lying one are

$$\frac{c_i^{(N_{\text{iter}})}}{c_1^{(N_{\text{iter}})}} = e^{-N_{\text{iter}} \tau (\varepsilon_i - \varepsilon_1)}, \qquad (11.8)$$

which tends to zero for $N_{\text{iter}} \to \infty$, because $\varepsilon_i > \varepsilon_1$. In conclusion, after a certain number of iterations, the contribution of eigenstates other than the ground state have been filtered out, and the trial vector becomes effectively the ground state, within the desired tolerance. There is, however, a problem with this procedure, and that is that after applying the exponential operator, the magnitudes of all the coefficients, including that of the lowest-lying state, are severely enhanced (for negative energies) or reduced (for positive energies). Therefore, the iterated vector is not normalized any longer, and in the long run it either explodes or collapses. This pitfall can be cured by normalizing the iterated vector according to $\tilde{\Psi}^{(n)} = \Psi^{(n)} / \|\Psi^{(n)}\|$. In addition, the lowest-lying eigenvalue can be estimated by

$$\varepsilon_1^{(n+1)} = -\frac{1}{\tau} \ln \left(\frac{c_1^{(n+1)}}{c_1^{(n)}} \right). \qquad (11.9)$$

This procedure can be generalized to determine several lowest-lying eigenvectors by operating on a set of orthonormal trial vectors, and making sure that they remain orthogonal at every iteration. When using this procedure in conjunction with a plane wave basis set, the coefficient τ in the exponential should be chosen small enough that the Trotter decomposition

$$e^{-\tau \mathbf{H}} = e^{-\tau \mathbf{T}} e^{-\tau \mathbf{V}} e^{-\tau^2 [\mathbf{T}, \mathbf{V}]/2} \approx e^{-\tau \mathbf{T}} e^{-\tau \mathbf{V}} \qquad (11.10)$$

is valid. In this case, the action of the potential and the kinetic operators can be computed independently in the most convenient representation (real or reciprocal space). Although very elegant, simple, and illustrative, this approach requires a large number of iterations to converge.

Electronic structure codes employ more efficient algorithms along the preceding lines. The main idea of the Davidson (1975) algorithm is to reduce the dimension of the matrix to be diagonalized by expanding the eigenvectors in some set of orthonormal vectors $\mathbf{v}_j$, with $j = 1, \ldots, k$. In this reduced basis, a set of n trial vectors Ψ_i can be written

$$\Psi_i = \sum_{j=1}^{k} \mathbf{v}_j\, a_{ji} = \mathbf{V}\mathbf{a}_i, \qquad (11.11)$$

where a_{ji} are expansion coefficients. Using the above, a reduced dimension matrix can be constructed as $\mathbf{H}_\mathrm{D} = \mathbf{V}^\dagger \mathbf{H} \mathbf{V}$, which satisfies the following eigenvalue equation:

$$\mathbf{H}_\mathrm{D}\,\mathbf{a}_i = E_i^{(k)}\,\mathbf{a}_i. \qquad (11.12)$$

The eigenvalues of $\mathbf{H}_\mathrm{D}$ have the property that they are all larger than the corresponding eigenvalue of $\mathbf{H}$ and, by increasing the size of the reduced basis to $m > k$, we have $E_i^{(k)} \geq E_i^{(m)} \geq \varepsilon_i$. These converge to ε_i for increasing size of the reduced basis.

This property can be exploited by constructing the following algorithm: first choose an initial set of vectors and build the matrix $\mathbf{V}$. Then operate with $\mathbf{H}$ on $\mathbf{V}$ to obtain the rectangular $M \times k$ matrix $\mathbf{V}_\mathrm{D} = \mathbf{H}\mathbf{V}$, and again with $\mathbf{V}^\dagger$ to form $\mathbf{H}_\mathrm{D} = \mathbf{V}^\dagger \mathbf{H} \mathbf{V}$. This is called the Davidson matrix, and it is a $k \times k$ square matrix. Diagonalization of $\mathbf{H}_\mathrm{D}$ using standard linear algebra library routines (remember that the dimension of the Davidson matrix is much smaller than the original one) produces the approximate eigenvalues E_i and eigenvectors Ψ_i^new. The quality of the eigenvectors is measured by the residual vectors $\mathbf{r}_i = (\mathbf{H} - E_i \mathbf{1})\Psi_i^\mathrm{new}$. If the norm of the residual $\|\mathbf{r}_i\|$ is smaller than a desired threshold, then the eigenvector is converged and not modified anymore. Otherwise, the iterative procedure must continue until all the required eigenvectors are converged. To continue, some additional orthogonal vectors are added to matrix $\mathbf{V}$, and the whole procedure is repeated, as the eigenvalues of $\mathbf{H}_\mathrm{D}$ converge to those of $\mathbf{H}$ for increasing dimension. If it happens that the size of the Davidson matrix becomes too large, then the lowest n basis vectors $\mathbf{v}_i$ are replaced by $\mathbf{V}\Psi_i^\mathrm{new}$, and the procedure is re-initialized.

The secret of the efficiency of the iterative algorithm lies in the choice of the expansion vectors. In the original Davidson method, vectors are added in the direction of the residuals, i.e. $\mathbf{w}_i = \mathbf{X}_i \mathbf{r}_i$, where the pre-conditioning matrix $\mathbf{X}_i = (\mathbf{H}_{ii} - E_i \mathbf{1})^{-1}$ selects prevalently the component of the residual in the direction of the desired eigenvector. Of course, there is space for improving the expansion algorithms. For example, the one implemented in the CPMD code (Marx and Hutter, 2000; CPMD Consortium, 2004) chooses

$$\mathbf{w}_i = \mathbf{X}_i \left[\mathbf{r}_i - \left(\frac{\Psi_i^{\text{new}\dagger} \mathbf{r}_i}{\Psi_l^{\text{new}\dagger} \mathbf{X}_l \mathbf{r}_l} \right) \Psi_i^{\text{new}} \right], \qquad (11.13)$$

and converges faster than the original Davidson prescription. The expansion vectors have to be orthogonalized to all the vectors already present in $\mathbf{V}$.

Another useful iterative diagonalization method is the celebrated Lanczos algorithm (Lanczos, 1950). This is basically a conjugated gradients procedure, which can be explained in the following way. Starting from a trial vector $\Psi^{(0)}$, a sequence of conjugated vectors is generated where every new vector in the sequence is orthogonal to all of the preceding ones. The first step, which is slightly different from the rest, is defined by

$$\Psi^{(1)} = \beta_1^{-1} \left(\mathbf{H} \Psi^{(0)} - \alpha_0 \Psi^{(0)} \right), \qquad (11.14)$$

with

$$\alpha_0 = \langle \Psi^{(0)} | \mathbf{H} | \Psi^{(0)} \rangle, \quad \beta_1 = \langle \Psi^{(1)} | \mathbf{H} | \Psi^{(0)} \rangle, \qquad (11.15)$$

while the general step in the recursive algorithm is given by

$$\Psi^{(n+1)} = \beta_{n+1}^{-1} \left(\mathbf{H} \Psi^{(n)} - \alpha_n \Psi^{(n)} - \beta_n \Psi^{(n-1)} \right), \qquad (11.16)$$

with

$$\alpha_n = \langle \Psi^{(n)} | \mathbf{H} | \Psi^{(n)} \rangle, \quad \beta_{n+1} = \langle \Psi^{(n+1)} | \mathbf{H} | \Psi^{(n)} \rangle. \qquad (11.17)$$

It is easy to see that, in the basis of the vectors generated by the Lanczos recursion, the Hamiltonian matrix $\mathbf{T}$ is tridiagonal,

$$\mathbf{T} = \begin{pmatrix} \alpha_0 & \beta_1 & 0 & 0 & \cdots & 0 \\ \beta_1 & \alpha_1 & \beta_2 & 0 & \cdots & 0 \\ 0 & \beta_2 & \alpha_2 & \beta_3 & \cdots & 0 \\ \vdots & \vdots & \vdots & \ddots & \vdots & \vdots \\ 0 & \cdots & 0 & \beta_{n-1} & \alpha_{n-1} & \beta_n \\ 0 & \cdots & 0 & 0 & \beta_n & \alpha_n \end{pmatrix}, \qquad (11.18)$$

and can be diagonalized by very efficient numerical algorithms implemented in linear algebra computer packages such as LAPACK (Anderson *et al.*, 1999). As in Davidson's method, the Lanczos algorithm generates a representation of the Hamiltonian matrix in a space of reduced dimensionality. For increasing dimension of the $\mathbf{T}$-matrix, its eigenvalues converge to ε_i and its eigenvectors to φ_i. A salient feature of the Lanczos algorithm is that the eigenvalues of $\mathbf{H}$ appear many times as degenerate eigenvalues of $\mathbf{T}$. Therefore, stable Lanczos algorithms have plenty of numerical subtleties, convergence checks, tolerances, etc., which are extensively described in the book by Cullum and Willoughby (2002). Once an initial Lanczos matrix has been generated, further refinements can be achieved by block recursion algorithms within the subspace of vectors associated with the individual eigenvalues (Pollard and Friesner, 1993). The block Lanczos algorithm is implemented in the FEMD code, which is included in the CPMD package (Alavi et al., 1995; CPMD Consortium, 2004). It actually operates with $e^{-\tau \mathbf{H}}$ in place of the Hamiltonian, with the advantage that all the eigenvalues are of similar magnitude (all close to 1), so that the eigenvalues all converge roughly at the same rate. This is sometimes a problem with $\mathbf{H}$, which has the eigenvalues more spread, and thus different convergence rates.

11.2 Self-consistency: mixing schemes

Reaching the self-consistent solution, i.e. $\rho_{\text{out}}(\mathbf{r}) = \rho_{\text{in}}(\mathbf{r})$ or $v_{\text{KS}}^{\text{out}}(\mathbf{r}) = v_{\text{KS}}^{\text{in}}(\mathbf{r})$, can be thought of as the problem of finding the fixed point of an equation of the type $f(x) = x$. The obvious strategy is to start from some guess x_0, and then iterate the equation $x_{n+1} = f(x_n)$. If this procedure converges, then the limiting value is the fixed point $\bar{x} = f(\bar{x})$, or self-consistent solution. The success of such a strategy strongly depends on the shape of the function $f(x)$. If the slope of $f(x)$ is too large and negative, then this simple iterative solution does not converge. This is precisely our case, because our function is

$$f(\mathbf{r}, \mathbf{r}') = \frac{\delta v_{\text{KS}}^{\text{out}}(\mathbf{r})}{\delta v_{\text{KS}}^{\text{in}}(\mathbf{r}')} = \int K(\mathbf{r}, \mathbf{r}'') \chi(\mathbf{r}'', \mathbf{r}') \, d\mathbf{r}'', \tag{11.19}$$

where $K(\mathbf{r}, \mathbf{r}'')$ is the linear response kernel

$$K(\mathbf{r}, \mathbf{r}'') = \frac{1}{|\mathbf{r} - \mathbf{r}''|} + \frac{\delta \mu_{\text{XC}}[\rho_{\text{out}}](\mathbf{r})}{\delta \rho_{\text{out}}(\mathbf{r}'')}, \tag{11.20}$$

and

$$\chi(\mathbf{r}'', \mathbf{r}') = \frac{\delta \rho_{\text{out}}(\mathbf{r}'')}{\delta v_{\text{KS}}^{\text{in}}(\mathbf{r}')} \tag{11.21}$$

is the response function of the electron gas. This response function is negative definite because increasing the value of the potential at point $\mathbf{r}'$ makes it less attractive, and then the electrons move to a more convenient region. Therefore, if $\mathbf{r}''$ is in the vicinity of $\mathbf{r}'$, the electronic density there will decrease, thus leading to a negative response. The slope may also be very large because the Coulomb kernel $1/|\mathbf{r} - \mathbf{r}'|$ in the reciprocal-space representation becomes $4\pi/G^2$, which diverges very strongly for small values of the wave vector G. A similar argument applies to the density response.

11.2 Self-consistency: mixing schemes

To overcome convergence problems when a simple *out–in* replacement procedure is ineffective, it is important to realize that the physical reason for the divergence is that large charge redistributions occur from one iteration to the next. These charge displacements can be damped out by mixing the input and output densities according to some appropriate prescription. The simplest strategy is *simple mixing*:

$$\rho_{\text{in}}^{(n+1)}(\mathbf{r}) = \alpha \rho_{\text{out}}^{(n)}(\mathbf{r}) + (1-\alpha)\rho_{\text{in}}^{(n)}(\mathbf{r}), \tag{11.22}$$

where α is an empirical parameter adjusted to minimize the number of iterations needed to achieve self-consistency. The smaller α, the more is retained from the input density.

In general this procedure is not satisfactory enough. It may work for simple cases with values of $\alpha \approx 0.4 - 0.6$, but usually smaller values have to be used to ensure convergence. Metallic and magnetic systems are notoriously difficult cases, where very small values of α – sometimes down to values of the order of 0.01 – are required in order to avoid the divergence of the iterative procedure. This means that only a minute fraction of the output density is used to construct the new input density, and a large number of iterations is required to achieve self-consistency.

The next natural step is to use information from the input and output densities of the preceding iterations. The simplest scheme along these lines consists of constructing modified input and output densities by mixing the two last steps (Anderson, 1965; Johnson, 1988):

$$\bar{\rho}_{\text{in}}^{(n)}(\mathbf{r}) = \beta \rho_{\text{in}}^{(n-1)}(\mathbf{r}) + (1-\beta)\rho_{\text{in}}^{(n)}(\mathbf{r}),$$
$$\bar{\rho}_{\text{out}}^{(n)}(\mathbf{r}) = \beta \rho_{\text{out}}^{(n-1)}(\mathbf{r}) + (1-\beta)\rho_{\text{out}}^{(n)}(\mathbf{r}), \tag{11.23}$$

and proposing a *guess* for the next iteration of the same form used in the simple mixing scheme:

$$\rho_{\text{in}}^{(n+1)}(\mathbf{r}) = \alpha \bar{\rho}_{\text{out}}^{(n)}(\mathbf{r}) + (1-\alpha)\bar{\rho}_{\text{in}}^{(n)}(\mathbf{r}), \tag{11.24}$$

where α is still an empirical mixing parameter. Parameter β, however, is determined by the condition that the *distance* between $\bar{\rho}_{\text{in}}^{(n)}$ and $\bar{\rho}_{\text{out}}^{(n)}$ is minimal, thus helping to accelerate the convergence of the self-consistent procedure. By minimizing the norm of $\Delta\bar{\rho}^{(n)} = \bar{\rho}_{\text{out}}^{(n)}(\mathbf{r}) - \bar{\rho}_{\text{in}}^{(n)}(\mathbf{r})$ with respect to β, it is easy to show that

$$\beta = \frac{\langle \Delta\rho^{(n)} | \Delta\rho^{(n)} - \Delta\rho^{(n-1)}\rangle}{\|\Delta\rho^{(n)} - \Delta\rho^{(n-1)}\|^2}, \tag{11.25}$$

where the brackets indicate the following scalar product:

$$\langle \Delta\rho^{(i)} | \Delta\rho^{(j)} \rangle = \int [\rho_{\text{out}}^{(i)}(\mathbf{r}) - \rho_{\text{in}}^{(i)}(\mathbf{r})][\rho_{\text{out}}^{(j)}(\mathbf{r}) - \rho_{\text{in}}^{(j)}(\mathbf{r})]\,d\mathbf{r}, \tag{11.26}$$

or its discrete version in terms of the density matrix, if an atom-centered basis set is used. This alternative is extremely simple and effective, allowing us to use values of α as large as 0.3 in the difficult cases alluded to above, and reducing the number of self-consistency iterations by a factor of ten or more. Still, some systems are reluctant to kneel in front of this approach, and more sophisticated schemes that generate a sequence of conjugated minimum search directions are required (Broyden, 1965; Johnson, 1988).

A simple generalization of the Anderson scheme to an arbitrary number of iterations, P, has been proposed under the name of direct inversion in iterative subspace (DIIS) (Pulay, 1980). The guess for the next iteration is constructed exactly as in (11.24), but the modified input and output densities are given by

$$\bar{\rho}_{\text{in}}^{(n)}(\mathbf{r}) = \sum_{i=1}^{P} \beta_i \rho_{\text{in}}^{(n-P+i)}(\mathbf{r}),$$

$$\bar{\rho}_{\text{out}}^{(n)}(\mathbf{r}) = \sum_{i=1}^{P} \beta_i \rho_{\text{out}}^{(n-P+i)}(\mathbf{r}), \quad (11.27)$$

under the normalization constraint that $\sum_{i=1}^{P} \beta_i = 1$. Minimization of the distance between $\bar{\rho}_{\text{in}}^{(n)}$ and $\bar{\rho}_{\text{out}}^{(n)}$ with respect to the P coefficients $\{\beta_i\}$ leads to a system of linear equations that can be put in the form of the following matrix equation:

$$\begin{pmatrix} C_{n-P+1,n-P+1} & \cdots & C_{n,n-P+1} & 1 \\ \vdots & \ddots & \vdots & \vdots \\ C_{n-P+1,n} & \cdots & C_{n,n} & 1 \\ 1 & \cdots & 1 & 0 \end{pmatrix} \begin{pmatrix} \beta_1 \\ \vdots \\ \beta_P \\ \lambda \end{pmatrix} = \begin{pmatrix} 0 \\ \vdots \\ 0 \\ 1 \end{pmatrix}, \quad (11.28)$$

with $C_{ij} = \langle \Delta \rho^{(i)} | \Delta \rho^{(j)} \rangle$ defined by (11.26). There is a limit in the number of iterations that can be mixed using the DIIS scheme, because after some iterations the linear system develops a linear dependency and the matrix becomes singular. In practice, it is observed that mixing more than four or five iterations does not improve the convergence properties of the algorithm anymore. Therefore, useful numbers of Pulay iterations are between three and five. This extrapolation procedure can be applied to the real-space density, as indicated above, but also to the density matrix, to the potential, and even to the Kohn–Sham or Hartree–Fock orbitals, although in these cases care must be taken to maintain the orbitals orthonormal (Hutter et al., 1994).

11.3 Direct minimization of the electronic energy functional

An alternative strategy to the diagonalization of the Hamiltonian matrix at fixed density is to minimize the energy functional directly. One of the advantages of

11.3 Direct minimization of the electronic energy functional

this procedure is that both diagonalization and self-consistency are attacked at once, while otherwise the diagonalization step has to be repeated several times after the mixing step. Direct minimization techniques have been pioneered by Car and Parrinello (1985) in their seminal work, later reviewed by Payne et al., (1992).

The simplest, but quite inefficient minimization method is *steepest descent* (SD). In this procedure the minimum is reached by a sequence of downhill steps in the direction of the local gradient of the functional with respect to the variational parameters, which in this case are the Kohn–Sham orbitals:

$$\varphi_i^{(n+1)}(\mathbf{r}) = \varphi_i^{(n)}(\mathbf{r}) - \Delta \frac{\delta \Omega_{\text{KS}}[\{\varphi_i^{(n)}(\mathbf{r})\}, \mathbf{R}]}{\delta \varphi_i^{(n)*}(\mathbf{r})}, \tag{11.29}$$

with

$$\Omega_{\text{KS}} = E_{\text{KS}} - \sum_{i=1}^{N}\sum_{j=1}^{N} \Lambda_{ij}(\langle \varphi_i | \varphi_j \rangle - \delta_{ij}), \tag{11.30}$$

and the Kohn–Sham energy functional E_{KS} is given by the usual expression. Δ above is a time-like parameter (the *time step*). The second term is to ensure the orthonormalization of the orbitals, with Λ_{ij} a set of Lagrange multipliers. The functional derivatives in (11.29) are given explicitly for the PPW method in Section 9.3.3. There, it can be seen that they correspond to the action of the Kohn–Sham Hamiltonian on the single-particle orbitals, supplemented with the orthonormalization constraint:

$$\varphi_i^{(n+1)}(\mathbf{r}) = \varphi_i^{(n)}(\mathbf{r}) - \Delta \left(\hat{\mathcal{H}}_{\text{KS}} \varphi_i^{(n)}(\mathbf{r}) - \sum_{j=1}^{N} \Lambda_{ij} \varphi_j^{(n)}(\mathbf{r}) \right). \tag{11.31}$$

The parameter Δ is adjusted to give the fastest possible convergence, i.e. the largest possible value that prevents the divergence of the SD procedure. In fact, the SD equations can be thought of as the discrete version of the first-order differential equation

$$\dot{\varphi}_i(\mathbf{r}) = -\frac{\delta \Omega_{\text{KS}}\{\varphi_i^{(n)}(\mathbf{r})\}}{\delta \varphi_i^{(n)*}(\mathbf{r})}, \tag{11.32}$$

whose solution decays exponentially in the time-like variable. If the time step is too large, then the discretization of this differential equation is not accurate enough, the generalized forces on the orbitals are given an excessive weight, they overshoot and the iterative process diverges. In most applications the value of a constant Δ is estimated by trial and error. Another alternative is to determine it dynamically by performing a line minimization along the direction of the gradient, and to choose $\Delta^{(n)}$ in such a way as to maximize the decrease in energy.

The steepest descent procedure is known to be inefficient (Press *et al.*, 1992). It may take hundreds, if not thousands, of steps to reach the minimum, because it is limited by the fact that the gradient at step $n+1$ is orthogonal *only* to the gradient at step n, and this re-introduces errors proportional to the previous gradient. It can be observed that the energy converges to the minimum exponentially with *time*. In the beginning the energy decrease is quite fast, but then it levels off and converges extremely slowly. In order to avoid this, each minimization step has to be carried out along a direction $\mathbf{d}^{(n)}$ that is *conjugated* to *all* the previous search directions, i.e. $\mathbf{d}^{(n)} \cdot \mathbf{G} \cdot \mathbf{d}^{(m)} = 0$, where $\mathbf{G}$ is the gradient operator. Algorithms that accomplish the above task are called *conjugate gradients* (CG). The above condition does not determine a unique CG algorithm, but a family of them, which differ in how the next search direction is chosen. One possible prescription is

$$\mathbf{d}^{(n)} = \mathbf{g}^{(n)} + \gamma^{(n)} \mathbf{d}^{(n-1)}, \tag{11.33}$$

with

$$\gamma^{(n)} = \frac{\mathbf{g}^{(n)} \cdot \mathbf{g}^{(n)}}{\mathbf{g}^{(n-1)} \cdot \mathbf{g}^{(n-1)}}, \tag{11.34}$$

where

$$\mathbf{g}^{(n)} = -\frac{\delta \Omega_{\text{KS}}\{\varphi_i^{(n)}(\mathbf{r})\}}{\delta \varphi_i^{(n)*}(\mathbf{r})}, \tag{11.35}$$

and $\gamma^{(1)} = 0$. Since each minimization step is independent of the previous ones, then the dimension of the search space is reduced by one at each iteration. In theory, a CG algorithm should reach the minimum of the target function in a number of steps equal to the dimension of the search space. In practice, however, the number of iterations can be significantly reduced from this value. It is interesting to notice that the Lanczos algorithm for matrix diagonalization described in Section 11.1 can also be viewed as a conjugated gradient minimization algorithm.

An alternative to the above relaxation dynamics is to perform an *annealing* procedure, which corresponds to a damped second order dynamics:

$$\mu \ddot{\varphi}_i(\mathbf{r}) + \eta \dot{\varphi}_i(\mathbf{r}) = -\frac{\delta \Omega_{\text{KS}}\{\varphi_i^{(n)}(\mathbf{r})\}}{\delta \varphi_i^{(n)*}(\mathbf{r})}, \tag{11.36}$$

with μ a mass-like coefficient and $\eta > 0$ a friction coefficient ensuring that the energy always decreases during the dynamical evolution of the orbitals. This second order set of differential equations can be integrated numerically using the following discretized algorithm:

$$\varphi_i^{(n+1)}(\mathbf{r}) = \frac{1}{1+\tilde{\Delta}(\eta,\mu)}\left[2\varphi_i^{(n)}(\mathbf{r}) - \left(1-\tilde{\Delta}(\eta,\mu)\right)\varphi_i^{(n-1)}(\mathbf{r})\right]$$

$$-\frac{1}{1+\tilde{\Delta}(\eta,\mu)}\frac{\Delta^2}{\mu}\left[\hat{\mathcal{H}}_{\mathrm{KS}}\,\varphi_i^{(n)}(\mathbf{r}) - \sum_{j=1}^{N}\Lambda_{ij}\varphi_j^{(n)}(\mathbf{r})\right] \quad (11.37)$$

with $\tilde{\Delta}(\eta,\mu) = \eta\Delta/2\mu$. The relaxation time for this frictional dynamics is $\tau_{\mathrm{ann}} = \mu/\eta$. It is easy to see that the steepest descent algorithm is recovered for $\tilde{\Delta}(\eta,\mu) = 1$, while $\tilde{\Delta}(\eta,\mu) = 0$ (no friction) corresponds to an undamped, conservative dynamics. This latter cannot be used for minimization purposes, but we shall see in the following section how it becomes useful in first-principles molecular dynamics simulations. Another possibility is to dynamically adjust the friction coefficient η so as to keep constant the *kinetic energy* of the orbitals, defined as $K_e = \mu\int|\dot{\varphi}_i(\mathbf{r})|^2 d\mathbf{r}$. This might be useful when the minimization process is difficult for a relaxation dynamics, e.g. because the landscape in the space of orbitals is too flat.

References

Alavi, A., Kohanoff, J., Parrinello, M., and Frenkel, D. (1995). Ab initio molecular dynamics with excited electrons. *Phys. Rev. Lett.* **73**, 2599–2602.

Anderson, D. G. (1965). Iterative procedures for non-linear integral equations. *Assoc. Comput. Mach.* **12**, 547–560.

Anderson, E., Bai, Z., Bischof, C., Blackford, S., Demmel, J., Dongarra, J., Croz, J. D., Greenbaum, A., Hammarling, S., McKenney, A., and Sorensen, D. (1999). *LAPACK Users' Guide*, 3rd edn. Philadelphia, PA, Society for Industrial and Applied Mathematics. www.netlib.org/lapack.

Blackford, L. S., Choi, J., Cleary, A., D'Azevedo, E., Demmel, J., Dhillon, I., Dongarra, J., Hammarling, S., Henry, G., Petitet, A., Stanley, K., Walker, D., and Whaley, R. C. (1997). *ScaLAPACK Users' Guide*. Philadelphia, PA, Society for Industrial and Applied Mathematics. www.netlib.org/scalapack.

Broyden, C. G. (1965). A class of methods for solving nonlinear simultaneous equations. *Math. Comput.* **19**, 577–593.

Car, R. and Parrinello, M. (1985). Unified approach for molecular dynamics and density-functional theory. *Phys. Rev. Lett.* **55**, 2471–2474.

CPMD Consortium (2004). *CPMD V3.9*. IBM Corp 1990-2004, MPI für Festkörperforschung Stuttgart. 1997-2001.

Cullum, J. K. and Willoughby, R. A. (2002). *Lanczos Algorithms for Large Symmetric Eigenvalue Computations, Vol. 1*. Philadelphia, Society for Industrial and Applied Mathematics.

Davidson, E. R. (1975). The iterative calculation of a few of the lowest eigenvalues and corresponding eigenvectors of large real-symmetric matrices. *J. Comput. Phys.* **17**, 87–94.

Hutter, J., Lüthi, H.-P., and Parrinello, M. (1994). Electronic structure optimisation in plane-wave-based density functional calculations by direct inversion in the iterative subspace. *Comput. Mater. Sci.* **2**, 244–248.

Johnson, D. D. (1988). Modified Broyden's method for accelerating convergence in self-consistent calculations. *Phys. Rev. B* **38**, 12807–12813.

Lanczos, C. (1950). An iteration method for the solution of the eigenvalue problem of linear differential and integral operators. *J. Res. Nat. Bur. Standards* **45**, 255–282.

Marx, D. and Hutter, J. (2000). *Ab-initio* molecular dynamics: theory and implementation. In *Modern Methods and Algorithms in Quantum Chemistry*, J. Grotendorst, ed. NIC, vol. 1. Jülich, John von Neumann Institute for Computing, 301–449.

Payne, M. C., Teter, M. P., Allan, D. C., Arias, T. A., and Joannopoulos, J. D. (1992). Iterative minimization techniques for *ab initio* total-energy calculations: molecular dynamics and conjugate gradients. *Rev. Mod. Phys.* **64**, 1045–1097.

Pollard, W. T. and Friesner, R. A. (1993). Efficient Fock matrix diagonalization by a Krylov-space method. *J. Chem. Phys.* **99**, 6742–6750.

Press, W. H., Teukolsky, S. A., Vetterling, W. T., and Flannery, B. P. (1992). *Numerical Recipes. The Art of Scientific Computing*. Cambridge, Cambridge University Press.

Pulay, P. (1980). Convergence acceleration of iterative sequences. The case of SCF iteration. *Chem. Phys. Lett.* **73**, 393–397.

12

First-principles molecular dynamics (Car–Parrinello)

At the end of Chapter 1 it was mentioned that the set of Newtonian equations of motion for the classical nuclear degrees of freedom,

$$M_I \frac{d^2 \mathbf{R}_I(t)}{dt^2} = -\frac{\partial}{\partial \mathbf{R}_I} \langle \Phi(\mathbf{R})|\hat{h}_e(\mathbf{R})|\Phi(\mathbf{R})\rangle - \frac{\partial V_{nn}(\mathbf{R})}{\partial \mathbf{R}_I}, \qquad (12.1)$$

where

$$\hat{h}_e(\mathbf{R}, \mathbf{r}) = -\frac{\hbar^2}{2m} \sum_{i=1}^{N} \nabla_i^2 + \frac{e^2}{2} \sum_{i=1}^{N} \sum_{j \neq i}^{N} \frac{1}{|\mathbf{r}_i - \mathbf{r}_j|} + \sum_{I=1}^{P} \sum_{i=1}^{N} v_{ext}(\mathbf{r}_i - \mathbf{R}_I) \qquad (12.2)$$

and

$$V_{nn}(\mathbf{R}) = \frac{e^2}{2} \sum_{I=1}^{P} \sum_{J \neq I}^{P} \frac{Z_I Z_J}{|\mathbf{R}_I - \mathbf{R}_J|}, \qquad (12.3)$$

can be integrated numerically to generate realistic physical trajectories in phase space. The fact that these trajectories are realistic is a consequence of the first-principles description of the acting forces, which is achieved at the expense of introducing the electronic component explicitly, within the adiabatic approximation. This avoids the bias that is necessarily introduced when the interatomic interactions are described through empirical potentials. Of course, the price is quite high, because within this scheme the electronic problem has to be solved every time step of the molecular dynamics (MD) integration, typically amounting to an increase in computational load of a factor of 1000 or more, with respect to classical simulations. Therefore, one must be very careful to analyze whether the problem under study really requires a first-principles description or not. As a rule of thumb, a first-principles description is necessary when the chemistry of the system plays an important role, e.g. when there is making and breaking of chemical bonds, changing environments, variable coordination, etc. If this is not the case, then better put the effort into constructing a classical force field, which can be obtained, e.g. by fitting the parameters of the potential to the results of a few suitably chosen first-principles calculations or MD simulations (Ercolessi and Adams, 1994). This allows for much faster and longer simulations of much larger samples, i.e. to a significant improvement in the statistics required to estimate thermodynamic quantities.

12.1 Density functional molecular dynamics

A feasible first-principles self-consistent MD approach can be obtained by solving the electronic problem for the ground state within the Kohn–Sham approach to density functional theory. In that case we have

$$\langle \Phi(\mathbf{R})|\hat{h}_e(\mathbf{R})|\Phi(\mathbf{R})\rangle = E_{KS}[\rho](\mathbf{R})$$

$$= T_R[\rho] + \frac{1}{2}\int \frac{\rho(\mathbf{r})\rho(\mathbf{r}')}{|\mathbf{r}-\mathbf{r}'|}d\mathbf{r}d\mathbf{r}' + E_{XC}[\rho]$$

$$+ \sum_{I=1}^{P}\int \rho(\mathbf{r})v_{ext}(\mathbf{r}-\mathbf{R}_I)d\mathbf{r} + \frac{1}{2}\sum_{I=1}^{P}\sum_{J\neq I}^{P}\frac{Z_I Z_J}{|\mathbf{R}_I-\mathbf{R}_J|}, \quad (12.4)$$

and the force on the nuclear coordinates is obtained by simple derivation, noting that only the last two terms introduce a non-vanishing contribution:

$$\mathbf{F}_I = -\frac{\partial E_{KS}[\rho](\mathbf{R})}{\partial \mathbf{R}_I}$$

$$= -\int \rho(\mathbf{r})\frac{\partial v_{ext}(\mathbf{r}-\mathbf{R}_I)}{\partial \mathbf{R}_I}d\mathbf{r} + \sum_{J\neq I}^{P}Z_I Z_J \frac{\mathbf{R}_I-\mathbf{R}_J}{|\mathbf{R}_I-\mathbf{R}_J|^3}. \quad (12.5)$$

Notice that for atom-centered truncated basis sets there is an additional force term due to the dependence of the basis functions on the nuclear coordinates (Pulay, 1969).

At this point, the straightforward procedure towards a computational scheme for a first principles molecular dynamics (FPMD) would be to keep the electronic subsystem *always* in the ground state compatible with the current nuclear configuration. Such a scheme can indeed be devised, but it has to be ensured that the electronic density is very well converged (in the sense of self-consistency) because, otherwise, a systematic perturbation (dragging) is being introduced into the nuclear dynamics (Remler and Madden, 1990). This kind of approach receives the name of *Born–Oppenheimer molecular dynamics* (BOMD), and is the approach embedded in most FPMD codes such as the plane wave-based VASP (Kresse and Furthmüller, 1996), PWSCF (Baroni *et al.*, 2005), and CASTEP (Segall *et al.*, 2002), and the local orbital-based codes SIESTA (Soler *et al.*, 2002), FP-LMTO (Methfessel, 1988), and some quantum chemistry codes (Estrín *et al.*, 1993).

Interestingly, these latter are very rarely prepared to run MD simulations (Parker *et al.*, 2003).

Nevertheless, even if the BOMD approach was theoretically proposed in the mid eighties (Bendt and Zunger, 1983), it was not the first one to be realized in practice. At that time the computational cost of a self-consistent electronic calculation was too high for the existing facilities, and researchers could not envisage a successful realization of the above plan, until Car and Parrinello (1985) introduced an alternative scheme for an FPMD which did not involve electronic self-consistency at every MD step. They were the first to show that FPMD was possible, and thus opened a completely new field in computational physics with an astonishing impact, not only in physics, but also in chemistry, materials science, and biochemistry.

Up to now we have used the terms first-principles and density functional as synonyms. It is important to remark that DFT is only one of the possible realizations of a first-principles calculation. One could also think of performing an FPMD simulation in which the electronic component is described using quantum chemistry methods, e.g., Hartree–Fock, MP2, or semiempirical (tight-binding) methods. The advantage of DFT is that its computational cost, at least within local or semi-local approximations like LDA or GGA, is significantly lower than Hartree–Fock, but it is accurate enough. In fact, Hartree–Fock molecular dynamics has been proposed and implemented to study the dynamics of a molecular system in an excited state, where DFT is not suitable (Hartke and Carter, 1992). To avoid confusion, we use the term density functional molecular dynamics (DFMD) for the class of DFT-based MD methods, BOMD for DFMD methods that diagonalize the electronic Hamiltonian at every time step, and Car–Parrinello molecular dynamics (CPMD) for methods that avoid the self-consistency step by allowing for a <u>fictitious dynamical evolution</u> of the single-particle orbitals.

12.2 The Car–Parrinello Lagrangian

As discussed in Section 11.3, the self-consistent solution of Kohn–Sham equations can also be formulated as a minimization problem in the manifold of the single-particle (orthogonal) orbitals of the non-interacting reference system, $\varphi_i(\mathbf{r})$. These can be considered as variational scalar fields that are represented numerically either on a discrete grid, in terms of PW coefficients in reciprocal space, by expanding them in atom-centered basis sets like STO, GTO, or numerical orbitals, or even in terms of the PW coefficients in the all-electron APW method (Soler and Williams, 1990). The self-consistent solution of Kohn–Sham equations, i.e. the multi-dimensional optimization of the Kohn–Sham energy functional with respect to the variational orbitals $\varphi_i(\mathbf{r})$, involves the optimization of a large number of parameters. Just as an example, imagine a system of eight Si atoms (four valence electrons) in a cubic box of side 10 bohr and an energy cutoff of 12 Ry. This implies that the number of Fourier coefficients is of the order of 5000 per orbital, thus amounting to about 90 000 variational coefficients for the 16 lowest-lying Kohn–Sham orbitals. Even if the Born–Oppenheimer surface is smooth, it is clear that optimizing so many parameters may be quite heavy from the computational point of view, especially taking into account that a tight convergence is required to have accurate forces on the nuclear degrees of freedom that are consistent with the energy. This is, in fact, a crucial condition to have energy conservation during MD simulations.

An alternative to this approach can be devised if the electronic density can be kept sufficiently close to the adiabatic density during the MD simulation. This plan can be accomplished by introducing a conservative, second order *fictitious* dynamics of the Kohn–Sham orbitals, as proposed and implemented by Car and Parrinello (1985). The CPMD scheme can be described as a dynamical system represented by the following Lagrangian:

$$\mathcal{L}_{\mathrm{CP}} = \frac{1}{2}\sum_{I=1}^{P} M_I \dot{\mathbf{R}}_I^2 + \mu \sum_{i=1}^{N} f_i \int |\dot{\varphi}_i(\mathbf{r})|^2 d\mathbf{r}$$

$$- E_{\mathrm{KS}}[\varphi_i(\mathbf{r})](\mathbf{R}) + \sum_{i=1}^{N} f_i \sum_{j=1}^{N} \Lambda_{ij} \left(\int \varphi_i^*(\mathbf{r})\varphi_j(\mathbf{r}) d\mathbf{r} - \delta_{ij} \right), \quad (12.6)$$

where the Kohn–Sham energy is given by

$$E_{\mathrm{KS}}[\varphi_i(\mathbf{r})](\mathbf{R}) = \sum_{i=1}^{N} f_i \int \varphi_i^*(\mathbf{r}) \hat{h}_1(\mathbf{R},\mathbf{r}) \varphi_i(\mathbf{r}) d\mathbf{r}$$

$$+ \frac{1}{2} \int \int \frac{\rho(\mathbf{r})\rho(\mathbf{r}')}{|\mathbf{r}-\mathbf{r}'|} d\mathbf{r} d\mathbf{r}' + E_{\mathrm{XC}}[\rho]$$

$$+ \frac{1}{2} \sum_{I=1}^{P} \sum_{J \neq I}^{P} \frac{Z_I Z_J}{|\mathbf{R}_I - \mathbf{R}_J|}. \quad (12.7)$$

The operator $\hat{h}_1$ in (12.7) is the one-electron Hamiltonian

$$\hat{h}_1(\mathbf{R},\mathbf{r}) = -\frac{\hbar^2}{2m}\nabla^2 + \sum_{I=1}^{P} v_{\mathrm{ext}}(\mathbf{r}-\mathbf{R}_I), \quad (12.8)$$

and the electronic density in terms of the dynamical orbitals is written

$$\rho(\mathbf{r}) = \sum_{i=1}^{N} f_i \int \varphi_i^*(\mathbf{r}) \varphi_i(\mathbf{r}) d\mathbf{r}. \quad (12.9)$$

The first term in (12.6) is the nuclear kinetic energy and the third term is the first-principles potential represented by the Kohn–Sham energy functional. The coefficients f_i are occupation numbers associated with the orbitals $\varphi_i(\mathbf{r})$. In the case of insulators or semiconductors they assume the values $f_i = 1$ for $i \leq N$ and $f_i = 0$ for $i > N$, where N is the number of electrons. For spin unpolarized systems the sums run up to the number of occupied states ($N/2$), and the occupation numbers become $f_i = 2$ for $i \leq N/2$. In the case of metals $f_i(\varepsilon) = (1 + e^{(\varepsilon - \mu)/k_B T_e})^{-1}$, where T_e is an electronic *temperature*. This latter is included not for fundamental reasons, since usually electronic Fermi temperatures are much higher than simulation temperatures. It is there to mimic Brillouin zone integration and to improve the convergence of the self-consistent procedure due to degeneracies at the Fermi level. If the electronic temperature is included as a physical variable, then the assumption implied by the use of the above expression is that energy exchanges between the electrons are so fast that the Fermi–Dirac equilibrium distribution is always verified (Mermin, 1965). This dynamics does not really take into account physical non-adiabatic processes.

In order to represent electronic densities that arise from a Slater determinant, the orbitals $\varphi_i(\mathbf{r})$, which from now on we shall call *dynamical Kohn–Sham orbitals* (DKSO), must be orthonormal. This is the motivation for the last term in the Lagrangian, which ensures the orthonormality of the DKSO at every step along the MD simulation. The Lagrange multipliers Λ_{ij} are determined in such a way as to verify this condition. Orthonormalization algorithms are presented in Section 12.4. In general, the DKSOs are different from the true Kohn–Sham orbitals that minimize E_{KS}.

Together with the orthonormalization constraints, the second term represents the most important innovation of Car and Parrinello. It corresponds to a *fictitious* kinetic energy

$$K_e[\varphi_i(\mathbf{r})] = \mu \sum_{i=1}^{N} f_i \int |\dot{\varphi}_i(\mathbf{r})|^2 d\mathbf{r} \tag{12.10}$$

associated with the time evolution of the DKSO (frequently also called *electronic degrees of freedom*, or simply *the electrons*). This does not imply that the electrons follow a real dynamical evolution. K_e is simply a convenient term that allows for a dynamical evolution of the orbitals independent of, although coupled to, that of the nuclear degrees of freedom. It is easy to visualize that the BOMD scheme mentioned above is recovered by neglecting the proper dynamics of the orbitals, i.e. by setting $\mu = 0$ in the second term in the CP Lagrangian.

12.3 The Car–Parrinello equations of motion

We then have a Lagrangian that depends on the nuclear coordinates $\mathbf{R}$ and on the electronic orbitals $\varphi_i(\mathbf{r})$. The Euler–Lagrange equations are obtained in the usual way of variational calculus:

$$\frac{d}{dt}\left(\frac{\partial \mathcal{L}_{CP}}{\partial \dot{\mathbf{R}}_I}\right) = -\frac{\partial \mathcal{L}_{CP}}{\partial \mathbf{R}_I}, \tag{12.11}$$

$$\frac{d}{dt}\left(\frac{\delta \mathcal{L}_{CP}}{\delta \dot{\varphi}_i^*(\mathbf{r})}\right) = -\frac{\delta \mathcal{L}_{CP}}{\delta \varphi_i^*(\mathbf{r})}. \tag{12.12}$$

The second equation involves functional derivatives because the orbitals are continuous scalar fields. In practice, however, these fields are represented in a basis (e.g. on a discrete real-space mesh or in terms of plane wave components), and the concept of functional derivation reduces to the partial derivation with respect to the components of the field in the basis. By performing the operations indicated above, we arrive at the Car–Parrinello equations of motion:

$$M_I \ddot{\mathbf{R}}_I = -\frac{\partial E_{KS}[\varphi_i(\mathbf{r})](\mathbf{R})}{\partial \mathbf{R}_I}, \tag{12.13}$$

$$\mu \ddot{\varphi}_i(\mathbf{r}, t) = -\frac{\delta E_{KS}[\varphi_i(\mathbf{r})](\mathbf{R})}{\delta \varphi_i^*(r)} + \sum_{j=1}^{N} \Lambda_{ij} \varphi_j(\mathbf{r}, t)$$

$$= -\hat{\mathcal{H}}_{KS} \varphi_i(\mathbf{r}, t) + \sum_{j=1}^{N} \Lambda_{ij} \varphi_j(\mathbf{r}, t), \tag{12.14}$$

where

$$\hat{\mathcal{H}}_{KS} = -\frac{\hbar^2}{2m}\nabla^2 + v_{ext}(\mathbf{r}, \mathbf{R}) + \int \frac{\rho(\mathbf{r}')}{|\mathbf{r} - \mathbf{r}'|} d\mathbf{r}' + \mu_{XC}[\rho] \tag{12.15}$$

is the Kohn–Sham Hamiltonian and Λ_{ij} are the Lagrange multipliers that ensure the orthonormality of the DKSO. Figure 12.1 shows schematically the various steps in a CPMD simulation.

If, for the time being, we ignore the left-hand side in Eq. (12.14), i.e. we focus on the stationary solution of this differential equation (vanishing second derivative), we recover the usual Kohn–Sham equations:

$$\hat{\mathcal{H}}_{KS} \varphi_i(\mathbf{r}) = \sum_{j=1}^{N} \Lambda_{ij} \varphi_j(\mathbf{r}). \tag{12.16}$$

The symmetric matrix Λ_{ij} can be diagonalized by means of a unitary transformation $\mathbf{U}$, because $\hat{\mathcal{H}}_{KS}$ is a hermitian operator:

$$\mathbf{U}^{-1} \Lambda \mathbf{U} = \epsilon \quad \text{and} \quad \Psi = \mathbf{U}^{-1} \varphi \tag{12.17}$$

12.3 The Car–Parrinello equations of motion

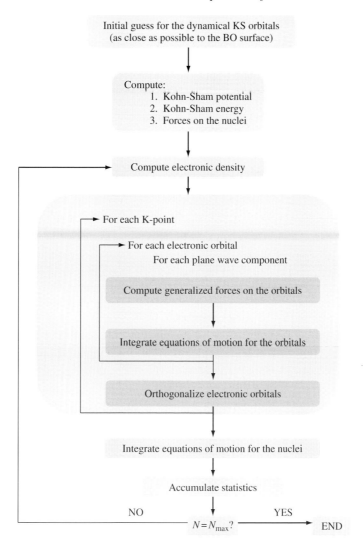

Fig. 12.1 Flow diagram of a first-principles molecular dynamics (CPMD) simulation. The statistics accumulated permits us to calculate thermodynamic quantities as time averages, assuming that the system behaves ergodically.

are the *canonical* Kohn–Sham orbitals, in analogy with the canonical orbitals in HF theory. Therefore, the Kohn–Sham equations can be re-written as

$$\left(\mathbf{U}^{-1}\mathbf{H}_{\text{KS}}\mathbf{U}\right)\left(\mathbf{U}^{-1}\varphi\right) = \tilde{\mathbf{H}}_{\text{KS}}\Psi = \epsilon\Psi = \left(\mathbf{U}^{-1}\Lambda\mathbf{U}\right)\left(\mathbf{U}^{-1}\Psi\right), \tag{12.18}$$

i.e.

$$\tilde{\mathbf{H}}_{\text{KS}}\Psi_i(\mathbf{r}) = \epsilon_i\Psi_i(\mathbf{r}), \tag{12.19}$$

where $\tilde{\mathbf{H}}_{KS}$ is the Kohn–Sham Hamiltonian matrix in the transformed basis. Therefore, the eigenvalues of matrix Λ can be interpreted as the single-particle eigenvalues associated with the Kohn–Sham Hamiltonian.

The above is not valid during the dynamical CP evolution, where the DKSOs are not solutions of the Kohn–Sham equations. Even if initially the DKSOs are *prepared* to be in the Born–Oppenheimer surface, i.e. they are solutions of the Kohn–Sham equations, during the nuclear evolution they will acquire some kinetic energy and will depart slightly from the ground state. The secret of a stable and meaningful CPMD simulation is that the DKSOs remain reasonably close to the Born–Oppenheimer surface during their dynamical evolution. If they do not, then the forces on the nuclear variables are poor and the dynamical evolution is unrealistic.

12.4 Orthonormalization

The exact integration of the above equations of motion for the orbitals is enough to ensure their mutual orthogonality, provided that initially they were orthogonal. The discretized numerical integration, however, introduces errors that induce a degradation of the orthogonality condition. Therefore, it is necessary to orthogonalize the orbitals every step along the iterative procedure. In the case of energy minimization, any orthonormalization algorithm, like Gram–Schmidt, can be used because there is no energy conservation constraint to respect. In a conservative second order dynamics, however, there is only one orthogonalization procedure that is consistent with the equations of motion, and it is given by the following expression for the Lagrange multipliers Λ_{ij} (Pastore *et al.*, 1991):

$$\Lambda_{ij} = \frac{(f_i + f_j)}{2} \int \varphi_j^*(\mathbf{r}) \hat{\mathcal{H}}_{KS} \varphi_i(\mathbf{r}) d\mathbf{r} - \mu \int \dot{\varphi}_j^*(\mathbf{r}) \dot{\varphi}_i(\mathbf{r}) d\mathbf{r}. \qquad (12.20)$$

The reason why there is only one correct expression is that, even if the orbitals are defined to within a unitary transformation (the electronic density is the only physically relevant quantity), the Lagrangian associated with the second order equations of motion includes a kinetic-like term that is not invariant upon unitary transformations.

An algorithm to enforce this type of holonomic constraint has been proposed in the context of geometrically constrained classical molecular dynamics under the name of SHAKE (Ryckaert *et al.*, 1977), and adapted to the orthogonalization of the dynamical orbitals (Car and Parrinello, 1988). The procedure consists of two steps: first, the equations of motion (12.12) are integrated according to a second order numerical integration algorithm such as that of Verlet (1967), but

12.4 Orthonormalization

without including the constraints. This leads to the following non-orthogonal, unnormalized updated orbitals:

$$\bar{\varphi}_i^{(n+1)}(\mathbf{r}) = 2\varphi_i^{(n)}(\mathbf{r}) - \varphi_i^{(n-1)}(\mathbf{r}) - \frac{\Delta^2}{\mu}\hat{\mathcal{H}}_{KS}\varphi_i^{(n)}(\mathbf{r}). \quad (12.21)$$

Then, a corrective action is applied to these orbitals in the following way:

$$\varphi_i^{(n+1)} = \bar{\varphi}_i^{(n+1)} + \sum_{j=1}^{N} x_{ij}^* \varphi_j^{(n)}, \quad (12.22)$$

where $\varphi_i^{(n)}$ are the orthonormal orbitals corresponding to the preceding iteration. The matrix $x_{ij}^* = \Delta^2 \Lambda_{ij}/\mu$ is determined by requesting that the updated orbitals remain orthonormal. This orthonormality condition leads to the following expression:

$$A_{ik} + \sum_{j=1}^{N} x_{ij} B_{jk} + \sum_{j=1}^{N} x_{kj}^* B_{ji}^* + \sum_{j=1}^{N} x_{ij} x_{kj}^* = \delta_{ik} \quad (12.23)$$

with

$$A_{ij} = \langle \bar{\varphi}_i^{(n+1)} | \bar{\varphi}_j^{(n+1)} \rangle, \quad B_{ij} = \langle \varphi_i^{(n)} | \bar{\varphi}_j^{(n+1)} \rangle. \quad (12.24)$$

In matrix notation, the above equation (12.23) reads

$$\mathbf{1} - \mathbf{A} = \mathbf{X}\mathbf{B} + \mathbf{B}^\dagger \mathbf{X}^\dagger + \mathbf{X}\mathbf{X}^\dagger, \quad (12.25)$$

where the dagger indicates the hermitian conjugation operation. The solution to this equation can be obtained by the following iterative procedure. The initial term in the recurrence is set to the zero-order approximation

$$\mathbf{X}_0 = \frac{1}{2}(\mathbf{1} - \mathbf{A}), \quad (12.26)$$

and then the iterative cycle proceeds according to the recursive formula:

$$\mathbf{X}_{k+1} = \frac{1}{2}\left[(\mathbf{1} - \mathbf{A}) + \mathbf{X}_k(\mathbf{1} - \mathbf{B}) + (\mathbf{1} - \mathbf{B}^\dagger)\mathbf{X}_k^\dagger - \mathbf{X}_k \mathbf{X}_k^\dagger\right]. \quad (12.27)$$

Actually, since the matrix of Lagrange multipliers is symmetric and real, $\mathbf{X}_k^\dagger = \mathbf{X}_k$. This algorithm is very stable and it typically converges in less than ten steps, unless the orbitals change too much from one iteration to the next, in which case it may not be able to converge at all. The longer the time step, the larger the number of iterations required to converge the iterative cycle. This imposes a restriction to the integration time step when using iterative orthogonalization, although in practice this is very rarely the limiting factor. Once the matrix of the constraints $\mathbf{X}$ has been determined, the new orthonormal orbitals are calculated according to (12.22).

It is important to remark that this orthonormalization procedure is not limited to the integration of second order equations of motion for the electronic orbitals, but it can also be applied to the evolution obtained from a first order relaxation dynamics such as steepest descent. In fact, the use of iterative orthonormalization accelerates convergence compared to other algorithms such as Gram–Schmidt.

12.5 Pre-conditioning

The maximum integration time step Δ of the equations of motion for the orbitals is determined by the fastest frequency present in the spectrum of the dynamical orbitals, which is roughly determined by the relation:

$$\omega^2_{i,\mathbf{G},\mathbf{G}'} \propto \frac{\delta^2 E_{\text{KS}}}{\delta C_i^*(\mathbf{G}) \delta C_i(\mathbf{G}')}, \qquad (12.28)$$

where $C_i(\mathbf{G})$ are the Fourier coefficients of the orbitals. In general, this tensor can be quite complicated to calculate, but it is easy to realize that the dominant contribution for large G vectors arises from the kinetic energy

$$\frac{\delta^2 T[\{C_i\}]}{\delta C_i^*(\mathbf{G}) \delta C_i(\mathbf{G}')} \approx \frac{G^2}{2\Omega} \delta_{\mathbf{G},\mathbf{G}'}, \qquad (12.29)$$

while the contribution from the potential energy is dominant for small wave vectors, e.g. for the Hartree potential it is given by

$$\frac{\delta^2 E_{\text{H}}[\{C_i\}]}{\delta C_i^*(\mathbf{G}') \delta C_i(\mathbf{G}'')} = \sum_{\mathbf{G}} \frac{4\pi}{G^2} C_i(\mathbf{G}'-\mathbf{G}) C_i^*(\mathbf{G}''+\mathbf{G}). \qquad (12.30)$$

The idea of pre-conditioning the integration of the equations of motion is based on the observation that for large wave vectors the eigenfrequencies are basically given by the kinetic kernel (12.29), which assumes a very simple form. In that limit, the force term in the equations of motion becomes

$$\frac{\Delta^2}{\mu} \hat{\mathcal{H}}_{\text{KS}} C_i(\mathbf{G}) \propto \frac{\Delta^2 G^2}{\mu} C_i(\mathbf{G}) \qquad (12.31)$$

and then a larger time step Δ_{prec} can be used if a G-dependent mass-like coefficient $\mu_{\text{prec}}(G) \propto G^2$ is assigned to the large G components of the orbitals. In fact, this procedure renormalizes down the frequency spectrum of the dynamical orbitals as $\tilde{\omega}_{i,\mathbf{G}} \propto \omega_{i,\mathbf{G}}/G$. This approach is valid as long as the kinetic kernel is more important than the potential one. When the potential terms become relevant, a different pre-conditioning factor should be adopted. Although sophisticated pre-conditioning schemes that use the G-dependence of the local potential can be devised, a simple algorithm consisting of using a constant mass-like term up to some cutoff vector G_{mass} and a G-dependent mass for larger values of G, i.e.

$$\mu_{\text{prec}}(G) = \begin{cases} \mu_0, & G \leq G_{\text{mass}}, \\ \mu_0 (G/G_{\text{mass}})^2 & G > G_{\text{mass}}, \end{cases} \quad (12.32)$$

already gives very satisfactory results at a negligible additional cost. In general, the value of G_{mass} depends on the particular system under study. Systems characterized by a free electron-like behavior will be able to use a small G_{mass}, while more tightly bound electrons will require a larger G_{mass}. Typical values of $E_{\text{mass}} = G_{\text{mass}}^2/2$ are between 1 and 4 Ry. The optimal value of E_{mass}, i.e. the one that maximizes the time step Δ_{prec}, can be rapidly obtained by a few trials.

12.6 Performance of CPMD

It has been shown that the Kohn–Sham orbitals minimize the energy associated with the Lagrangian $\mathcal{L}_{\text{CP}}$ given by (12.6), at fixed nuclear configuration. These orbitals define the electronic density univocally. If the integration of the second order equations of motion (12.12) is started from the minimum in the Born–Oppenheimer surface (or close to the minimum), then the time evolution of the dynamical orbitals will consist of small oscillations around the Born–Oppenheimer surface.

In general, the coupled dynamics of nuclei and orbitals is different from the true nuclear dynamics, where the electrons are in the instantaneous ground state, as realized in the BOMD scheme. Since the Lagrangian includes the dynamical evolution of the electronic orbitals, the conserved quantity in a conservative CPMD dynamics is $E_{\text{cons}} = K_e + K_n + E_{\text{KS}}$, with K_e the fictitious electronic kinetic energy given by (12.10) and $K_n = 1/2 \sum_I M_I \dot{\mathbf{R}}_I^2$ the true kinetic energy of the nuclear degrees of freedom. A necessary condition for the electronic orbitals to perform small oscillations around the instantaneous ground state is, then, that $K_e \ll K_n$. This, however, is not sufficient, because the system can depart from the ground state while still keeping a small K_e at the expense of transferring energy to the nuclear degrees of freedom. Therefore, a second condition for a meaningful CPMD simulation is that energy exchanges between nuclear and electronic degrees of freedom are kept to a minimum.

The dynamics of the electronic orbitals contains two types of component: one is due to their own dynamics, which is controlled by the *fictitious mass* μ, and the other arises through E_{KS} as a consequence of the dragging force exerted by the nuclei in their motion. The latter guides the average trajectory of the orbitals, while the former introduces oscillations on top of the main component. This behavior is shown in the lower panel in Fig. 12.2(a), where the kinetic energy of the orbitals for a vibrating N_2 molecule is plotted as a function of time. In the upper panel of Fig. 12.2(a) is shown the time evolution of the *ab initio* potential given by the Kohn–Sham energy E_{KS} (thin dashed line). The sum of this latter and the nuclear kinetic energy, $E_{tot} = E_{KS} + K_n$, is the quantity that would be conserved in a BOMD simulation, where the electronic orbitals are always on the Born–Oppenheimer surface (solid line). The Lagrangian constant of motion, i.e. the conserved quantity in a CPMD simulation, is $E_{cons} = E_{tot} + K_e$ (thick dashed line). The amplitude of the oscillations has been exaggerated for the sake of visualization. In practice it is much smaller than what is shown here.

On the one hand, the fictitious mass μ controls the energy transfer between orbitals and nuclei, which is proportional to $\mu^{-1/2}$. This unphysical transfer appears because the DKSOs are now treated as dynamical variables, exactly as the nuclear coordinates, and it is known that degrees of freedom of similar

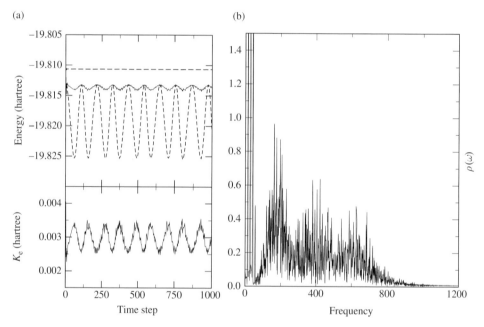

Fig. 12.2 (a) Lower panel: the fictitious kinetic energy K_e of the dynamical Kohn–Sham orbitals for a vibrating nitrogen dimer, obtained using the CPMD pseudopotential plane wave code. Norm-conserving pseudopotentials were used, and the PW cutoff was set to 60 Ry. The N_2 molecule was placed in a cubic cell of 10 bohr side. Upper panel: the potential energy E_{KS} (thin dashed line), the total energy $E_{tot} = E_{KS} + K_n$ (solid line), and the Lagrangian constant of motion $E_{cons} = E_{KS} + K_n + K_e$ (thick dashed line). (b) Frequency spectrum of K_e obtained by Fourier transformation. A fictitious mass $\mu = 400$ a.u. and a time step $\Delta = 3$ a.u. have been used.

frequencies tend to mix together. On the other hand, μ also controls the integration time step, which is proportional to $\mu^{1/2}$. A compromise is then required to have the time scales of the two motions well separated in order to keep the energy transfer at reasonably low values, while preserving an integration time step sufficiently large for the dynamics to sample the interesting time scales of the problem. A frequency analysis of the above time evolution is shown in Fig. 12.2(b), where the frequency of the dragging nuclear motion (the high peak on the far left) is observed to be well below the broad band of proper frequencies of the electronic degrees of freedom.

A careful analysis of the dynamical behavior of the CP equations of motion has been carried out by Pastore *et al.* (1991), who analyzed different model situations. The conclusion was that the CP dynamics is stable, unless one of the following situations occurs.

(i) The energy gap between occupied and empty single-particle states opens and closes periodically as a consequence of the nuclear evolution. In that case, K_e receives a *kick* every time the gap closes, and the magnitude of the oscillations increases at regular periods. This happens, e.g., in some molecules where the HOMO and the LUMO are close, and the energetic ordering changes with the interatomic distance.

(ii) The energy gap is sufficiently small that the frequencies associated with nuclear vibrations and electronic variables are within the same range. Then, what happens is a continuous energy transfer between the two sets of degrees of freedom until both are equilibrated at the same temperature. Sometimes this behavior can be cured by decreasing the fictitious mass μ. In other difficult cases, such as metallic systems or open-shell molecules, a smearing of the Fermi surface or the introduction of fractional occupation can help solve the problem, but the ultimate solution is BOMD. Interestingly, a stable CPMD dynamics can be obtained by evolving the whole electronic shell with fractional occupation up to the next significant gap in the spectrum (Kohanoff and Hansen, 1996; Marzari *et al.*, 1997).

(iii) Two single-particle energy levels cross, one becoming occupied and the other empty. In this case K_e remains small until the level crossing, and then it starts acquiring energy at the expense of the nuclear motion, which becomes damped. This type of behavior may be observed during the migration of a defect in a bulk solid.

Perturbation theory in classical mechanics indicates that motions with two different types of frequency appear when the system is displaced away from the Born–Oppenheimer surface (Pastore *et al.*, 1991):

$$\omega_{ij}^{(1)} = \sqrt{\frac{f_j(\varepsilon_i^\star - \varepsilon_j)}{\mu}}, \qquad (12.33)$$

$$\omega_{ij}^{(2)} = \sqrt{\frac{(f_j - f_i)(\varepsilon_i - \varepsilon_j)}{2\mu}}, \qquad (12.34)$$

where ε_i^* are the energies of the empty single-particle eigenstates. The most important electronic frequency is the lowest one, because it mixes more efficiently with the nuclear frequencies. The condition to avoid the mixing of the two sets of degrees of freedom is that $\Omega_{max} \ll \sqrt{2E_g/\mu}$, where Ω_{max} is the highest nuclear frequency and $E_g = \varepsilon_i^*(\min) - \varepsilon_j(\max)$ is the gap in the single-particle spectrum. If the above relation is not satisfied, one can always decrease μ in order to reduce the energy transfer between electronic and nuclear variables, at the expense of reducing the integration time step.

If there is energy transfer between the two sets of degrees of freedom, an alternative to decreasing μ is to couple the two sets of variables to separate thermal reservoirs, e.g. using a Nosè thermostating procedure (Nosè, 1984). The thermostats are set to maintain, individually, the kinetic energies of the electronic and nuclear components fluctuating close to a pre-fixed requested value, by exchanging energy with an additional dynamical variable that describes the thermal bath. In this way, a two-temperature regime can be maintained for long integration times (Blöchl and Parrinello, 1992). This procedure, however, has to be carefully monitored because there is no guarantee that the thermostat will keep the electronic orbitals close to their ground state in the Born–Oppenheimer surface. They can actually drift away, while the fluctuations are kept under control by the thermostat.

By comparing the forces on the nuclear degrees of freedom computed according to the CPMD prescription with the exact BOMD forces obtained by self-consistent minimization, it can be observed that the CPMD forces exhibit a high frequency component due to the oscillations of the electronic orbitals, and a second, lower frequency related to the nuclear oscillations. The BOMD forces do not exhibit the first component. It can be shown that in a successful CPMD simulation, the departures of the CPMD forces from the BOMD ones are averaged out during the time evolution, i.e. the dynamical force oscillates around the correct BOMD force. This is the secret of the success of the Car–Parrinello method.

It has been mentioned that the nuclei drag the electronic orbitals along their motion. In fact, the opposite effect also takes place, namely that the coupling of electronic and nuclear variables has the effect of damping the nuclear dynamics. This can be thought of as if the nuclei got *dressed* with the mass of the dynamical orbitals. In the atomic limit this mass renormalization effect can be estimated as $\tilde{M} = M + 4\mu K_e/3$ (Blöchl and Parrinello, 1992), or $\tilde{\omega} = \omega(1 + 4\mu K_e/3M)^{1/2}$, where $\tilde{\omega}$ is the renormalized frequency and K_e is the kinetic energy of the electronic orbitals per atom, expressed in hartree. This correction can be important in the case of light atoms (H, He, and first row elements). For instance, in carbon compounds this correction can amount to 6% of the calculated frequencies (Kohanoff et al., 1992). It can also have a substantial effect in thermodynamic properties such as melting points, provided that the electronic density is dragged non-rigidly. These issues have been thoroughly analyzed by Tangney and Scandolo (2002).

Reviews on the Car–Parrinello method and recent developments can be found in several references (Remler and Madden, 1990; Galli and Parrinello, 1991; Marx and Hutter, 2000).

References

Baroni, S., de Gironcoli, S., dal Corso, A., and Giannozzi, P. (2005). Plane-wave self-consistent field. www.pwscf.org.

Bendt, P. and Zunger, A. (1983). Simultaneous relaxation of nuclear geometries and electronic charge densities in electronic structure theories. *Phys. Rev. Lett.* **50**, 1684–1688.

Blöchl, P. and Parrinello, M. (1992). Adiabaticity in first-principles molecular dynamics. *Phys. Rev. B* **45**, 9413–9416.

Car, R. and Parrinello, M. (1985). Unified approach for molecular dynamics and density-functional theory. *Phys. Rev. Lett.* **55**, 2471–2474.

Car, R. and Parrinello, M. (1988). The unified approach for molecular dynamics and density functional theory. In *Proceedings of NATO Workshop: Simple Molecular Systems at Very High Density*, A. Polian, P. Loubeyre, and N. Boccara, eds. NATO ASI. New York, Plenum Press, 455–476.

Ercolessi, F. and Adams, J. (1994). Interatomic potentials from 1st-principles calculations – the force-matching method. *Europhys. Lett.* **26**, 583–588.

Estrín, D. A., Corongiu, G., and Clementi, E. (1993). Structure and dynamics of molecular systems using density functional theory. In *METECC, Methods and Techniques in Computational Chemistry*. Cagliari, Stef, Chapter 12, 541–567.

Galli, G. and Parrinello, M. (1991). Ab initio molecular dynamics: principles and practical implementation. In *Computer Simulations in Material Science*, M. Meyer and V. Pontikis, eds. Dordrecht, Kluwer, 283–304.

Hartke, B. and Carter, E. A. (1992). Spin eigenstate-dependent Hartree–Fock molecular dynamics. *Chem. Phys. Lett.* **189**, 358–362.

Kohanoff, J. and Hansen, J.-P. (1996). Statistical properties of the dense hydrogen plasma: an *ab initio* molecular dynamics investigation. *Phys. Rev. E* **54**, 768–781.

Kohanoff, J., Andreoni, W., and Parrinello, M. (1992). Zero-point-motion effects on the structure of C_{60}. *Phys. Rev. B* **46**, 4371–4373.

Kresse, G. and Furthmüller, J. (1996). Efficient iterative schemes for *ab initio* total-energy calculations using a plane-wave basis set. *Phys. Rev. B* **54**, 11169–11186. cms.mpi.univie.ac.at/vasp.

Marx, D. and Hutter, J. (2000). *Ab-initio* molecular dynamics: theory and implementation. In *Modern Methods and Algorithms in Quantum Chemistry*, J. Grotendorst, ed. NIC, vol. 1. Jülich, John von Neumann Institute for Computing, 301–449.

Marzari, N., Vanderbilt, D., and Payne, M. C. (1997). Ensemble density-functional theory for *ab initio* molecular dynamics of metals and finite-temperature insulators. *Phys. Rev. Lett.* **79**, 1337–1340.

Mermin, N. D. (1965). Thermal properties of the inhomogeneous electron gas. *Phys. Rev.* **137**, A1441–A1443.

Methfessel, M. (1988). Elastic constants and phonon frequencies of Si calculated by a fast full-potential linear-muffin-tin-orbital method. *Phys. Rev. B* **38**, 1537–1540.

Nosè, S. (1984). A molecular dynamics method for simulations in the canonical ensemble. *Mol. Phys.* **52**, 255–268.

Parker, C. L., Ventura, O. N., Burt, S. K., and Cachau, R. E. (2003). DYNGA: a general purpose QM-MM-MD program. I. Application to water. *Mol. Phys.* **101**, 2659–2668.

Pastore, G., Smargiassi, E., and Buda, F. (1991). Theory of *ab initio* molecular-dynamics calculations. *Phys. Rev. A* **44**, 6334–6347.

Pulay, P. (1969). *Ab initio* calculation of force constants and equilibrium geometries in polyatomic molecules. I. Theory. *Mol. Phys.* **17**, 197–204.

Remler, D. K. and Madden, P. A. (1990). Molecular dynamics without effective potentials via the Car–Parrinello approach. *Mol. Phys.* **70**, 921–966.

Ryckaert, J. P., Ciccotti, G., and Berendsen, H. J. C. (1977). Numerical integration of the Cartesian equations of motion of a system with constraints: molecular dynamics of n-alkanes. *J. Comput. Phys.* **23**, 327–341.

Segall, M. D., Lindan, P. J. D., Pickard, C. J. *et al.* (2002). First-principles simulation: ideas, illustrations and the CASTEP code. *J. Phys. Condens. Matt.* **14**, 2717–2744.

Soler, J. M. and Williams, A. R. (1990). Augmented-plane-wave forces. *Phys. Rev. B* **42**, 9728–9731.

Soler, J. M., Artacho, E., Gale, J. D. *et al.* (2002). The SIESTA method for *ab initio* order-N materials simulation. *J. Phys. Condens. Matter* **14**, 2745–2780. www.siesta.org.

Tangney, P. and Scandolo, S. (2002). How well do Car–Parrinello simulations reproduce the Born–Oppenheimer surface? Theory and examples. *J. Chem. Phys.* **116**, 14–24.

Verlet, L. (1967). Computer "experiments" on classical fluids. I. Thermodynamical properties of Lennard-Jones molecules. *Phys. Rev.* **159**, 98–103.

Index

Note: Main references are given in bold.

ab initio molecular dynamics (*see* FPMD)
adiabatic
   approximation, **6–10**, 11
      molecules, 6–7
      solids, 7, 10
   basis, 9
   connection, **66–67**
   dynamics, 9
   electronic eigenstates, 8
all-electron methods, **125–126**
   augmentation spheres, 126
   augmented partial waves, 126
      linearization, 126
   matching conditions, 126
anharmonicity, 12
aperiodic systems, **139–140**
   long-range electrostatics, 139
   modified Poisson equation, 139–140
   neutral, charged and polar systems, 139
   vacuum region, 139
APW method, 143, 179, **205–206**, **222–223**
   basis functions, **205**
   logarithmic derivatives, 206
   matching conditions, 205
   matrix elements, **223**
   scattering phase shifts, 206
ASA, **221–222**
   canonical bands, 222
atom-centered basis sets (*see* local orbital methods)
atomic calculation
   all-electron, 154
   relativistic corrections, **156–157**
      Dirac's equation, **156**
      minor and major components, 156
      spin–orbit coupling, 156–157
atomic orbitals (AO), 189, **191–192**
   Laplace equation, 192
   Racah normalization, 192
   real spherical harmonics, 192
   spherical harmonics, 192

atomic pseudopotentials (*see* pseudopotentials), **143–176**
atomic sphere approximation (*see* ASA)
augmented plane wave method (*see* APW)
average density approximation (ADA), 94

B3LYP functional, 96
band energy, **276**, 296
basis functions (or orbitals), 125, 178
   augmentation, 204
      interstitials, 205
   energy-dependent, 125, 178
   energy-independent, 125, 179, 209–211
   orthogonality, 190
   orthogonalization (Löwdin), 190
   PBC-adapted, 181
   periodic systems (solids), 181–182
   screening transformation (LMTO), 190
   shape, 191
basis set, 38, **178–213**
   Λ-functions, one-range expansion, 211
   atom-centered, 188, **189–202**
      atomic orbitals (AO, *see* atomic orbitals)
      molecular orbitals (MO, *see* molecular sorbitals)
      size, 40, 126, 178, 185, **198–199**
         correlation-consistent (cc-pVDZ, aug-cc-pVTZ), 199
         diffuse functions (+), 199
         multiple-ζ, **198**, 211
         polarization functions ($*$), 198–199
         split valence, 198
         tempered (even- and well-), 199
   atomic orbitals, 193
      fireball (*see also* SIESTA), 193–194
         localization radius and energy shift, 193
         optimization, 194
         pseudo-atomic orbitals (PAO), 193
         strict and soft localization, 193–194
   augmented, 127, **202–213**
      atom-centered, **206–211**
         logarithmic derivatives, 207

basis set (cont.)
    one-center expansion, 207
        partial wave expansion, 207
        structure constants, 207, 210
    atomic spheres (*see also* muffin-tin spheres), 203
        cellular method (Wigner–Seitz), 203
        interstitials, 203
        linearization (*see* linear methods)
        muffin-tin orbitals (*see* MTO)
        secular equation, 204
        spherical waves (ASW), 211, 220
    classes, **127–128**
    completeness, 190
    contracted Gaussians (CGTO), 200–201
        core orbitals, 200
        cusp, 200
        split valence (3-21G, 6-31G*), 201, 260
    convergence, 127, 190
    expansion coefficients, 178
    extended, 127
    floating, **188–189**
        Gaussians, 189
        B-splines, 189
        finite differences, 189
        Lagrange polynomials, 189
        PW with local metric tensor, 188–189
        spherical waves, 189
        wavelets, 189
    Gaussian-type orbitals (*see* GTO)
    generalized eigenvalue equation, **179**, **181**
    Hamiltonian matrix, **179**, **182**
    hydrogenic and Λ-functions, **194–195**
        Laguerre polynomials, 194
        orthogonality (single exponent), 194
        structure matrix, 194
    linear dependence, 191
    localized, 127
    mixed, 127, **202**
        completeness, 202
        GTO+PW, 202
    multi-center expansion, 190–191
    numerical, **192–194**
    over-completeness, 185
    overlap matrix, **180**, **182**
    plane waves (*see* PW)
    Pulay forces (*see* Pulay forces)
    Roothaan–Hall equations (*see* Roothaan–Hall)
    Slater-type orbitals (*see* STO)
    superposition error (BSSE), **201–202**
        counterpoise correction, 202
    truncation, 185, 190
Bloch sums, 210
Bloch's theorem, **129**, 132, 133, 141, 181, 182, 274
bond energy, **282**, 296
    second moments approximation (SMA), 297
bond order potentials (BOP), 296
    continued fraction expansion, 297
    empirical, 296
    recursion method, 297
bond orders, **249**, 282, 296

Born–Oppenheimer, 6, 11
Bragg reflection, 130
Bravais lattice, 128
Brillouin zone (BZ), **129–133**, 181
    refolding, 130
    sampling **133–136**
        dispersion relation, 130
        grid centering and de-centering, 134
        irreducible wedge, 134
        k-vectors (**k**-points), 134–135
        metals, 135
        multiplicity factor, 134
        reciprocal lattice vectors, **129**, 133, 183
        reduced zone diagram, 130–132
        special **k**-points, 135
Brillouin's theorem, 40

CADPAC, 260
Car–Parrinello method, **325–337**
    canonical Kohn–Sham orbitals, 328–330
    constants of motion, 333–334
    damping of the nuclear dynamics, 336–337
    dynamical Kohn–Sham orbitals (DKSO), 327
    energy transfer, 334–335
    equations of motion, **328–330**
    fictitious dynamics, 326
    fictitious kinetic energy, **327**, 333
    flow diagram, 329
    Lagrange multipliers, 326, 327, 330
    Lagrangian, **325–327**
    orthonormalization, **330–332**
        iterative (SHAKE), 330 332
    performance, **333–337**
    pre-conditioning, **332–333**
    stability, 335–336
CASTEP, 324
chemical potential, 55, 63
CI (configuration interaction), **39–41**, 47
    CISD, 40, 46, 48
    CISDT, 40, 48
    CISDTQ, 40
    equations, 39
    excited states, 39
    full, 40, 48
    matrix elements, 39–40
        symmetry group, 40
    single-reference, 40
    truncated expansion, 40
classical force fields, 270, **298–300**
    angular forces, 298–299
    bond order (Tersoff, Brenner), 298
    electrostatics, 299
        distributed mutipoles, 299
        ESP and RESP, 299
    polarizable, 298, 300
    repulsion-dispersion potentials (Lennard-Jones), 300
    steric constraints, 300
    torsional potential, 299
    two-body potentials, 298, 299

classical nuclei approximation, **10–13**
closed shell, 36, 38, 61, 180
　atoms, 108
　molecules, 123
Colle–Salvetti electronic correlation, 88, 104–105
complete active space self-consistent field (CASSCF), 41, 47, 72
　active and inactive spaces, 41
　CASMP2, 47
condensed phases, **128–141**
configuration interaction (*see* CI)
core Hamiltonian, **28**, 250
correlated wave function, 38
correlating function, 49
　Jastrow-type factor (*see* Jastrow correlation factor)
correlation energy (*also* term), 37, 45–46, 59, 66, 75, 105–106
　density, 78
　coupling constant average, 75, 106
correlation, 5, **18**, 45
　dynamical, 83, 108
　hole, 25, **69**
　non-dynamical, 72
　potential, 55
Coulomb operator, 35, 180
coupled clusters (CC), **47–48**
　CCSD, 48
　truncated expansions, 47
CPMD (*see* Car–Parrinello molecular dynamics)
CRYSTAL, 260
crystalline solids, 128
　basis, 128
　conventional cell, 128
　energy bands (*see* energy bands)
　lattice vectors, 128
　primitive lattice vectors (*also* unit vectors), 128, 129
　supercell, 133
　Wigner–Seitz cell, 128

Debye–Hückel, 17–19
　screening length, 19
decoherence, 11
density functional perturbation theory (DFPT, *see also* GL theory), 105, 248
density functional theory (*see* DFT)
density matrix, **23**, 239, 276, 296
　ground state, 60
　one-body, **23**, 59, 68, 249
　two-body, **23**
DFT, 26, 51, **56–72**
　constrained search formulation (Levy), 58
　ensemble, 59
　ground and excited states, 58–59
　ground state energy, 57
　time-dependent (TDDFT), 59, 71

DGAUSS, 254, 260
diagonalization, 311, **312–316**
　block recursion algorithms, 316
　Davidson algorithm, **314–315**
　direct (LAPACK), 312
　FEMD, 316
　iterative methods, 40, **312–316**
　Lanczos algorithm, **315–316**, 320
　Trotter decomposition, 313
dielectric function, 21
differential overlap, 287
dipole–dipole interaction, 108, 188
direct correlation function (*see also* pair correlation function), 23–24
direct minimization of the energy functional, **318–321**
　conjugated gradients, **320**
　damped second order dynamics, **320–321**
　relaxation dynamics, 320
　steepest descent, **319**
disordered systems (liquid and amorphous), 140–141
　PBC and Brillouin zone sampling, 141
　polar fluids, hydrogen-bonding, 141
dissociation limit, 46
Dmol, 193
dynamical mean field theory (DMFT), 113
Dyson equation (*see* Green's function)

EAM, 298
　atomic embedding function, 297
effective core potential (ECP), 126, 144, 288
effective medium theory, 298
Ehrenfest theorem, 12
electron affinity, 35, 71, 292
electron–electron interaction, 5, 8, 22–25, 29
electron–hole pairs, 10, 21
electron–nuclear interaction, 8, 15, 22, **124–127**
electron–phonon interaction, 10
electronic density, **23**, 51, **123**, 238
　periodic systems (Brillouin zone average), **135**
electronic density of states (EDOS), **275**, 296, 297
electronic eigenstates
　diagonal correction, 11
　mixing, 6
electronic excitations, 10, 21, 71
　ΔSCF method, 71
　in DFT, 62
　Slater determinants
　　double excitations (doubles, D), 38, 45,47
　　single, triple, and quadruple excitations (S, T, Q), 38
　Slater's transition state method, 71
electronic levels in atoms, 124–125
　core electrons, 124–126, 144, 174
　frozen orbitals, 124, 125
　semi-core electrons, 125, 144, 155, 175
　valence electrons, 124, 125, 144, 174
electronic orbits, 3

electronic structure methods, **217–266**
  APW, LAPW, and PAW, **222–226**
  atom-centered basis sets (*see* local orbital methods), **248–259**
  augmented basis sets, **220–228**
    full potential, 221–222
    non-muffin-tin potential, 221
  Gaussians (*see* Gaussian basis sets), **259–266**
  LCMTO and LMTO, **227–228**
  pseudopotential plane wave (see PPW method), **228–248**
electronic wave function, 5
embedded atom model (*see* EAM)
empirical force fields, **299–300**
  AMBER, 299
  CHARMM, 299
  GROMOS, 299
  OPLS, 299
  UFF, 299
energy bands, **130–132**, 274
  degeneracy, 130
  dispersion relation, 130, 132
  metallic and insulating behavior, 132
  perturbation theory (nearly free electron model), 131, 143
energy functional, 62
energy gap, 130–132
energy scale
  electronic motion, 6
  nuclear vibrations, 7
  rotational motion, 7
exact exchange (EXX), **97–102**, 105
  functional, 98
  Hartree–Fock vs DFT, 98
exchange
  energy (*also* term), 75
  energy density, 78
  exact (*see* exact exchange)
  hole, 25, 69
  operator, **35**, 124, 180
exchange-correlation
  energy, **25–27**, 238
  functionals, **26**, 62
    relativistic effects, 157–158
    summary, **113–114**
  hole, 69, 78
  linear response kernel, 109, 316
  modified energy functional, 62, **66–67**, 69
  non-locality, 94, 110
  potential, **63**
    discontinuity, 84
explicitly correlated methods, **48–49**
  Hylleraas method and function, 48–49
  R12 and CI-R12 methods, 49
extended Hückel theory, **286–287**, 289, 291, 292, 294
EXX (*see* exact exchange)

fast Fourier transform (FFT), 186, 229–230, 233, 238, 240
Fermi energy, 20, 64, 136, 275
Fermi gas, 20

Fermi momentum, 20, 68
Fermi surface, 135
  sampling (metals), **136–139**
    Fermi energy (or level), 136
    k-point density, 137
    occupation numbers, 136
    self-consistency convergence, 136
    smearing, 137
      Fermi function, 137
      Mermin functional, 137
    tetrahedron method, 136
Fermi temperature, 10
FFT (*see* fast Fourier transform)
first-principles molecular dynamics (FPMD, *see also* ab initio MD)
FLAPW method, 224
  matrix elements, 224–225
  warped muffin-tin, 225
Fock
  exchange energy (*see also* exact exchange), 97, 100, 105
  matrix eigenvalues, 44
  operator, **35**, 36, 44, 251
FP-LMTO method, **228**, 324
  interstitial potential, 228
FPMD, 13, **323–337**
  Born–Oppenheimer MD (BOMD), 324
  Car–Parrinello MD (CPMD, *see also* Car–Parrinello), 325
  density functional MD (DFMD), 324–325
fractional occupation, 41, 72
Friedel oscillations, 21
frontier orbitals, 72
full-potential LMTO (*see* FP-LMTO)
functional, 53
  derivation, **53–55**
  universal, 58, 61

GAMESS, 260, 266
Gaunt coefficients, 210, 225
GAUSSIAN, 260, 266
Gaussian basis sets, **259–266**
  Coulomb integrals
    Boys function, 262, 263
    Gauss–Rys quadratures, 264
    integral transforms, 261–263
    numerical integration, 264
  electron–nuclear interaction, 261–262
  kinetic integrals, 261
  multipole expansion, 265–266
  Obara–Saika recurrence relations, 264–265
  overlap integrals, 260
  periodic boundary conditions (PBC), 260
  two-electron integrals, 262–264
  vertical and horizontal recurrence relations, 265
generalized gradient approximation (*see* GGA)
generalized Kohn–Sham theory, 96–97
  semiconductor band gaps, 96
geometry optimization, 13, 217, 226, 243

GGA, **86–91**
  BLYP functional, 88
  Langreth–Mehl functional, 88
  PBE functional, 88–90
  trends, **90–91**
    asymptotic limits, 90
    binding and atomic energies, 90
    bond lengths and angles, 90
    hydrogen-bonding, 90
    self-interaction, 91
    semiconductors and metals, 90
GL theory, 105–107
  atomization energies, 107
  GL2, 107
glue potentials, 298
Görling–Levy theory (*see* GL theory)
Gouy–Chapman theory, 17
gradient corrections, 26
gradient expansions, **85–94**
  second-order density response, 86
Green's function, 20, 297
  Dyson's equation, **110**, 112
  GW approximation (GW *or* GWA), 59, **111**
    dielectric function, 112
      RPA and plasmon pole
        approximation, 112
    dynamically screened Coulomb
      interaction (W), 111
    polarization function, 111
    self-consistent, 112
  KKR method, 218
  many-body, 71, **110–112**
    non-interacting, 110–111
    relation to pair correlation
      function, 110
    self-energy, 110, 111, 113
    vertex corrections, 112
ground state energy, 22
ground state wave function, 22
GTO, **196–198**
  Cartesian Gaussians (CGTO), 196–197
    primitive (PGTO), 197, 200
    redundancy, 197
  exponents, 196
  Gaussian product formula (rule),
    197–198, 260
  optimization for DFT, 201
  polar Gaussians, 196

Hamiltonian
  electronic, 8, 15
  general, 4
Harris functional, **280–281**
Hartree
  energy (term), 24, 59, 75
  equation, **30**
  potential, **20–21**, 37
Hartree approximation (HSCF), 25, **28–31**, 37
  time-dependent, 11
Hartree–Fock
  approximation (HF), 25–26, **31–37**, 44
    time-dependent, 11

canonical orbital representation, **35**, 124
  energy, **34**, 36, 45, 249, 251
  equations, **35**, **124**
    basis set (*see* Roothaan–Hall equations)
  left–right correlation, 83
  restricted (RHF), 36, 180
  spin unrestricted (SUHF), 102, 180, 251
    basis set (Pople–Nesbet), 180
  theory (*see also* Hartree–Fock approximation),
    25–26, 59
  wave function (*see also* Slater determinant), 37
Heitler–London, 4
Hellmann–Feynman theorem, **12**, 188
Helmholtz equation, 208
HF-KS, 96–97
  B3LYP, **96**, 260
Hohenberg–Kohn theorem, **56–58**, 61
homogeneous electron gas (*also* uniform electron
    gas), 16, 22, 68, 75
  correlation energy
    Ceperley–Alder, 76
    Perdew–Zunger (PZ), 76
    Vosko–Wilk–Nusair (VWN), 76
    Wigner, 52
  exchange energy (Slater–Dirac), 52, 76
  Fermi energy, 52
  fully spin-polarized, 68, **76–77**
  gap in the excitation spectrum, 93
  kinetic energy (Thomas–Fermi), 52
  pair correlation function, 78
  spin-unpolarized, 68, **76–77**
hopping (*see* tight-binding)
Hubbard, U., 113
hybrid Hartree–Fock–Kohn–Sham methods (*see*
    HF-KS)
hybrid QM-MM methods (*see* QM-MM)

inhomogeneous electron gas, 25
integrals
  Coulomb, 30, **33**
  exchange, **34**
  four-center, 40, 191, 253, 278, 288, 295
  hopping (*see* tight-binding)
  one-center, 278
  one-electron, 252–253
  three-center, 191, 253, 254, 274,
    278, 288, 295
  two-electron, 29, 253
  two-center, 274, 278
    parameterization, 278–279
inter-electronic distance (r_s), 76
ionization energy (potential), 35, 71, 291, 294

Janak's theorem, 71, 72
Jastrow correlation factor, 49, 103, 104

k-point refolding, **137–139**
  spurious geometrical distortions, 138
  supercell vs Brillouin zone sampling, 138
kinetic
  correlation, **65–68**
  energy functional, 59, 238

kinetic (cont.)
    correlation contribution, 60
    non-interacting electrons, 60, 62, 75
    operator
        electronic, 8, 15, 22, 60
        nuclear, 8
KKR method (Korringa–Kohn–Rostoker), 207, **218–220**, 227
    integral equation, 218
    logarithmic derivatives, 220
    secular equation, 210, **219–220**
    structure constants, 220
    variational principle, 219
KLI (*see* OEP-KLI)
Kohn–Sham
    eigenvalues, 71
    energy, 251
    equations, **61**, **64**, **123**
        periodic systems (solids), **135**
    functional, 62, 65
    Hamiltonian, 251
    orbitals, **62**, 68, 106
    potential, **62**
    spin orbitals, 64
    theory, **59–65**
        observations, **70–72**
    unrestricted, 251
Koopmans' theorem, **35**, 71
Krieger–Li–Iafrate approximation
    (*see* OEP-KLI)

Landau–Zener, 8
LAPW method, 212, **223–225**
    empirical Wigner–Seitz rule, 224
    full potential (*see* FLAPW method)
    lattice harmonics, 224–225
    linearization energy, 224
    matrix elements, 223–224
LCAO, **191**, 211
LCMTO method, 179, 205, 209, **227**
LDA, 52, **77–79**
    beyond, **85–114**
    exchange-correlation
        energy density, 78
        energy functional, **78**, 81
        hole, 77
            spherical average, 80–81
            sum rule, 80
        potential, 81
    improvements, **84–85**
        density inhomogeneities, 85
        Hubbard-type on-site correlations, 85
        non-locality, 85
LDA+DMFT, 113
LDA+U, 112–113
LDA-LSDA
    limitations, **82–84**
        atomic electron densities, 82
        dissociation limit and ionization
            energies 83
        excitation energies, 84
        hydrogen bonds, 83
        metallic surfaces, 83
        molecular dissociation and left–right
            correlation, 83
        self-interaction cancellation, 82, 83
        semiconductor band gaps, 83
        strong Coulomb correlations, 84
        van der Waals interactions, 83
    performance, **80–81**
    trends, **82**
        bond lengths and angles, 82
        dielectric properties, 82
        ionization potentials, 82
        vibrational frequencies, 82
Lieb–Oxford bound, 89, 92
Lindhard approximation (*see* RPA)
linear combination of atomic orbitals
    (*see* LCAO)
linear combination of muffin tin orbitals (*see*
    LCMTO method)
linear methods, **211–213**
    linearized APW (*see* LAPW method)
    linearized MTO (*see* LMTO method)
    NMTO, 213
    reference energy, 212
linear scaling (*see also* order-N), 270
LMTO method, 213, **227–228**
    ASA, **227**
        combined correction, 227
        empty spheres, 228
    MTA, 227
local density approximation
    (*see* LDA)
local density of states (LDOS), 297
    moments, 297
local orbital methods (based on atom-centered
    basis sets), **248–259**
    core Hamiltonian, 250
    density matrix, 239
    Fock matrix elements, 251
    forces, **257–259**
        force theorems, 259
        Helmann–Feynman theorem, 257
        Pulay forces, 258
    Hartree term, **253–254**
        atom-centered expansion, 254–254
        auxiliary basis sets, 253–254
    Kohn–Sham equations, **251–252**
    Kohn–Sham matrix elements, **251**
    multi-pole expansion, **255–257**
        fast multi-pole method (FMM), 257
        Hartree energy, 256
        moments of the charge distribution, 256
    one-electron integrals, **252–253**
        kinetic term, 252
        nuclear attraction, 252–253
    overlap distribution (differential overlap),
        256, 287
    overlap integrals, 249
    response functions, 259
    Roothaan–Hall equations, 251, 252
    spin density matrices, 251

structure constants, 256
total energy, 249–252
two-electron integrals, 253
local spin density approximation
    (*see* LSDA)
localized Hartree–Fock (LHF), 102
logarithmic derivatives, 126, 149–150,
    206–209, 220
LSDA, **79–80**
    exchange-correlation
        energy density, 79
        energy functional, 79
    interpolation function
        von Barth–Hedin (vBH), 80
        Vosko–Wilk–Nusair (VWN), 80

magnetic systems, 61, 64
magnetization density, 64, 79, 124
many-body
    classical potentials, **296–298**
    perturbation theory (MBPT),
        105–108, 110
    problem
        classical, 16
        quantum, 16, 105
Mermin functional (grand potential), 137
meta-GGA (meta-generalized gradient
    approximation), **91–94**, 95
    kinetic energy density, 91
    limitations, 93–94
    PKZB functional, 91
    self-interaction, 92
    TPSS functional, 93
    trends, 93
molecular dynamics (MD), 217,
    226, 243
molecular orbitals (MO), **189**, 191
Møller–Plesset theory (MP), **44–47**, 105
    2n+1-theorem, 45
    MP2 approximation, **45**, 96, 109
    MP3, **45**
    MP4, **46**, 48
MTO, **207–209**, 220
    envelope function, 208, 209
    logarithmic derivatives and matching conditions,
        208–209
    one-range addition theorem, 209
    partial waves and phase shifts, 208
    spherical Bessel, Neumann, and Hankel
        functions, 208
    structure constants, 210
    tail cancellation, 209–210, 213
muffin tin
    approximation (MTA), 207,
        219, 221–222
    orbitals (*see* MTO)
    potential, 219, 221
    radius, 203, 219
    spheres, 203, 220
Mulliken charges, 284, 291
multi-configuration methods (MCSCF), 41
multi-determinantal wave function, 38

multi-reference methods, 42, 47, 72
    configuration interaction (MRCI), 42, 72
    MRCISD, 42

non-adiabatic
    couplings, 9
    dynamics, 9
non-bonding interactions, 108
non-collinear magnetism, 65
non-interacting
    chemical potential, 63
    electrons, 60
    equivalent system, 60
    Hamiltonian, 63
    kinetic energy (*see* kinetic
        energy functional)
    reference potential, **61–63**, 66
    reference system, **61**, 66–67, 71, 106
non-radiative transitions, 8–9
nuclear
    excitations, 10
    spin, 4
    wave function, 8
nuclear–nuclear (inter-nuclear) interaction,
    8, 15
NWChem, 260

occupation numbers, 123, 136
OEP, **98–100**
    EXX, 100
        -LDA and -GGA correlation, 103
        long-range tail cancellation, 103
        RPA correlation, 104
    integral equation, 99–100
    KLI approximation, 101–102, 109
        -ASA, 101
        -GL2, 109
    KLICS functional, **104**, 107
    linear response function, 99
    local exchange potential, 100, 106
    -MBPT(2), 107
    non-interacting Green's function, 99
    properties, 102
        dissociation limit, image potential, 102
        HOMO-LUMO gap, 102
        negatively charged ions, Rydberg series, 102
        self-interaction, 102
        unoccupied states, 102
one-electron
    energies, 33
    wave function (orbitals), 5, **28**
open shell, 61, 64, 180
    molecules, 124
optimized effective potential (*see* OEP)
orbital-dependent
    correlation functionals, 101, **103–105**
        Colle–Salvetti, 104–105
    functionals, 98, 100
orbital-free DFT, 56, **304–307**
    atomic shell structure, 304, 306
    kinetic energy

orbital-free DFT (cont.)
  GGA and meta-GGA, 306
  gradient expansion, 304, 305
  WDA, 306
  linear response, 306
  non-local kinetic energy functionals, 304–307
  second-order response, 307
order-N, 270
orthogonalized plane wave method (OPW), **143–144**, 225

pair correlation function, **18–24**, 67–68
  coupling constant average, 67–68
partial waves
  expansion, 179
  one-center expansion, 219
Pauli's exclusion principle, 25, 31, 34, 36
PAW method, **225–226**
  projector functions, 226
periodic boundary conditions (PBC), **133**, 135, 181
perturbation theory
  general, **42–43**
  quantum many-body, **43–48**
  re-summations, 47–48, **107–108**
phase shifts (scattering), 149, 153, 206
plane wave method (see PW)
plasma frequency, 17, 21
plasma oscillations, 21
plasmon, 10
plasmon pole approximation, 112
Poisson–Boltzmann, 19
Poisson's equation, 17–18, 20
polarons, 8
post-Hartree–Fock methods, 26, **37–49**, 51
potential energy surface (PES), 12, 15
PPW method, **228–248**
  auxiliary (pseudo-ionic) charge distribution, 235, **237**
  charged systems and background energy, 240
  energy functional, **230–238**
    atomic structure factor, 233
    charge neutrality, 231–232
    Ewald sums, **235–238**
    exchange-correlation energy, **238**
    ionic self-interaction, 236
    kinetic energy, **238**
    local electrostatic energy, **230–235, 237**
    minimum image convention, 236
    non-local pseudopotential energy, **238**
  flow diagram of a calculation, 244
  forces on the nuclear coordinates, **243–245**
  Fourier transforms, **232–234**, 239
  generalized forces on the electronic orbitals, **242–243**
  Kohn–Sham potential, 229, **238–242**
    exchange-correlation term, **240**
    gradient of the density, 240
    Kleinman–Bylander separable form, 241
    local electrostatic term, **239**
    non-local pseudopotential term, **240–242**
      real-space implementation, 241–242

potential reference (zero), 239–240
response functions, **247–248**
stress tensor, **245–247**
principle of integer preference, 102
projected augmented waves method (see PAW method)
promotion energy, 283
pseudopotential, **126–127**, 143
  angular momentum-dependent (see pseudopotential, non-local)
  construction, **148–154**
  cutoff (or core) radius, **149–151**, 171
  empirical, 148
  examples, **160–164**
  frozen core approximation, 144
  generation, **154–155**
  generalized norm-conservation, **159–160**, 169
  generalized separable forms, 168–169, 174
  ion–electron interaction, 148
  ionic cores (or ions), 126
  local, 127, 146
  local component, **146–147**, 231
  multi-reference projectors, 169
  non-linear core corrections (NLCC), **155–156**, 162–163, 175
    core pseudization radius, 156
  non-local, 126, **145–146**
  norm-conserving, **149–154**, 174
    Bachelet–Hamann–Schlüter (BHS), 151
    Friedel sum rule, 149
    Hamann–Schlüter–Chiang (HSC), 150
    Kerker, 151
    LQPH, 153
    RRKJ, 153–154
    Troullier–Martins (TM), 151–153, 160
  practical aspects, **174–176**
  pseudo-wave function, 126, 127, 145
  quality and validation, 147, 175
  reference electronic configuration, 150, 154, **158–159**, 174
  reference energy, 150
  relativistic, **156**, 175
  scalar relativistic, **157**, 175
  separable form, **164–169**
    choice of the local component, 166–168
    ghost states, 167–168
    Kleinman–Bylander (KB), 166–167
    projection function, 166
  smoothness, 151–153, 169, 174
  test configurations, 158, 161–164
  theory, **144–148**
  transferability, 148, 150, 158, 169, 171, 175
  ultrasoft (Vanderbilt, US, or VUS), **169–174**, 226
    core pseudization, 172
    generalized eigenvalue equation, 171–173
    non-local overlap operator, 171
  unscreening, 154
pseudopotential plane wave method (see PPW method)
Pulay forces, 188

PW, 143, **182–188**
  advantages, 187–188
  charged systems and compensating background, 188
  convergence, 187
  disadvantages, 188
  eigenvalue equation, 184
  energy cutoff, **184–187**
  Fourier coefficients, 184
  Fourier transform and Fourier series, 182–183
  matrix elements, 184
  number of PW components, 186–187
  PBC and Brillouin zone sampling, **183**
  real-space grid, 185–186
PWSCF, 324

QM-MM methods, **300–304**
  forces on the MM nuclei, 303
  forces on the QM nuclei, 302
  frozen ghost orbitals, 303
  interaction energy, 302
  interface, 301
  macromolecules (enzymes), 300
  modified Kohn–Sham Hamiltonian, 302
  molecules in solution, 300
  pseudopotentials, 303
  reaction field, 303–304
    ONIOM, 304
    Onsager, 304
    shape of the cavity, 304
  total energy, 301
quantum Monte Carlo, 49, 76, 95, 112
quantum phase coherence, 10
QUICKSTEP, 253–254

random phase approximation (*see* RPA)
Rayleigh–Ritz variational principle, 56, 57
reference potential (*see* non-interacting reference potential)
Roothaan–Hall equations, **180**, 251, 294
RPA, **21**, 44, 80, 112
  Gell-Mann and Brueckner, 76

scattering integral equation, 20–21
Schrödinger equation, 4, 9
  nuclear, 11
  time-independent, 4, 8
screened exchange (sX), 96
screening, 17
  length, 17, 19
  linear, 19, 21
  nuclear charge, 126
self-consistency, 20, 31, 63, 311
  Anderson mixing, **317–318**
  direct inversion in iterative subspace (DIIS), **318**
  linear response kernel, 316
  mixing schemes, **316–318**
  simple mixing, **317**
self-consistent field (SCF), 28, **32**
self-interaction, 31, 34, 37, 69
self-interaction correction (*see* SIC)

semi-empirical methods, 254, **286–294**
  complete neglect of differential overlap (CNDO), **290–292**
    CNDO/1, 291
    CNDO/2, **292**, 295
  core–core repulsion, **289**, **293–294**
  diatomic differential overlap, 287
  differential overlap (overlap distribution), 287
  energy, 289, 291
  extended Hückel theory (*see* extended Hückel theory), 286–287
  Fock matrix, 286, 288, 291, 292, 296
  intermediate neglect of differential overlap (INDO and MINDO), **292**
  monoatomic differential overlap, 287
  neglect of diatomic differential overlap (NDDO), **288**, 296
    AM1, **294**, 296
    MNDO, **293**
    PM3, **294**, 296
  neglect of differential overlap, 287–289
  overlap matrix, 286, 291
  parameterization, 289, 292
  relation to tight-binding methods, 294–296
  resonance integrals, **290**, 296
  rotational invariance, 289, 290
  zero differential overlap (ZDO), 287
shell model, 298
SIC (self-interaction correction), 70, 100
  -OEP, 100
  pseudopotentials, 70
SIESTA, 193, 254, 324
simplified electronic structure methods, **270–307**
  general features, 271
  hybrid QM-MM (*see* QM-MM)
  orbital-free DFT (*see* orbital-free DFT)
  semi-empirical methods (*see* semi-empirical methods)
  tight-binding (*see* tight binding)
Slater determinant, 5, 11, 25, **32**, 61, 72
Slater–Koster integrals, 279
Slater-type orbitals (*see* STO)
spin density functional theory (SDFT), **64–65**
spin orbitals, 32, 38
spin-adapted configurations, 40
spin–orbit coupling, 29, 124
spin-polarization density (*also* spin density), 64
spin-polarization vector, 65
spin-polarized systems, **63–65**, 66, 124
spin-unpolarized insulators, 123
STO, 48, **195–196**
  basis set generation, 195–196
  completeness, 195
  exponents, 195–196
  linear combination of GTO (*see* STO-kG basis set)
  linear combinations, 196
STO-kG basis set, **200**
Strong correlations, 112–113
  Hubbard model, 112
  Mott–Hubbard insulators, 113
structure constants (matrix), 194, 207, 210, 220, 256

surfaces and interfaces
    charged, 140
    slab geometry, 140

thermal
    distribution (Boltzmann), 19
    wavelength, 10
Thomas–Fermi
    approximation (*also* theory, TF), **51–52**
    screening length (wave number), 20, 89
Thomas–Fermi–Dirac theory (TFD), **52–53**
    equation, **55**
    functional, **53**, 304
tight binding, 254, **272–286**
    ab initio, **277–283**
    band model, 282
    bond model, 282
        binding energy, 283
        cohesive energy, 282
        promotion energy, 283
    Coulomb repulsion (on-site and inter-site), **284–285**, 295
    electronic density of states, 275
    empirical tight binding, **273–277**, 294
    energy (*see also* band energy, bond energy), 276
    energy bands, 274–275
    generalized eigenvalue problem, 276
    Hamiltonian, 273–275
    Hamiltonian matrix elements, 278–280
    hopping integrals (terms), 273–275, 278, 288, 294, 296
    Kohn–Sham Hamiltonian, 277
    local charge neutrality, 282–283, 285
    moments of the eigenvalue distribution, 275
    multi-band models, 276
    non-self-consistent, 281–283
    on-site energy, **273–275**, 278, 294
    orthogonal, 284–285
    overlap matrix, 276, 279
    parameterization, 278–280
    polarizable, 286
    relation to semi-empirical methods, 294–296
    repulsive potential (energy), 277, 282
    self-consistent (SCTB), **283–286**, 295
        charge transfer (SCCT), **284–285**, 295
        multipolar expansion, 285–286

van der Waals interactions, 26, 83, **108–109**
    long-range tail, 109
    PES, binding energies and bond lengths, 109
VASP, 324
von Weiszäcker kinetic energy density, **92**, **305**
v-representability, 71, 72

weakly interacting fragments, 46, 108
weighted density approximation (WDA), **94–96**
    pair correlation function, 94
    self-interaction cancellation, 95
Wigner–Seitz
    cell (*also* unit cell), 128, 207
    radius, 221

Yukawa potential, 19, 96